Integrated Transport

Travel is an essential part of everyday life and today most journeys are multimodal. It is the total travel experience that counts and integrated transport must reduce the inconvenience of transfers between modes. Most research and many publications on transport policy advocate sustainable transport, but the priority given to integration has been negligible. Yet integration is one of the most important means to advance sustainable transport and sustainability more generally.

While integrated transport systems are seen to be an ideal, there is a failure to make the transition from policy to practice. The authors argue that the achievement of sustainable transport is still a dream, as an integrated transport policy is a prerequisite for a sustainable transport system. It is only when the two concepts of sustainability and integration operate in the same direction and in a positive way that real progress can be made.

In this book, transportation experts from across the world have addressed the questions about what is integration, why is it so important and why is it so hard to achieve? The book provides an in-depth analysis of these issues and it aims to provide a better understanding of the subject, about what should be strived for, about what is realistic to expect, and about how to move forward towards a more integrated provision of transport infrastructure, services and management.

Moshe Givoni is a Senior Researcher at the Transport Studies Unit (TSU) which is part of the School of Geography and the Environment (SoGE) at Oxford University. He is also a Research Fellow at Wolfson College. Before joining Oxford he was a Marie Curie Fellow at the Department of Spatial Economics, Free University Amsterdam. He gained his PhD at the Bartlett School of Planning, University College London, and his academic background also includes degrees in Economics and Geography (BA) and Business Administration (MBA) from Tel-Aviv University.

David Banister is Professor of Transport Studies at the University of Oxford and Director of the Transport Studies Unit. He is also currently Director of the Environmental Change Institute in the School of Geography and the Environment at the University of Oxford. Until 2006, he was Professor of Transport Planning at University College London.

Integrated Transport

From Policy to Practice

Moshe Givoni and David Banister

Routledge
Taylor & Francis Group
LONDON AND NEW YORK

First published 2010
by Routledge
2 Park Square, Milton Park, Abingdon, Oxon, OX14 4RN

Simultaneously published in the USA and Canada
by Routledge
270 Madison Avenue, New York, NY 10016

Routledge is an imprint of the Taylor & Francis Group, an informa business

Typeset in Univers by
Pindar NZ, Auckland, New Zealand
Printed and bound in Great Britain by
TJ International Ltd, Padstow, Cornwall

British Library Cataloguing in Publication Data
A catalogue record for this book is available from the British Library

Library of Congress Cataloging-in-Publication Data
Integrated transport : from policy to practice / Moshe Givoni and David Banister.
p. cm.
Papers from a two day meeting held in Sept. 2008 in support of activities of the Network for European Communications and Transport Activities Research.
Includes bibliographical references and index.
1. Transportation and state—Congresses. 2. Transportation—Planning—Congresses. 3. Transportation engineering—Congresses. I. Givoni, Moshe. II. Banister, David.
HE193.I58 2010
388—dc22 2009048669

ISBN10: 0–415–54893–4 (hbk)
ISBN10: 0–203–85088–2 (ebk)

ISBN13: 978–0-415–54893–9 (hbk)
ISBN13: 978–0-203–85088–6 (ebk)

To our families

Contents

Notes on contributors

Julian Allen is a Senior Research Fellow in the Transport Department at the University of Westminster. His current research interests are urban freight transport, the impact of manufacturing and retailing techniques on logistics and transportation systems, and the history of freight transport.

Karen Anderton is currently undertaking a DPhil at Oxford University's Transport Studies Unit. She is an independent climate change consultant with seven years' experience working on policy research and has managed several international projects with various public, private and non-profit organizations. She has an MSc (with Distinction) in Environment and Development.

Peter Bakker is a Senior Researcher at KiM Netherlands Institute for Transport Policy Analysis in The Hague, an independent institute within the Netherlands Ministry of Transport. He specialized in public transport and taxi studies during a 20-year professional career. He has a Masters in Human Geography and a Bachelors in Passenger Transport.

David Banister is Professor of Transport Studies at the University of Oxford and Director of the Transport Studies Unit. He is also currently Director of the Environmental Change Institute in the School of Geography and the Environment at the University of Oxford. Until 2006, he was Professor of Transport Planning at University College London.

Maria Teresa Borzacchiello is a Post-Doctoral Fellow at the Institute for Environment and Sustainability of the European Commission Joint Research Centre in Italy. In 2008 she gained her PhD in 'Engineering of Hydraulics, Land-use and Transportation Systems' at the University of Napoli Federico II, Italy. Her main research interests cover the evaluation of the impacts of transportation systems and spatial data infrastructures for transportation.

Michael Browne is Professor of Logistics at the University of Westminster. He has directed research projects on many aspects of freight transport and logistics, including: research on the energy use implications of global sourcing, potential benefits from improved city logistics strategies, and forecasting future trends in logistics. He chairs the Central London Freight Quality Partnership.

Biagio Ciuffo is a Post-Doctoral Fellow at the Institute for Environment and Sustainability of the European Commission Joint Research Centre in Italy. In 2008 he gained his PhD in 'Engineering of Hydraulics, Land-use and Transportation Systems' at the University of Napoli Federico II. Socio-economic analysis and dynamic modelling of transportation networks have been his major research topics.

Eran Feitelson is a Professor in Geography and in Public Policy and former head of the Federmann School of Public Policy and Government at the Hebrew University of Jerusalem, as well as a past chair of the Department of Geography. He has a PhD from Johns Hopkins University and a Masters in Geography and in Economics from the Hebrew University of Jerusalem. Currently he is a Visiting Professor at Oxford University. His research interests include sustainable development, transport–environment interfaces, transboundary water management, land use policies and politics, and the role of planning in advancing sustainability notions.

Josef (Yossi) Gamlieli is a PhD candidate at the Federmann School of Public Policy and Government at the Hebrew University of Jerusalem. He holds a Masters degree in Geography, Environmental Management and Planning from the Hebrew University. Yossi has 15 years' experience in environment and transport planning in both the public and private sectors in Israel.

Moshe Givoni is a Senior Researcher at the Transport Studies Unit (TSU) which is part of the School of Geography and the Environment (SoGE) at Oxford University. He is also a Research Fellow at Wolfson College. Before joining Oxford he was a Marie Curie Fellow at the Department of Spatial Economics, Free University Amsterdam. He gained his PhD at the Bartlett School of Planning, University College London, and his academic background also includes degrees in Economics and Geography (BA) and Business Administration (MBA) from Tel-Aviv University.

Peter Headicar is a Reader in Transport Planning in the Department of Planning at Oxford Brookes University. His main research interests are in the relationships between land use planning and travel behaviour, and transport planning policy and practice.

Robin Hickman is an Associate Director and transport planner at Halcrow Group Ltd, and a Research Fellow at the Transport Studies Unit, University of Oxford. He is a specialist on transport and climate change issues, and integrated transport and urban planning strategies.

Torben Holvad is Economic Advisor at the European Railway Agency, France, Senior Research Associate at the Transport Studies Unit, University of Oxford, and External Associate Professor at the Department of Transport, Danish Technical University. He obtained his Economics degrees from Copenhagen University (MSc) and the European University Institute in Florence (PhD).

Neil Hoose is an independent consultant with over 20 years' experience in the Intelligent Transport Systems sector, ranging from systems design and implementation to policy analysis. Neil is a Visiting Professor at the Centre for Transport Studies, Imperial College London and a Fellow of the Institution of Highways and Transportation.

Mark Koetse obtained his PhD in economics from VU University Amsterdam in 2006. His dissertation focused on meta-analysis, and contains both applications of this research method in the field of investment behaviour and simulation studies on methodological issues. Since January 2006 he has been employed as a post-doc researcher at the Department of Spatial Economics at VU University Amsterdam. His current research interests lie in the fields of transport economics, economics of climate change, environmental economics and valuation, and meta-analysis.

Carl Koopmans is Research Director at SEO Economic Research and Professor of Economics at VU University, both in Amsterdam. He holds a degree in Econometrics and a PhD in Economics. After research into labour, housing, energy and the environment, he has specialized in transport and cost-benefit analysis.

Tom van Lier is a Research Associate in the research group MOSI-Transport and Logistics (Vrije Universiteit Brussel) led by Professor Cathy Macharis. His PhD work focuses on sustainable logistics and calculations of external costs of transport. He is currently also a teaching assistant for the courses transport and logistics management, operations research and operations management.

Cathy Macharis is a Professor at Vrije Universiteit Brussel. She teaches courses in operations and logistics management, as well as in transport and sustainable mobility. Her research group MOSI-Transport and Logistics focuses on establishing linkages between advanced operations, research methodologies and impact assessment. She has been involved in several national and European research projects dealing with topics such as the location of intermodal terminals, assessment of policy measures in the field of logistics and sustainable mobility.

Peter Nijkamp is Professor in Regional and Urban Economics and in Economic Geography at VU University Amsterdam. His research interests cover quantitative plan evaluation, regional and urban modelling, multicriteria analysis, transport systems analysis, mathematical systems modelling, technological innovation, entrepreneurship, environmental and resource management, and sustainable development. He is past president of the Netherlands Research Council (NWO).

Jos van Ommeren is Associate Professor in Economics at VU University Amsterdam and Research Fellow of the Tinbergen Institute. His current research interests include on-street and employer-provided parking, transport-related fringe benefits, commuting, job search, spatial economics, housing economics and transport economics.

John Parkin worked in consulting engineering before joining academia to pursue interests in transport system design innovation. His research includes monitoring and evaluating sustainable transport interventions, and investigating perceptions of risk and effort. He has worked on schemes from conception to construction and provides training for practitioners.

Ethem Pekin is a Research Associate at Vrije Universiteit Brussel. He is a member of the research group MOSI-Transport and Logistics. His main research interests are intermodal freight transportation, location analysis and GIS applications. His doctoral research further develops a location analysis model for intermodal terminals to assess policy measures.

John Preston is Professor of Rail Transport and Director of the Transportation Research Group at the School of Civil Engineering and the Environment, University of Southampton. He has over 25 years' experience of transport research and teaching, has held over 100 research grants and contracts, and has over 200 publications.

Vincenzo Punzo has been Assistant Professor at the Department of Transportation Engineering of the University of Napoli Federico II since 2005. Dynamic simulation of traffic flow has been his major research topic. He is author of 30 refereed papers in journals and international conference proceedings. Currently he is leading the European COST Action 'Methods and Tools for Supporting the Use, Calibration and Validation of Traffic Simulation Models'.

Aura Reggiani is Full Professor of Economic Policy at the University of Bologna, Italy. She is a specialist in spatial and transport economics and modelling, with a particular focus on the study on network evolution and complexity. She has a long list of international publications in her field of expertise and she is currently the President of NECTAR (Network on European Communications and Transport Activity Research).

Piet Rietveld is Professor in Transport Economics at VU University Amsterdam and a Fellow of both the Tinbergen Institute and the RSAI. He has been active in the fields of travel behaviour, transport policy, transport and spatial development, and the valuation of transport externalities. Recently the theme of how climate change impacts on transport has become one of his research topics.

Muhammad Sabir has been working as a PhD researcher at VU University Amsterdam since 2006. His research work focuses on the impacts of weather and climate change on road transport and travel behaviour. Prior to his PhD, he worked as a research officer in the Pakistani Planning Commission.

Catherine Seaborn is a Senior Consultant and transport planner with Halcrow Group Ltd. Her research interests cover both technical and policy aspects of transport planning with emphasis on the interrelationships between transportation systems and the urban realm. She holds Masters degrees in City Planning and in Transportation from the Massachusetts Institute of Technology.

Dominic Stead is Associate Professor at Delft University of Technology. Much of his research and teaching is comparative in nature and focuses on issues of governance. He has experience of a wide range of research projects related to spatial planning and transport policy, including the EU-funded ESPON, Interreg and Framework Programme projects. He has published widely in international books and journals.

Yusak O. Susilo is a Senior Lecturer in Transport and Spatial Planning at the Centre of Transport and Society (CTS) of the University of the West of England, Bristol. His main research interest is in understanding the way individuals compose their daily travel patterns and the interaction of such patterns with changes in activity location, urban form and socio-demographic factors. He received his doctoral degree from the Department of Urban Management, Kyoto University, Japan. Before joining CTS he was a research fellow at the Delft University of Technology.

Joseph S. Szyliowicz is a Professor at the Josef Korbel School of International Studies, University of Denver. He founded the Intermodal Transportation Institute and served as its Director for ten years. He has been a member of the US delegation to APEC's Transportation Working Group, has served on various international conferences and research projects, and has published extensively on sustainability and security issues.

Vincenzo Torrieri is Full Professor at the Department of Transportation Engineering at the University of Napoli Federico II. Besides his long academic experience, since 2003 he has been Director of the TEST research company (Napoli, Italy) whose mission is to provide engineering services for the qualification of advanced transport vehicles and transport systems, as well as to support spin-offs and perform technology transfer activities.

Allan Woodburn is a Senior Lecturer in Freight and Logistics in the Transport Studies Department at the University of Westminster, London. He is involved in a wide range of teaching, research and consultancy activities in the field of freight transport, both within the UK and internationally. Allan completed his doctorate, which examined the role of rail freight within the supply chain, in 2000. Since then, his main research interests have been rail freight policy, planning and operations, focusing specifically on efficiency and sustainability issues.

Luca Zamparini is Associate Professor of Economics at the University of Salento, Italy. His current research interests involve transport economics, in particular monetary estimations of the value of time and of the value of reliability, and tourism economics, with a particular focus on the relationships between seasonality of demand and labour market dynamics.

Pierre Zembri is Professor of Transport Geography at the University of Cergy-Pontoise, France, and Director of the Research Unit MRTE (Mobility, Networks, Territories and Environment). His recent research works focus on the relationship between transport networks (airlines, railways, public transport) and territories in a deregulated context.

Preface

In September 2008 as part of the academic activities of the Network for European Communications and Transport Activities Research (NECTAR – http://www.nectar-eu.org) a meeting of Cluster 1 on 'Networks' was organized by the Transport Studies Unit, in the School of Geography and the Environment at the University of Oxford.

About 20 academic experts attended the two-day meeting which focused on debating the theme of Integrated Transport. The results of that meeting, together with some additional contributions, are brought together in this book on *Integrated Transport: From Policy to Practice*.

We would like to thank all who have participated in the Cluster meeting and are especially grateful to all the contributors to this book and for their help in bringing the material to publication. We would also like to thank Aura Reggiani for her work as the President of NECTAR.

NECTAR is a European-based scientific association established in 1992 which emerged from the European Science Foundation Network. Its primary objective is to foster research collaboration and exchange of information between experts in the field of transport, communication and mobility from all European countries and the rest of the world. It is a multidisciplinary social science network that brings together a wide variety of perspectives on transport and communication problems and their impacts on society in an international perspective.

The Editors,
Moshe Givoni
David Banister
Transport Studies Unit
School of Geography and the Environment
University of Oxford

Chapter 1

The need for integration in transport policy and practice

Moshe Givoni and David Banister

1. Introduction

In almost every piece of research and publication on transport policy a combination of the words 'transport' and 'sustainability' can be found. Similarly, next to these words a combination of the words 'transport' and 'integration' will normally also be present. While the inclusion of the term 'sustainability' in the transport policy discourse can be traced back to about 20 years ago, following the publication of the *Brundtland Report* (WCED, 1987), 'integration' has a much longer history in transport planning. Since sustainability became the dominant paradigm in the transport policy and research discourse, priority given to integration has been almost silent. Integration, even if it is not explicitly recognized, is probably still one of the most important means to advance sustainable transport and sustainability more generally. The concepts of sustainability and integration need to be promoted as being complementary in transport policy and practice.

Although integration might be easier to define and agree upon than sustainability, there are many elements of integration that need to be pursued. There are also many barriers to overcome in achieving integration in practice. This book argues that it is very difficult to turn integrated transport policy to integrated transport practice, and that as a consequence, sustainable transport is still only an ideal. It is only when these two concepts operate in the same direction in a positive way that real progress can be made.

This book addresses the questions: what is integration, why is it so important, why is it so hard to achieve, and are we never likely to get to a truly integrated transport system or even just a truly integrated and coherent transport policy. It provides an in-depth analysis of these issues and it aims to provide a better understanding of the subject – what should be strived for, what is realistic to expect, and how to move forward towards a more integrated provision of transport infrastructure, services and management.

This introductory chapter sets the background for the in-depth consideration of integrated transport. In Section 2 a short overview of the need for integration is provided, and this is followed in Section 3 by some of the definitions of integrated transport commonly used in the literature. Section 4 describes the structure and content of the book, and the last section provides some further thoughts before the individual chapters take over.

2. What to integrate and why?

Integration is normally required where a system is made up of several parts and the elements need to complement each other so that the whole system can work more efficiently. Integration in this respect can be considered as physical, operational or managerial. The same approach can be applied to the transport network where the physical network consists of many sub-networks and a large variety of users, operators and governing institutions. All these elements need to be integrated to provide an efficient transport system that serves the transport needs of society at minimal (environmental) cost. Whether this integration is possible and how it can be achieved is returned to later.

The transport system is often described as the blood system of society and especially that of the economy. This is indeed a good description, as virtually every economic activity and most social activities involve the transport of people and goods from one place to another in some form or another. Alongside transport's contribution to society, through its social and economic benefits, transport activities also entail a cost to society, mainly in the form of negative environmental impacts. The contribution of transport to climate change is probably the most important of these negative impacts, but other impacts such as air pollution are also substantial. Getting the 'right' mix of these components is a key element of sustainability. Without dwelling on the different definitions of sustainability, a transport system that can generate more or the same socio-economic benefits but at a lower (environmental) cost should be considered more sustainable.

As the transport system has grown and developed over time, and as new modes of transport have been introduced, specialization has taken place in which those involved with the supply, operation and management of the transport system have tended to focus on one or a limited sub-set of the transport system components. The current organization of the transport system became focused around a specific network (road, rail, air) and individual modes of transport. This specialization in turn and over time has become embedded in the institutions set in place to control and manage the transport system. From the supply aspect, the concept of the whole journey from the origin to the destination is often forgotten, as attention concentrates on the main section of the journey (for example the rail journey or the flight). So the need for multimodal travel to get from one place to another in most cases is often overlooked, and individual journeys are considered in a simple way as a single mode when many journeys involve more than one mode of travel.

This process or specialization (by mode) has taken place on the supply side of the transport system, but of course not on the demand side (i.e. from users' perspectives). People and goods still need to get from one place to another and in what they perceive to be the 'best' possibility available to them. In other words, from the demand perspective the choice of mode (and therefore network) is made based on the overall journey, considering all the elements that are involved in travelling from the origin to the destination. In this assessment, the weakest part of the journey, or the most difficult one, is often outside the main mode used (e.g. getting to or from the rail station).

This mismatch between the priority of users to just 'get there' (they or their goods) and the priority of suppliers to meet this need, but only for a specific part of the journey and often only after a choice of mode and network has been made, means that sustainability objectives have been compromised. This specialization did not only take place within the transport system, but is generic to many sectors and other elements and systems which are part of the everyday socio-economic activity. For example, transport often plays a major role in determining the levels of demand for health and education services (at a particular location), but this important element in the overall decision is often ignored by the authorities in terms of where they choose to locate hospitals and schools.

The lack of a coherent approach to the supply and organization of the transport system results in two adverse effects which compromise the goals of sustainability as defined above. First, too much travel is being generated to achieve a certain level of socio-economic benefit or welfare, and this travel results in higher (environmental) costs. Second, the amount of transport or travel activities that is necessary to support the desired level of welfare, or desired level of activities that entail transport in some form, results in higher than necessary (environmental) costs. This outcome also results from the best use not being made of the available capacity on the transport system, as transport is not provided and consumed in the most efficient way. To make transport more sustainable, the amount of transport or travel needs to be reduced, and the level of transport activity that is still needed must be undertaken in the most efficient way.

One of the implications of the mismatch described above between the need to simply 'get there', and the focus of the supply on just one part of the transport system results in the car being seen as being more attractive than other modes of transport with respect to travel time and convenience. When wanting to travel people see just one transport network that offers different travel/transport possibilities to get from one place to another, and they will normally opt to use the least costly option (in terms of time, money, convenience and reliability). Where integration of transport sub-networks is absent on the supply side, it is not surprising that the private car is usually seen as being the more attractive choice, as it involves the use of only one of the transport sub-networks and it provides door-to-door transport. There are two downsides to this. First, in most cases the car is much less advantageous in terms of emissions of both greenhouse gases and air pollutants, a feature which at present is not part of the decision of mode and network choice. Second, car transport normally requires more capacity (for a certain level of demand), especially given that many car journeys are used to carry only one person. When the capacity to meet that demand is not supplied, congestion appears with further detrimental effects on the quality of transport and sustainability. This book will argue that better integration, in different forms and at different levels, is probably the most important requirement to change mode choice and to make all modes of transport more attractive.

Although integrated transport as a policy goal has a long history, this subject has not been central to the debate. Indeed integration has probably always been one element in public policy with respect to any matter and sector. As Pressman and Wildavsky (1984, p. 133, cited in Chapter 2 in this volume) write, 'no suggestion for reform is more common than "what we need is more co-ordination"'. Even when integration did get the attention and priority it deserved, the results in practice were usually far from satisfactory, with the transport system being as fragmented (at the different levels) and as unimodal-focused as before, and usually with more infrastructure in place.

Two notable examples are the UK and EU official transport policies in recent years. In 1998 the UK government published a new transport White Paper called 'A new deal for transport: better for everyone', which is also known as the Integrated Transport White Paper, as it was based on the notion of integrated transport (DETR, 1998). One recommendation of the White Paper was for the creation of the Commission for Integrated Transport, which was supposed to bring together all stakeholders in the transport system. A few years later the UK government translated the integrated transport policy envisaged in the White Paper into a concrete plan known as the Ten Year Plan. Through better integration the White Paper aimed to reduce car use and increase public transport use. The minister responsible for transport at the time famously said, 'I will have failed, if in five years time there are not many [more] people using public transport and far fewer journeys by car. It's a tall order but I urge you to hold me to it'. Even before the ten-year period ended, it was apparent that very little modal shift had taken place from car to other modes, and that the transport system in 2010 is as disintegrated as it was in 2000 (Docherty and Shaw, 2008). Somewhat similarly to the UK, in 2001 the EU published a Transport White Paper at the heart of which was also the aim to achieve modal shift partly through a more integrated transport system (CEC, 2001), but there has been little subsequent evidence of modal shift.

There seem to be three main challenges to greater integration. First, the supply of transport needs to be integrated through realizing the role of the transport system to provide transport from place of origin (the house or place of production) to destination (work or shop). Transport should not be seen from bus stop to bus terminal or from one airport to another, as passengers rarely start or end their journeys there. Second, there is a need to integrate transport considerations into the decision-making (at different levels) concerning the location of all the activities that generate demand for transport in one way or another. A third level of integration is necessary if the two others are to get real priority, and that is integration of the institutions responsible for the transport network and activity, and those responsible or involved in other activities which require some transport.

These three levels of integration encompass integration of numerous elements within and outside the transport system and integration at different levels and in many forms.

Many of these myriad components of integration are considered in the different chapters of the book, and they are also captured in many definitions of integration. Some of these definitions are now presented and discussed.

3. Some definitions of integrated transport

There are many different definitions that can help in understanding the integrated transport system or integrated transport policy. At the basic level, integration will refer to integration of the different elements within the transport system. This would cover integration within one of the transport system sub-networks (e.g. the rail network). At a higher level this will refer to integration of several of these sub-networks (e.g. integrating rail and urban bus services). Integration within the transport network often relates to the terms 'multimodal' and 'intermodal', which are used interchangeably, but in general reflect the use of more than one mode of transport within one journey (of passengers or goods) and/or the consideration of more than one mode of transport (e.g. in transport policy). In relation to these terms, the EU recently adopted the term 'co-modality', which is defined as 'the efficient use of different modes on their own and in combination, [which] will result in an optimal and sustainable utilization of resources' (CEC, 2006, p. 4). Integration within the transport system strongly suggests a focus on the interchange or transfer element of a journey that consists of a chain of trips.

In this book, integration within the transport system is considered to be essential. But to achieve it, integration at other levels is required and thus a broader definition of integration is needed. Such an approach to integration is necessary to promote efficient transport, to ensure that when transport activity is undertaken the best use is made of the available capacity and that the most appropriate modes are used, and to ensure consistency with the broader sustainability goals. At the same time, a wider definition of integration is required if the aim is to reduce the need for travel.

There are several definitions for integration adopted in the literature and used in different chapters of this book. These are summarized in Table 1.1 and cover the meaning of integrated transport in this book. Most cover, or are related to, the paper by Eggenberger and Partidário (2000), where five different forms of integration are suggested in the context of 'dealing with development planning and assessment, and its impact on our living space' (p. 204). These five forms of integration are:

- *Substantive* – The integration of physical or biophysical issues with social and economic issues.
- *Methodological* – The integration of environmental, economic and social (impact) assessment approaches.
- *Procedural* – The integration of environmental, social, economic planning/assessment, spatial planning and EIA (Environmental Impact Assessment).
- *Institutional* – The provision of capacities to cope with the emerging issues and duties.

- *Policy* – The integration of 'sustainable development' as overall guiding principle in planning and EIA.

The different definitions for integrated transport that are grouped in Table 1.1 have much in common. They all include the functional or the basic level of integration described above, namely integration at the level of the actual transport and travel activity. All the definitions are then extended to include policy (of transport and also of other related sectors) and institutions. They also give some attention to the triple objectives of social, economic and environment goals, which are captured in the term sustainability, thus reflecting some of the complexity inherent in integration.

The forms of integration suggested by Eggenberger and Partidário (2000) and the broad definitions of integration given in Table 1.1 are all useful in the context of this book. They are, therefore, used to define the meaning of integration and are comprehensive enough to not merit a new definition in this book. If there is a limitation in the above definitions, as is the case with the many definitions given to the term sustainability (which, explicitly or implicitly, is embedded in the above definitions), it is that they are very hard to operationalize or to interpret when it comes to actual planning, development and governance of the transport system. It is for this reason that the book mainly focuses on this transition from policy to practice, which seems to fail again and again, no matter how well integrated the transport policy is. Perhaps this is a core weakness with respect to transport integration, namely that such a multifaceted concept is difficult to define and even harder to implement. This might suggest that a more simple starting point would be to solely concentrate on the transport definition, rather than the richer more complex ideas covering intersectoral perspectives, and the institutional and organizational constraints within which integrated transport operates. We return to this issue in the concluding chapter of this book.

4. Analysing integrated transport in this book

Even in a large book, such as this, is it is not possible to cover all aspects of integrated transport. The approach taken has been to cover the main issues first (in Part 1) and then further examine them and extend the range of issues covered through a series of case studies (Parts 2–4). The book addresses both the concept of integrated transport policy and the challenge of turning integrated transport policy into an integrated transport system. It also comments on what should be integrated, building on the above definitions, and it addresses the challenges in implementing this policy. The structure of the book takes the reader through this transition from policy to practice.

Part 1 contains six chapters, and it starts with a conceptual analysis of integrated transport policy (Chapter 2), taking a broad EU-wide perspective. This is followed by an analysis of the need to integrate transport planning and spatial planning (what is more commonly referred to as integrated land use and transport) in Chapter 3. Here the focus is more at the national level, but it demonstrates both the potential for integration and the problems in making real progress through the 11 themes identified for better integration. These operate at all levels of decision

Table 1.1 Some definitions of integration used in the book

Term	*Definition*	*Source*
Key dimensions in the notion of integrated transport policy	*Horizontal Dimension*: concerns the integration of policies between the various agencies or sectors involved in policy making, or even between departments within the same agency. *Vertical Dimension*: concerns integration between different tiers of government. *Spatial (or Intra-jurisdictional) Dimension*: integration between the same sector in geographically adjacent agencies. *Temporal Dimension*: the integration of policy documents with different production dates or time horizons and/or the sequencing of policy measures. *Modal Dimension*: including policy packaging, coordinated timetabling, common ticketing, multimodal travel information and planning and the interoperability of transport equipment.	See Chapter 2. Based on: Potter and Skinner (2000), Stead (2003), Underdal (1980) and Vigar and Stead (2003)
Integration (in public transport context)	The organizational process through which the planning and delivery of elements of the transport system are brought together, across modes, sectors, operators and institutions, with the aim of increasing net social benefits.	NEA *et al.* (2003)
Integration ladder	8. Integration of policy measures 7. Integration of policy sectors 6. Institutional and administrative integration 5. The integration of environmental issues in transport policy making 4. Integration with social objectives 3. Integration with market needs 2. Modal integration 1. Physical and operational integration of public transport	Hull (2005) (Hull considered horizontal integration 1–3 as achievable, but vertical integration 4–8 as much harder to achieve)
Integrated Transport (as a scalar, starting with the lower end of the scale)	*Functional or Modal Integration*: a policy enabling different travel modes to complement each other, making multimodal journeys easier. *Transport and Planning Integration*: land use and transport are closely linked, as patterns of land use and facilities are direct influences on travel generation. *Social Integration*: at this level, all those who have a stake in transport have their needs considered. *Environmental, Economic and Transport Policy Integration*: essentially, all of the above policies are combined in a holistic way.	Potter and Skinner (2000)

making, but they need to be addressed and implemented in a comprehensive 'packaging' approach. Chapter 4 deals with another important, but different, level of integration – that of institutions, both within the transport sector and between transport and other sectors. Integration is now widely accepted as a prerequisite for an efficient and better transport system, both for the passenger and freight transport networks. Yet, for a long time integration was only really seen as crucial, and therefore pursued, with respect to freight transport. Chapter 5 presents the concept and practice of integration in freight transport. The main reason why integration was traditionally pursued more vigorously and much more successfully in freight transport, compared to passenger transport, is probably because commercial forces play a greater role in the freight sector. In other words, 'the market' works better in freight transport. The last two chapters in Part 1 examine in turn the value of reliability in transport, which is one of the main motivations for physical integration of the different sub-networks of the transport system (Chapter 6). The challenges in evaluating integrated transport policies, when trying to determine if investments aimed at integrating the transport system (mainly at the operational level) are worthwhile and give more effective outcomes than other investments, are covered in more detail in Chapter 7.

Part 2 contains four chapters which address issues related to the application of integrated transport policy. Chapter 8 builds on the discussion of integrating transport and spatial planning by examining how far people are likely to travel, by certain modes, to participate in different activities, considering especially the duration of the activities they plan to engage in. The analysis clearly shows how travel behaviour is influenced by where activities people want to engage in are located, reinforcing the need to integrate land use and transport planning. For a long period of time, walking and cycling were not perceived as modes of transport, or at least not as modes that deserve planning and infrastructure consideration. This has now changed and Chapter 9 discusses in detail the aspects of planning that are required in the creation of walking and cycling networks and their integration with other, mainly public transport, networks. At the same time, the chapter strongly emphasizes that integration in this respect does not mean walking and cycling networks should be considered as one network, but that it is essential to separately plan each of these networks and often physically separate them from one another and from the motorized road network. Information and Communications Technology (ICT) is increasingly used to facilitate transport activities in different ways (traffic flow management being one notable example), and there is a real opportunity to achieve greater integration of services by exploiting recent developments in information technology. But this potential is mainly harnessed in separately improving different transport networks. Even within one network, like the road network, the power of ICT to generate useful information to improve the transport system is used by individual elements of the network and is not shared (e.g. between the fleet operator and the road network manager). At present, there is a lack of integration which is not related to technology and the shortcoming of this and the barriers for the integration of ICT resources across all stakeholders in the transport network is examined in Chapter 10. To end this part, Chapter 11 provides further evidence of the benefits of and the need for integration across transport networks by showing that improving the journey to the rail station can impact on the satisfaction with the rail journey, and as a consequence can potentially increase rail use.

In Part 3, the book turns to address the issue of the evaluation of integrated transport policies, which was introduced in Chapter 7. Attention is given to different aspects of measuring the elements of, and benefits from, integration. Other aspects of integration are also addressed in the five chapters in this part. Chapter 12 describes the results of several studies that evaluated the benefits of integrated transport schemes. The use of cost-benefit analysis and benefit-cost ratio to judge the level of benefits is explored. Another tool to assist in decision making is introduced in Chapter 13, namely the use of Geographic Information System (GIS). GIS is used to visualize the results of different policies (namely subsidies to rail and waterways freight transport) to promote intermodal freight transport in Belgium. From an environmental perspective the use of railways and waterways to transport freight is better than using the roads, and policies to promote modal shift away from road use mainly by internalizing (fully or partly) the external costs of freight transport are examined in this chapter. Chapter 14 uses the same evaluation framework as in Chapter 12 to discuss the benefits from investments in integrating rail services with other rail services and other modes of transport. This chapter also further supports the analysis presented in Chapter 11 on integrating the access/egress journey to/from the railway station with the actual rail journey. In Chapter 15, there is another application of evaluation techniques, which considers a proposed new connection of the Island of Venice to the mainland, with the objective of achieving a modal shift and better integration. Weather can affect the journey experience (for passengers) in various ways, and this is explored in Chapter 16. For public transport use the effect is more profound due to the need to wait for the service and transfer between services, often at places with no protection from the weather. Weather, especially, rain, snow and strong winds, is also likely to slow the journey and this can have an impact on mode choice and the use of public transport. Changes in the weather conditions influences the speed (and hence travel time) of a journey by public transport.

The final part of the book presents three case studies analysing in detail integrated transport policy in practice. Each case explores the unsuccessful practice of integrated transport policy at three different spatial levels. Chapter 17 tells the remarkable story of a new town planned from scratch with integrated transport policy in mind, and where all the conditions for successful implementation of it were favourable, only to achieve quite disappointing results. Chapter 18 emphasizes the difficulties in integrating the provision of different public transport services across different regions and municipalities. Chapter 19 discusses integration (or intermodalism as it is referred to) at a national level, looking at the challenges in promoting integrated transport in the US. Finally, Chapter 20 provides some concluding remarks and attempts to bring together some of the main themes discussed in the previous chapters.

5. Final remarks

The approach taken in this book is to see integrated transport as consisting of many levels, from the physical integration of the network (from the supplier or user perspective), through to the integration of transport into other policy objectives, and its integration with other sectors of governance and sectors of the economy. Integration in this sense relates to aspects of the planning, provision and management of the transport network and transport services, and to the ease with which the various modes or parts of the network are combined to make up the overall journey. The argument this book is making is that integration at all these levels is essential for an effective supply and use of the transport system. It is also essential for the transport system to meet the socio-economic needs of society, needs which include limiting the damage from transport infrastructure and activities to the environment.

Despite the apparent need for integrated transport, as recurrently highlighted in contemporary transport policy statements, there is little evidence of successful outcomes. This is probably obvious for any user of the transport system. The aim of this book is to investigate the nature of integration and what needs to be integrated, and then to identify the most important aspects of integration that will result in turn in a more sustainable transport system. The reasons for the lack of integration are crucial to the discussion, since the need for integration has been long recognized but, it is clear, it has never been successfully implemented. The book thus also aims to address this issue by giving attention in the various chapters to why the transport system is (still) not integrated. We return to this subject in the concluding chapter and discuss whether we can expect the transport system to be fully integrated at some point in the future.

Too much thinking has gone into looking at modes and networks as separate forms of transport. True integration requires new thinking about transport and different methodological approaches. As a start, it requires addressing transport as one system and this is the starting point of this book.

References

Buchanan C. and Crow G. (1974) An integrated transport system. *Journal of the Royal Society of Arts*, 2, February, pp. 117–128.

Commission of the European Communities (CEC) (2001) *European Transport Policy for 2010: time to decide*. Commission of the European Communities, COM(2001)370, September, Brussels.

—— (2006) Keep Europe moving: sustainable mobility for our continent. *Mid-term Review of the European Commission's 2001 Transport White Paper*. Commission of the European Communities, COM(2006)314 final, June, Brussels.

Department of the Environment, Transport and the Regions (DETR) (1998) A new deal for transport: better for everyone. *The Government's White Paper on the Future of Transport*. July, DETR.

Docherty I. and Shaw J. (2008) *Traffic Jam: ten years of 'sustainable' transport in the UK*. Bristol, UK: The Policy Press.

Eggenberger M. and Partidário M. (2000) Development of a framework to assist the integration of environmental, social and economic issues in spatial planning. *Impact Assessment and Project Appraisal*, 18(3): 201–207.

Hull A. (2005) Integrated transport planning in the UK: from concept to reality. *Journal of Transport Geography*, 13: 318–328.

NEA, OGM and TSU (2003) *Integration and Regulatory Structures in Public Transport*. Final Report. DGTREN, Brussels.

Potter S. and Skinner M.J. (2000) On transport integration: A contribution to better understanding. *Futures*, 32: 275–287.

Pressman J.L. and Wildavsky A. (1984) *Implementation*. 2nd edn. Berkeley, CA: University of California Press.

Stead D. (2003) Transport and land use planning policy in Britain – really joined up? *International Social Science Journal*, 55(2): 333–347.

Underdal A. (1980) Integrated marine policy: What? Why? How? *Marine Policy*, 4(3): 159–169.

Vigar G. and Stead D. (2003) Implementing transport policy in local government. In I. Docherty and J. Shaw (eds) *New Deal for Transport? The UK's Struggle with the Sustainable Transport Agenda*, London: Blackwell, pp. 51–72.

World Commission on Environment and Development (WCED) (1987) *Our Common Future*. Oxford: Oxford University Press.

Part 1

The main issues in integrated transport

Chapter 2

Integrated transport policy

A conceptual analysis

Dominic Stead

Introduction

Few policy makers or policy documents miss the chance to call for more integrated transport policy (alongside other calls for more efficient, more sustainable, safer, higher quality transport and the like). Detailed examination of policy documents and closer questioning of policy makers soon reveals however that integrated transport can mean different things to different people and that there is no widely accepted definition of what integrated transport policy means (Potter and Skinner, 2000). The same is also true for the notion of joined-up or integrated government in general: Cowell and Martin (2003), for example, argue that probing beyond the rhetoric reveals a range of diverse meanings and manifestations behind the concept. Moreover, clear statements from the academic or policy literature about what integrated transport is or might look like, or how it might be achieved, are scarce. According to Peters (1998, p. 296) for example, 'policy coordination is a term used with almost universal approbation but less often defined'. What is nevertheless clear is that the notion of integrated transport policy has a number of key dimensions. First, there is a *horizontal* dimension, which concerns the integration of policies between the various agencies or sectors involved in policy making, or even between departments within the same agency. Second, there is also a *vertical* dimension which concerns integration between different tiers of government, from the European and national level down to the local level (Vigar and Stead, 2003). These are two of the most commonly encountered dimensions but are by no means the only dimensions of policy integration: various others can be distinguished. There is a *spatial* (or intra-jurisdictional) dimension of policy integration (Stead, 2003), concerning integration between the same sector in geographically adjacent agencies (e.g. the integration of policy between the agencies responsible for transport policy in neighbouring public authorities). There is also a *temporal* dimension concerning the integration of policy documents with different production dates or time horizons and/or the sequencing of policy measures (Stead, 2003; Underdal, 1980). In the specific case of transport there is also a very important *modal* dimension concerning the integration of transport policy (Potter and Skinner, 2000), which can encompass a wide variety of issues including policy packaging, coordinated timetabling, common ticketing, multimodal travel information and planning, and the interoperability of transport equipment. Many or all of these different dimensions of policy integration tend to be conflated in policy arenas (Cowell and Martin, 2003) and tensions may exist between these different dimensions.

Demands for policy integration (and coordination) are certainly not new. Pressman and Wildavsky (1984), for example, note that 'no suggestion for reform is more common than "what we need is more co-ordination"' (p. 133). The *Brundtland Report*, the 1987 report from the United Nations, World Commission on Environment and Development, which laid the groundwork for the 1992 Earth Summit, notes the tendency for institutions to be 'independent, fragmented, and working to relatively narrow mandates with closed decision processes' (WCED, 1987). Peters (1998, p. 295) observes that ever since governing structures began to be differentiated into departments and ministries, there have been complaints or concerns about the lack of integration: that one organization does not know what another is doing, and that programmes or policies are contradictory or redundant, or both.

This chapter focuses on the conceptualization of transport policy integration, a notion which is broader in scope and more multidimensional than integrated transport (the latter has primarily only a modal and a horizontal dimension). This chapter looks specifically at the European dimension of transport policy integration: it examines the ideas underlying the concept and the different forms of policy integration, and it outlines various examples of mechanisms that promote policy integration (particularly in relation to horizontal policy integration). The paper draws on a combination of three main sources: (i) a survey of national policy makers involved in transport, health and environment issues carried out in 2005/2006 (Stead 2008); (ii) two UNECE workshops with national government officials and stakeholders in 2006 (Stead, 2008); and (iii) the wider literature on policy integration.

European policy context[1]

The idea of policy integration has featured in European policies for some time, especially in relation to the horizontal dimension (see above), where the integration of environment and transport policy received significant attention from the 1980s onwards (Geerlings and Stead, 2003), although this attention has arguably waned in recent years since the 'Cardiff Process'[2] was sidelined by the Lisbon Agenda (see for example Jordan *et al.*, 2008). The idea of integrating environmental considerations into other EU sectoral policies such as transport appeared in various European environmental action programmes and the integration principle has had legislative force in the European Community since the 1986 Single European Act. Since then, the integration principle has been strengthened by the Maastricht and Amsterdam Treaties (signed in 1992 and 1997 respectively). The latter recognizes integration as a key means of promoting sustainable development and requires environmental protection requirements to be 'integrated into the definition and implementation of other Community policies'.

Various European policy documents call for an integrated approach to policy making. These include the 1990 *Green Paper on the Urban Environment* (CEC, 1990); the 2001 Sustainable Development Strategy (CEC, 2001a), which calls for further integration of environmental concerns into sectoral policies; the 2004 *Thematic Strategy on the Urban Environment* (CEC, 2004a), which advocates an 'integrated approach to the management of the urban environment'; and the 2005 review of the Sustainable Development Strategy (CEC, 2005).

More specific to transport policy, the 2001 European Transport White Paper (CEC, 2001b) makes frequent reference to integration, as well as mentioning the concepts of 'intermodality' (linking different modes of transport) and 'interoperability' (compatibility of different transport infrastructures and/or networks), which also imply policy integration. The White Paper recognizes that transport policy alone is not sufficient to tackle current transport problems and advocates an *integrated* approach with other areas of policy making, such as economic policy, land use planning policy, social and education policy, and competition policy. It also highlights the need to integrate environmental considerations into transport policy (and other Community policies). More recently, the European Commission's 2009 Communication on the Future of Transport Policy (CEC, 2009) contains numerous references to integrated transport, mostly using the term to refer to integration between transport modes.

The 2002 report of the European Conference of Ministers of Transport (ECMT) concludes that 'sustainability requires that policy making for urban travel be viewed in a holistic sense' and that 'planning for transport, land-use and the environment no longer be undertaken in isolation one from the other' (ECMT, 2002: 33): clear references to the horizontal dimension of policy integration (on integrating spatial planning and transport see also Chapter 3 in this volume). Further references to the horizontal dimension can also be found elsewhere in statements such as: 'lack of a co-ordinated planning process for all transport (road and public transport), land use, and environmental considerations [at the local or regional level] can lead to a segmented approach to policy making, preventing the development and implementation of comprehensive, integrated plans addressing all related aspects of urban travel' (p. 21). The report also argues policy coordination (and integration) is essential to ensure that packages of complementary policies, rather than single measures, are implemented (possible reference to the temporal dimension of policy integration). Reference to the spatial dimension of policy integration is also evident in statements such as: 'lack of co-ordination on urban travel and land-use policy among constituent municipalities in a metropolitan area can lead to serious organizational problems and inefficiencies in, for example, the provision of public transport services' (p. 21).

Concepts of policy integration and types of mechanisms to promote policy integration

Policy integration is advocated for a variety of reasons (see for example EEA, 2005a; Stead and de Jong, 2006). The benefits associated with better integration include greater consistency between policies in different sectors (horizontal) and at different levels of decision making (vertical); less duplication in the policy-making process, providing more focus to the achievement of a government's overall goals rather than the achievement of narrower sector-oriented goals; helping to promote innovation in policy development and implementation; and encouraging greater understanding of the effects of policies on other sectors.

Integrated policy making implies going beyond mere coordination of policies and encompasses joint work among sectors, creating synergies between policies, sharing goals for their formulation and responsibility for their implementation (Stead *et al.*, 2004; Stead and Meijers, 2009). Definitions of policy integration are far from simple or consistent, particularly since a number of related terms exist in the literature such as coherence, consistency, collaboration, cooperation, coordination, and integration (see also Stead *et al.*, 2004). Shannon and Schmidt (2002, p. 17) define policy integration as 'an activity that links policy actors, organizations, and networks across sector boundaries'. According to Underdal (1980, p. 162), one of the early writers on the concept of policy integration, a policy is integrated 'when the consequences for that policy are recognized as decision premises, aggregated into an overall evaluation and incorporated at all policy levels and into all government agencies involved in its execution'. Underdal identifies three criteria that should be met in order for a policy to be described as integrated: (i) comprehensiveness – the recognition of a broader scope of policy consequences in terms of time, space, actors and issues; (ii) aggregation – policy alternatives are evaluated from an overall perspective; and (iii) consistency – policy penetrates all policy levels and all government agencies. More recently, Eggenberger and Partidário (2000) identified five forms of integration (that are relevant to spatial planning): substantive; methodological; procedural; institutional; and political (Figure 2.1).

Many mechanisms can help to promote it but no single mechanism alone can deliver policy integration. The 1996 OECD report on policy coherence (OECD, 1996), closely related to the concept of policy integration, provides a thorough overview of the subject and identifies eight basic 'tools of policy coherence' (i.e. ways of promoting policy integration) relating to: political leadership; strategic framework/ priorities; information management and analysis; policy coordination structures; policy coordination processes; policy-budget coordination; policy implementation and administrative culture (Figure 2.2). The 2000 United Kingdom Cabinet Office report on the management of cross-cutting policies and services identifies a range of different ways of promoting cross-cutting policies or joint working activities (United Kingdom Cabinet Office, 2000). These range from new or merged organizational structures and budgets to simple sharing of information or appropriate mechanisms for consultation between departments. The report also identifies a number of distinct forms of cross-cutting interventions and joint working arrangements, such as organizational change; merged structures and budgets; joint teams (virtual or real); shared budgets; joint management arrangements; shared objectives and performance indicators; and consultation and information sharing. Ling (2002), in an analysis of activities taking place under the name of joined-up government, identifies four distinct ways of achieving more integrated policy in practice: new types of

1. Substantive
   - The integration of physical or biophysical issues with social and economic issues
   - The integration of emerging issues such as health, risks, biodiversity, climate change and so on
   - The (appropriate) integration of global and local issues
2. Methodological
   - The integration of environmental, economic and social (impact) assessment approaches such as cumulative assessment, risk assessment, technological assessment, cost-benefit analysis, multicriteria analysis
   - The integration of the different applications, and experiences with the use of particular tools such as GIS (geographical information system)
   - The integration and clarification of (sector) terminologies (including the element of 'strategic')
3. Procedural
   - The integration of environmental, social, economic planning/assessment, spatial planning and EIA (environmental impact assessment)
   - The integration of sector approval/licensing processes, spatial planning and EIA
   - The adoption of coordination, cooperation and subsidiarity as guiding principles for (governmental) planning at different levels of decision making
   - The integration of affected stakeholders (public, private, NGO) in the decision-making process
   - The integration of professionals in a truly interdisciplinary team
4. Institutional
   - The provision of capacities to cope with the emerging issues and duties
   - The definition of a governmental organization to ensure integration
   - The exchange of information and possibilities of interventions between different sectors
   - The definition of leading and participating agencies and their respective duties and responsibilities
5. Policy
   - The integration of 'sustainable development' as an overall guiding principle
   - The integration of sector regulations
   - The integration of sector strategies
   - The timing and provisions for political interventions
   - Accountability of government

ure 2.1
e forms of
egration
rce: Adapted
n Eggenberger
Partidário,
0

organization (e.g. culture and values, information and training); new accountabilities and incentives (e.g. shared outcome targets and performance measures); new ways of delivering services (e.g. joint consultation and involvement); and new ways of working across organizations (e.g. shared leadership, pooled budgets, merged structures and joint teams). Meanwhile, the European Environment Agency (EEA) reviews progress in implementing policy integration in European member states and at the European level as a whole under five headings: political commitment and strategic vision; administrative culture and practices; assessment and consultation;

The experience of OECD countries, distilled into a handful of practical lessons, has led to the identification of the following basic tools of coherence. These are organizational concepts which, translated into appropriate structures, processes and methods of work, have proved conducive to higher degrees of policy coherence in governments from different political and administrative traditions. Some may seem, at first glance, deceptively obvious. However, experience shows that successfully putting them into practice requires painstaking experimentation and careful adaptation to the legal, administrative and political requirements of each national system.

- Commitment by the political leadership is a necessary precondition to coherence, and a tool to enhance it
- Establishing a strategic policy framework helps ensure that individual policies are consistent with the government's goals and priorities
- Decision makers need advice based on a clear definition and good analysis of issues, with explicit indications of possible inconsistencies
- The existence of a central overview and coordination capacity is essential to ensure horizontal consistency among policies
- Mechanisms to anticipate, detect and resolve policy conflicts early in the process help identify inconsistencies and reduce incoherence
- The decision-making process must be organized to achieve an effective reconciliation between policy priorities and budgetary imperatives
- Implementation procedures and monitoring mechanisms must be designed to ensure that policies can be adjusted in the light of progress, new information, and changing circumstances
- An administrative culture that promotes cross-sector cooperation and a systematic dialogue between different policy communities contributes to the strengthening of policy coherence

Figure 2.2 OECD tools of policy coherence
Source: OECD, 1996

policy instruments; and monitoring and review mechanisms (EEA, 2005a). In relation to administrative culture and practices, the EEA evaluates progress in implementing policy integration under four different headings: planning, budgetary and audit processes; internal management regimes; strategic coordination; and coordination and communication mechanisms (EEA, 2005b). These various examples serve to illustrate that there are many ways of classifying mechanisms to promote policy integration, often with substantial overlap between classification systems.

Examples of policy integration mechanisms

A range of mechanisms are available (and used in various parts of Europe) to promote transport policy integration. Selected examples are illustrated in this section, most of which primarily support horizontal policy integration. Many of the mechanisms highlighted are specifically designed to promote policy integration whereas others contribute to policy integration but are primarily designed to achieve other goals. Some of the mechanisms are specific to the transport policy sector but many are not, which is of course to be expected since many mechanisms are meant to cut across different sectors. Examples of mechanisms for policy integration are presented here according to the five interrelated forms of policy integration identified by Eggenberger and Partidário (2000): *substantive, methodological, procedural, institutional* and *political* (Figure 2.1).

Substantive

Substantive mechanisms of policy integration are understood here (according to the author's interpretation of Eggenberger and Partidário's five forms of integration in Figure 2.1) to include strategy documents that promote the concept or principles of policy integration and programming documents (setting out investment plans, for example). Many of these are not specific to the transport sector but include close attention to it: examples include sustainable development strategies; energy and/or climate change strategies; and environmental action programmes. In addition to these documents are other documents that are specific to the transport sector, such as transport and environment strategies or strategies for transport integration.

National sustainable development plans or strategies in various European countries refer to transport policy integration when considering the relation between transport and the environment (Stead, 2008). The same is also true for National Environmental Action Programmes (NEAPs), which were produced across Central and Eastern Europe as part of the implementation process of the international Environmental Action Programme for Central and Eastern Europe (UNECE, 2003). Attention to transport policy integration can also be found in energy and climate change strategies. Examples include Finland's energy and climate change strategy and Ireland's climate change strategy (Stead, 2008). Some of the above documents contain specific objectives for transport and other policy areas such as health and environment (e.g. the German sustainable development strategy).

A smaller number of strategy documents more specific to the transport sector can be found which refer to the concept or principles of policy integration. Integrated transport and environment strategies, for example, have been developed and implemented in various European countries, some of which focus on policy integration (EEA, 2004). In Finland, environmental guidelines for the transport sector up to 2010 (Finnish Ministry of Transport and Communications, 2005) have been developed which contain goals and targets across nine key issues including the integration of environmental aspects into transport planning. The document also identified key measures, monitoring requirements, research needs and key responsibilities for each of these nine key issues. At the European level, a few policy and strategy documents devoted to the integration of environmental concerns into transport policy were developed as part of the Cardiff Process (e.g. European Commission, 2003; CEC, 2004b).

Methodological

The mechanisms to promote policy integration that are contained in this section primarily concern monitoring and assessment techniques. These include ex-ante policy and project impact assessment (see also Chapter 7 in this volume), such as strategic environmental assessment (SEA), regulatory impact assessment (RIA), cost-benefit analysis (CBA), social impact assessment (SIA) and environmental risk assessment, as well as ex-post assessment techniques to a more limited extent.

Governments around the world have applied a variety of ex-ante assessment processes to support policy making. SEA, for example, seeks to identify and assess the environmental impacts of plans and programmes (and less frequently, policies) and identify ways of remediating these impacts. RIA on the other hand seeks to identify the costs and benefits of a new regulation and evaluate the performance of existing regulations in the same policy area (Kirkpatrick and Parker, 2004). These ex-ante processes can help to promote policy integration by focusing attention on the wider impacts (especially the environmental impacts) of plans and programmes while under development. The effectiveness of these instruments can often be enhanced by giving the public and NGOs a substantial say in the evaluation process and the drafting of evaluation criteria relevant to them (see procedural forms of policy integration, p. 22).

Since SEA is a legal requirement for development of transport plans at the regional and local level within the European Union, frequent examples can be found across the whole of Europe. However, the content and procedures of SEA differ widely due to national differences in the application of the European SEA Directive (2001/42/EC). In Germany, for example, an environmental risk assessment is required as part of the strategic environmental assessment process, whereas this is not part of the process in all European countries. Meanwhile, regulatory impact assessment has been applied in most OECD countries (Kirkpatrick and Parker, 2004) and represents one way of monitoring and assessing the cross-sectoral impacts of policies and programmes (Figure 2.3).

Procedural

A wide variety of procedural ways to promote policy integration can be identified. These include budgetary procedures (e.g. linking governmental planning cycles and budgets with the delivery of overarching sustainable development strategies or sectoral environmental integration strategies); processes for consultation and participation in policy making (as a way of considering the impacts of policy for other sectors and/or levels of government); and career rotation for policy makers, joint policy teams and environmental management systems (as a way of stimulating environmental awareness in government operations and policies).

The internal management planning and budgetary mechanisms of government can be used as a top-down tool to promote cross-cutting objectives, including environmental objectives (EEA, 2005b). However, despite the potential for doing so, few countries have exploited opportunities to link governmental planning cycles and budgets with the delivery of overarching sustainable development strategies or sectoral environmental integration strategies (EEA, 2005a). At the EU level, efforts were made by the European Parliament and selected Commission departments in the 1990s to use the budgetary process to integrate environmental considerations into key policy areas, notably regional development (EEA, 2005b). One consequence was that provisions were inserted into the EU budget, making expenditure conditional on certain environmental requirements. Environmental

Regulatory impact assessment is a policy tool to assess the costs, benefits and risks of any proposed regulation. The assessment incorporates aspects of sustainable development appraisal and health impact assessment. Regulatory impact assessments must be completed for all proposed policy changes, whether European or domestic, which could affect the public or private sectors, charities, the voluntary sector or small businesses. The regulatory impact assessment includes:

- an identification of the policy objectives
- an identification and quantification of the scale of the issue being addressed
- a description of possible options, including the key risks associated with the options, and how these can be mitigated
- an inventory of who is affected, including the business sectors and groups on which there may be a disproportionate impact
- a comparison of the benefits and costs for each option
- a summary of who or what sectors bear the costs and benefits of each option
- a summary of any unintended consequences and indirect costs
- a summary of the impacts, including the impact of each option on small firms and any measures for helping them comply
- an outline of how to communicate the changes that each option would bring
- an indication of how the policy will be monitored
- a summary of the results of the consultation exercise
- an implementation and delivery plan for the recommended option
- detailed plans for post-implementation review
- a recommendation of the preferred option, including the analysis of costs and benefits

gure 2.3 ·gulatory ·pact sessment in ɜ UK
urce: United ŋgdom Cabinet fice, 2006

auditing of policies is carried out in several countries and can help to identify the environmental implications of government spending plans. Examples can be found in Portugal (where environmental auditors in the transport and the agriculture ministries are required to carry out assessments of the potential environmental impacts of sectoral policies), and in the UK (where the Parliamentary Environmental Audit Committee annually assesses the ex-ante and ex-post environmental implications of departmental spending plans) (EEA, 2005a).

Intergovernmental consultation and public consultation on proposed policy is a way of considering the impacts of policy on other sectors and/or levels of government, and examples can be found across most of Europe. However, consultation processes are more common and better established in some countries than in others. Some countries have a more active civil society familiar with expressing their views, whereas citizens and interest groups in other countries are weaker or more passive or feel unable to criticize plans or government interventions. Despite the fact that Portugal does not have a long tradition of public participation, significant steps have recently been made to involve non-governmental organizations (NGOs) and other stakeholders in policy making over recent years (Figure 2.4), which has meant that the environmental impacts of policies have been brought more strongly into policy-making processes (OECD, 2001a).

The rotation of civil servants between ministries can offer a means of strengthening relationships and enhancing coordination and communication. In some countries, multidisciplinary professionals are highly valued and the rotation of civil servants is actively encouraged (e.g. the UK). In other countries, however, specialization and sectoralization of professions is considered much more desirable (see for example Lodge, 2003). The specialization and sectoralization has its advantages and disadvantages. On the one hand, it can for example contribute to long-term stability of personnel within administrations and retain their expertise, but on the other hand it may stifle strategic thinking, creativity or innovation in policy development as a result of the fragmentation of tasks and policies and a lack of a broader view by policy makers.

Instruments or tools to promote greater coordination and communication between departments can range from proactive formal exchange of information to informal *ad hoc* exchange. The extent of vertical communication and coordination varies across Europe (and can be affected by factors such as the centralization or federalization of government): Scandinavian countries are considered to be relatively strong in this regard (Niestroy, 2005). Interdepartmental or interagency teams responsible for policy implementation represent another way of promoting more integrated policy. Interdepartmental or cross-sectoral workshops and training activities can help to promote intersectoral communication and information exchange between professions, departments and/or levels of government. Within an administration, environmental management systems can stimulate greater general environmental awareness about the administration's own operations (e.g. procurement and housekeeping) as well as its policies (OECD, 2001b).

Given Belgium's federal structure and devolution of powers, several structures have been created to promote consultation and cooperation between the different levels of power and to ensure consistency in the action of the federal state and its entities. Sixteen interministerial conferences, each related to a particular policy area, have been created. These interministerial conferences are specialist committees whose members are the ministers concerned from the different levels of governments. The interministerial conferences provide the opportunity to hold joint meetings between various groupings, such as environment and transport. Interministerial conferences have no formal decision-making power: their purpose is to ensure smooth and efficient consultation and discussion between the different levels of government involved in a particular policy area. Permanent or temporary working groups can be established under interministerial conferences in order to examine an issue in detail. Such working groups can include representatives from different levels of government as well as experts and/or members of parliament.

Figure 2.4 Joint interministeri conferences Belgium
Source: Belgia Federal Public Service for Health, Food Chain Safety a Environment, 2006

Institutional

Institutional structures have a strong influence on policy integration (Peters, 1998; Stead, 2008). Support and endorsement from the highest level of government is also essential. A central steering role can help to coordinate policies from different departments (e.g. the prime minister's office in Finland). Interdepartmental committees, commissions, task forces, working groups and steering groups can all help to bring members of different ministries together and promote horizontal cooperation between departments and/or sectors (Stead, 2008). They can also help to involve a wide range of actors (including from outside government) in policy making. New high-level cross-cutting departments, units or committees have been introduced in many European countries to oversee and promote policy integration (especially in relation to environmental policy integration) across the government. The establishment of sustainable development or environment units is now common across OECD countries (EEA, 2005a). In Belgium, each Federal Public Service (ministry) has a unit ('cell') responsible for sustainable development to ensure that all federal sectoral policies comply with the federal plans for sustainable development (Belgian Interdepartmental Commission for Sustainable Development, 2004). These cells are required to carry out ex-ante assessments of the impact of all important government decisions in terms of sustainable development. Jacob and Volkery (2004) suggest that the influence of such units on the overall policy direction has however been rather limited in many of their respective departments.

Intersectoral programmes involving cooperation between several departments for developing and/or implementing policy (e.g. Belgium's joint interministerial conferences – see Figure 2.4) are another important example of institutional structures to promote policy integration. Many of these bodies were originally created in order to prepare or implement national sustainable development strategies or similar frameworks (EEA, 2005b). The development of institutional structures at the European level to promote greater policy integration has been relatively limited (ibid.). The General Affairs and External Relations Council, for example, was given a coordination role for environmental policy integration under the Cardiff Process but its role was limited.

Sustainable development or environmental advisory councils have also been established by governments across Europe to follow up national sustainable development strategies. Membership of the councils varies considerably, some taking the form of stakeholder councils and others acting more as scientific/expert councils (EEA, 2005b), but always involve a combination of governmental and non-governmental experts (Macrory and Niestroy, 2004). The main role of stakeholder councils has been building consensus among stakeholders and communicating issues on sustainable development or the environment to the public. The main role of scientific/expert councils on the other hand has been assessing government policies and identifying recommendations for future policies. All councils are expected to deliver independent advice to their governments, although some are more independent than others (EEA, 2005b).

Political

The interrelationships between the five forms of policy integration discussed in this section are complex, especially in relation to the fifth form of policy integration – political factors. The substantive, methodological, procedural and institutional forms of policy integration (described above) are all clearly shaped by political decisions. Each of these different forms of policy integration influences policy decisions and the extent to which intersectoral or intergovernmental issues are taken into account. In addition, political decisions influence the extent to which policies are interpreted, implemented and integrated. In a wider political sense (beyond party politics), the influence of key individuals (politicians, policy makers and other stakeholders) and their personal networks is also very important to policy integration (EEA 2005b; Stead 2008). A number of examples are discussed in the following section.

The influence of these political (and institutional) factors is very much apparent in the long-term historical development of European common transport policy. Various attempts to develop a comprehensive set of European policies across different transport sectors from the late 1950s onwards 'were blocked at every turn', according to Stevens (2004, p. 60) due to very different political ideologies of member states of the European Community. The proposals were considered far too extensive and liberal for some member states, while other member states judged them to be far too incompatible with their existing situation (Kerwer and Teutsch, 2001). As a result, the development of the European common transport policy (particularly up to the mid-1980s) was slow, incremental and piecemeal (i.e. modally based). According to Stevens (2004, p. 60), European transport policy in the 1980s was 'a patchwork of ... mildly useful mono-modal initiatives'. This situation has since changed to some degree, although my own summary of the most recent European transport policy strategy (from 2001) at the time it was published was that it is still far from integrated across modes. My opinion was (and still is) that the strategy 'reads like a collection of separate contributions from different sections within the European Commission which have been written in the absence of a common vision for European transport policy' (Stead 2001, pp. 417–18).

Recent progress towards policy integration

Recent trends and developments in policy integration are illustrated here drawing primarily on discussions with government practitioners at workshops on policy integration organized by the United Nations Economic Commission for Europe (UNECE) and the World Health Organization (WHO) Regional Office for Europe (see Stead, 2008 for details). The focus of the discussion here is on the horizontal integration of the transport, health and environment policy sectors.

In many cases, policy goals, targets and instruments in the transport, health and environment sectors are often still quite sectoral in nature and determined at the departmental level. Inconsistencies between sectoral policies still arise as a result. Accountability for policies and their effects are departmental in nature, which means that policy measures are usually primarily designed to fulfil departmental goals and targets rather than to address wider objectives outside the department. Thus, transport policies typically primarily address issues such as congestion or providing and managing infrastructure. There are, however, signs of integration in some national and local transport policies across Europe, which are now starting to consider environmental issues such as air quality or noise, and health issues such as road safety. Some would argue, however, that these are still rather narrow considerations of environment and health (physical activity is rarely part of transport policy, for example). Various differences between professional culture and practice (e.g. vocabulary, education, procedures, priorities) in government departments exist that also pose barriers to developing and implementing policies that address transport, health and environment concerns.

Although formal networks between the transport, health and environment sectors exist in many countries, many are perceived as lacking real power or influence. Some networks between the sectors are considered to be paying lip service rather than properly contributing to policy integration. Formal networks between departments are usually poorly resourced. Political interest in these networks is often low, especially since it is usually difficult to demonstrate tangible results from them. In many cases, there is no coordinating body, agency or structure to manage the horizontal relationships between sectors. While rotation of personnel has its advantages for promoting better understanding between policy sectors, especially if personnel move between different departments, there is also a downside: frequent rotation can lead to loss of continuity in an organization (low 'institutional memory') and a lack of long-term expertise in a department. In some countries, particularly in some of the new EU member states and South Eastern Europe, there is no tradition of working or even communicating with other policy sectors. There is, thus, little connectivity between sectors in some of these countries. In some instances, there is even the feeling that the sectors are in competition and are actually antagonists. Departments may even have to compete against each other for funding, which creates a poor climate for mutual cooperation or trust between sectors.

In general, relations between the transport, health and environment sectors are being given more recognition and support than in the past. Despite the formal relationships, however, there is often the feeling that the power or influence of one sector on another is low (see above). There is often the feeling that, however formalized and established the relationships between the sectors, the bottom line for many decisions is not so much the contribution to wider environmental or health objectives but the contribution to economic growth. As well as the internal relationships between departments, it also needs to be recognized that there are formal and informal relationships between these departments, with a range of public and private sector organizations that exert their influence and shape policy. In the case of transport for example, there are formal relationships between public transport providers (which can be public or private) and infrastructure operators (e.g. track companies, port authorities), as well as informal relationships with construction companies (e.g. infrastructure building and maintenance firms) and signalling companies. Each of the sectors has its own set of constituents with their own separate priorities and interests. In addition, there are vested interests from outside government departments which exert powerful influences over policy agendas and policy decisions (see for example Hamer, 1987; Stevens, 2004).

Inertia is an important consideration in any debate about policy reform. There is almost always strong resistance to change, both institutionally and individually, if new structures, methods, procedures, instruments or practices are proposed. This is very much the case for policy integration. It is difficult to sell the idea of policy integration to the extent that changes in structures, methods, procedures, instruments or practices can be justified. Policy integration is a very intangible concept. It is also difficult to identify incentives to promote intersectoral working and the achievement of multisectoral policy objectives in government. Consequently, policy integration is on everyone's wish list but is no one's top priority. In addition, education and training is frequently mentioned as a cause of professional differences between sectors and a reason for a lack of uniformity between government departments (see above). Professional differences are arguably becoming less important: multidisciplinary professional education and training (and recruitment) is more common. So too is staff mobility within and between departments. Nevertheless, it is also the case that some roles within government are more specialized than ever before.

Conclusions

Although there is increasing attention being given to the issue of transport policy integration, the concept remains fuzzy for many policy makers, somewhat analogous to the concept of sustainable development. As in the case of sustainable development, there is widespread consensus at all levels of policy making that it is a good idea, but a more limited understanding about what exactly it is or precisely how to achieve or monitor it (Stead *et al.*, 2004; Stead and Meijers, 2009). The concept of transport policy integration is very much multidimensional: horizontal, vertical, spatial, temporal and modal dimensions can all be identified. Despite ongoing efforts to break down the traditional silo (i.e. sectoral or departmental) mentality, integrated transport policy remains illusive.

Policy integration has a number of potential benefits such as promoting synergies (win-win solutions) between sectors or tiers of government, promoting consistency between policies in different sectors, and reducing duplication in the policy-making process. The need for policy integration, especially in relation to transport policy, is becoming increasingly recognized. However, achieving policy integration is no easy task, particularly since the number of actors and the amount of information involved in the policy-making process is generally increasing (for examples at local, regional and national levels (see Feitelson and Gamlieli, Chapter 17, Zembri, Chapter 18, and Szyliowicz, Chapter 19).

Resources and support (political, professional and public support) are two necessary preconditions for the promotion of policy integration and the introduction of the various mechanisms outlined above. Allocating resources according to overall goals, rather than government departments, will facilitate their successful implementation. Making joint funds available for intersectoral or intergovernmental cooperation and joint projects is also important for policy integration. In addition, policy integration needs to be supported by incentives or rewards to promote joint working and increase the motivation of professionals to think and work across disciplines and overcome the rigidity of existing working practices. Training, education and information management can play an important role here.

The institutional basis for intersectoral or intergovernmental working is crucial for policy integration: organizational structures and administrative responsibilities must not only permit but also encourage joint working (see also Chapter 4 in this volume). Overall government objectives cutting across ministerial or departmental boundaries should be defined (including long-term objectives as well as short- and medium-term objectives). In order for policy integration to be recognized as a priority, the costs and benefits need to be closely examined and explained to politicians, professionals and the public alike in simple, specific terms (e.g. cost savings, lives affected). Formal requirements for project and policy assessment (e.g. strategic environmental assessment and health impact assessment) and evaluation (e.g. ex-post evaluation) can help to increase integration between the transport, health and environment sectors. Formal agreements and policy commitments made at international forums (e.g. charters, directives and regulations) can increase attention to policy integration at the national, regional and local levels.

Some evidence of progress towards more integrated transport policy can be found but various barriers remain (Stead, 2008). Clearly, perfect policy integration is not possible, as Metcalfe (1994), for example, recognizes, but it can nevertheless be improved. While a range of factors can help to promote policy integration, there is of course no single solution. More importantly, policy integration needs to be recognized as a means to an end rather than an end in itself. Thus, policy integration should be seen in terms of its contribution to achieving practical action on the ground that is more integrated and that simultaneously fulfils the goals of different policy sectors or tiers of government.

Notes

1 For an account of 'integration' in the US see Chapter 19 in this volume.

2 The Cardiff Process is the name given to a process launched by European heads of state and government (The European Council) at their meeting in Cardiff in June 1998, requiring environmental considerations to be integrated into the policies, programmes and projects across a range of sectors (putting article 6 of the EC Treaty into practice).

References

Belgian Federal Public Service for Health, Food Chain Safety and Environment (2006) *Fourth National Communication on Climate Change*, Brussels: Federal Public Service Health, Food Chain Safety and Environment. Available online at: http://unfccc.int/national_reports/annex_i_natcom/submitted_natcom/items/3625.php (accessed 30 September 2009).

Belgian Interdepartmental Commission for Sustainable Development (2004) *Federal Plan for Sustainable Development 2004–2008*, Brussels: Interdepartmental Commission for Sustainable Development. Available online at: http://www.icdo.be/NL/publicaties/federale_plannen/2e_plan (accessed 30 September 2009).

Commission of the European Communities – CEC (1990) *Communication from the Commission to the Council and Parliament. Green Paper on the Urban Environment. COM(90)218 final*, Luxembourg: Office for Official Publications of the European Communities. Available online at: http://ec.europa.eu/environment/urban/policy_initiatives.htm (accessed 30 September 2009).

—— (2001a) *A Sustainable Europe for a Better World: A European Union Strategy for Sustainable Development. Communication of the European Commission. COM(2001)264 final*, Luxembourg: Office for Official Publications of the European Communities. Available online at: http://ec.europa.eu/environment/eussd (accessed 30 September 2009).

—— (2001b) *European Transport Policy for 2020: Time to Decide. COM(2001)370 final*, Luxembourg: Office for Official Publications of the European Communities. Available online at: http://ec.europa.eu/transport/strategies/2001_white_paper_en.htm (accessed 30 September 2009).

—— (2004a) *Communication from the Commission to the Council, the European Parliament, the European Economic and Social Committee and the Committee of the Regions. Towards a Thematic Strategy on the Urban Environment. COM(2004)60 final*, Luxembourg: Office for Official Publications of the European Communities. Available online at: http://ec.europa.eu/environment/urban/thematic_strategy.htm (accessed 30 September 2009).

—— (2004b) *Integrating Environmental Considerations into Other Policy Areas – A Stocktaking of the Cardiff Process. Commission Working Document COM(2004)394 final*, Luxemburg: Office for Official Publications of the European Communities. Available online at: http://ec.europa.eu/research/environment/newsanddoc/off_docs_en.htm (accessed 30 September 2009).

—— (2005) *Communication from the Commission to the Council and the European Parliament. The 2005 Review of the EU Sustainable Development Strategy: Initial Stocktaking and Future Orientations. COM(2005)37 final*, Luxemburg: Office for Official Publications of the European Communities. Available online at: http://ec.europa.eu/environment/eussd (accessed 30 September 2009).

—— (2009) *Communication from the Commission. A Sustainable Future for Transport: Towards an Integrated, Technology-led and User Friendly System. COM(2009)279/4 final*, Luxemburg: Office for Official Publications of the European Communities. Available online at: http://ec.europa.eu/transport/strategies/2009_future_of_transport_en.htm (accessed 30 September 2009).

Council of the European Communities (2001). Directive 2001/42/EC of the European Parliament and of the Council of 27 June 2001 on the assessment of the effects of certain plans and programmes on the environment. *Official Journal of the European Communities* L 197 (21.7.2001), 30–37.

Cowell, R. and Martin, S. (2003) 'The joy of joining up: Modes of integrating the local government modernisation agenda', *Environment and Planning C: Government and Policy* 21, 2: 159–179.

Eggenberger, M. and Partidário, M. (2000) 'Development of a framework to assist the integration of environmental, social and economic issues in spatial planning', *Impact Assessment and Project Appraisal* 18, 3: 201–207.

European Commission (2003) *Integration of Environment into Transport Policy – From Strategies to Good Practice*, Luxemburg: Office for Official Publications of the European Communities. Available online at: http://ec.europa.eu/environment/archives/gpc/index.htm (accessed 30 September 2009).

European Conference of Ministers on Transport – ECMT (2002) *Implementing Sustainable Urban Travel Policies. Final Report*, Paris: OECD.

European Environment Agency – EEA (2004) *Indicator Fact Sheet. TERM 2004 35 – Integrated Transport and Environment Strategies in the European Union*, Copenhagen: EEA. Available online at: http://www.eea.europa.eu/themes/transport/indicators (accessed 30 September 2009).

—— (2005a) *Environmental Policy Integration in Europe. Administrative Culture and Practices EEA Technical Report No 5/2005*, Copenhagen: EEA. Available online at: http://www.eea.europa.eu/publications (accessed 30 September 2009).

—— (2005b) *Environmental Policy Integration in Europe. State of Play and an Evaluation Framework. EEA Technical Report No 2/2005*, Copenhagen: EEA. Available online at: http://www.eea.europa.eu/publications (accessed 30 September 2009).

Finnish Ministry of Transport and Communications (2005) *Environmental Guidelines for the Transport Sector until 2010*. Helsinki: Ministry of Transport and Communications. Available online at: http://www.lvm.fi/web/en/publication/view/820811 (accessed 30 September 2009).

Geerlings, H. and Stead, D. (2003) 'The integration of land use planning, transport and environment in European policy and research', *Transport Policy* 10, 3: 187–196.

Hamer, M. (1987) *Wheels Within Wheels: A Study of the Road Lobby*, London: Routledge.

Jacob, K. and Volkery, A. (2004) 'Institutions and instruments for government self-regulation: environmental policy integration in a cross-country perspective', *Journal of Comparative Policy Analysis* 6, 3: 291–309.

Jordan, A., Schout, A. and Unfried, M. (2008) 'The European Union', in A. Jordan and A. Lenschow (eds) *Innovation in Environmental Policy? Integrating the Environment for Sustainability*, Cheltenham: Edward Elgar, pp. 159–179.

Kerwer, D. and Teutsch, M. (2001) 'Transport policy in the European Union', in A. Héritier, D. Kerwer, C. Knill, D. Lehmkuhl, T. Teutsch and A.C. Douillet (eds) *Differential Europe. The European Union Impact on National Policymaking*, Lanham, MD/Boulder, CO/New York, NY/Oxford: Rowman and Littlefield, pp. 23–56.

Kirkpatrick, C. and Parker, D. (2004) 'Regulatory impact assessment and regulatory governance in developing countries', *Public Administration and Development* 24, 4: 333–344.

Ling, T. (2002) 'Delivering joined-up government in the UK: dimensions, issues and problems', *Public Administration* 80, 4: 615–642.

Lodge, M. (2003) 'Institutional choice and policy transfer: reforming British and German railway regulation', *Governance* 16, 2: 159–178.

Macrory, R. and Niestroy, I. (2004) 'Emerging transnational policy networks: the European environmental advisory councils', in N.J. Vig and M.G. Faure (eds) *Green Giants? Environmental Policies of the United States and the European Union*, Cambridge, MA: MIT Press, pp. 305–328.

Metcalfe, L. (1994) 'International policy coordination and public management reform', *International Review of Administrative Sciences* 60, 2: 271–290.

Niestroy, I. (2005) *Sustaining Sustainability. A Benchmark Study on National Strategies Towards Sustainable Development and the Impact of Councils in Nine Eu Member States*. Utrecht: Uitgeverij Lemma.

Organisation for Economic Cooperation and Development – OECD (1996) 'Building policy coherence: tools and tensions'. *OECD Public Management Occasional Paper No. 12*. Paris: OECD.

—— (2001a) *Environmental Performance Reviews: Portugal*. Paris: OECD.

—— (2001b) *Environmental Performance Reviews: Achievements in OECD Countries*. Paris: OECD.

Peters, B.G. (1998) 'Managing horizontal government: the politics of coordination', *Public Administration* 76, 2: 295–311.

Potter, S. and Skinner, M.J. (2000) 'On transport integration: a contribution to better understanding', *Futures* 32, 3: 275–287.

Pressman, J.L. and Wildavsky, A. (1984) *Implementation*. 2nd edn. Berkeley, CA: University of California Press.

Shannon, M.A. and Schmidt, C.H. (2002) 'Theoretical approaches to understanding intersectoral policy integration', in I. Tikkanen, P. Glück and H. Pajuoja (eds) *Cross-Sectoral Policy Impacts on Forests, EFI Proceedings No. 46*, Joensuu: European Forest Institute. Available online at: http://www.efi.int/portal/virtual_library/publications/proceedings (accessed 30 September 2009).

Stead, D. (2001) 'The European Transport White Paper', *European Journal of Transport and Infrastructure Research* 1, 4: 415–418.

—— (2003) 'Transport and land use planning policy in Britain – really joined up?', *International Social Science Journal* 55, 2: 333–347.

—— (2008) 'Institutional aspects of integrating transport, environment and health policies', *Transport Policy* 15, 3: 139–148.

Stead, D. and de Jong, M. (2006) *Practical Guidance on Institutional Arrangements for Integrated Policy and Decision-Making, United Nations Economic Commission for Europe and World Health Organization Regional Office for Europe Report ECE/AC.21/2006/7 – EUR/06/THEPEPST/7,*

Geneva/Rome: UNECE/WHO-Europe. Available online at: http://www.thepep.org/en/commitee/committee_fourth.htm (accessed 30 September 2009).

Stead, D., Geerlings, H. and Meijers, E. (eds) (2004) *Policy Integration in Practice: The Integration of Land Use Planning, Transport and Environmental Policy-Making in Denmark, England and Germany*. Delft: Delft University Press.

Stead, D. and Meijers, E. (2009) 'Spatial planning and policy integration: concepts, facilitators and inhibitors', *Planning Theory and Practice* 10, 3: 317–332.

Stevens, H. (2004) *Transport Policy in the European Union*, Basingstoke: Palgrave Macmillan.

Underdal, A. (1980) 'Integrated marine policy: What? Why? How?', *Marine Policy* 4, 3: 159–169.

United Kingdom Cabinet Office (2000) *Wiring it Up. Whitehall's Management of Cross-cutting Policies and Services*, London: The Stationery Office. Available online at: http://www.cabinetoffice.gov.uk/strategy/publications.aspx (accessed 30 September 2009).

—— (2006) *The Tools to Deliver Better Regulation. Revising the Regulatory Impact Assessment: A Consultation*, London: Cabinet Office. Available online at: http://www.berr.gov.uk/files/file44543.pdf (accessed 30 September 2009).

United Nations Economic Commission for Europe – UNECE (2003) *Environmental Policy in Transition: Ten Years of UNECE Environmental Performance Reviews*, New York: United Nations.

Vigar, G. and Stead, D. (2003) 'Implementing transport policy in local government', in I. Docherty and J. Shaw (eds) *New Deal for Transport? The UK's Struggle with the Sustainable Transport Agenda*, London: Blackwell, pp. 51–72.

World Commission on Environment and Development – WCED (1987) *Our Common Future*. Oxford: Oxford University Press.

Chapter 3

Planning for sustainable travel

Integrating spatial planning and transport

Robin Hickman, Catherine Seaborn, Peter Headicar and David Banister

1. Introduction

The effective integration of spatial planning and transport has proved difficult to achieve, as the location and form of development does not necessarily promote sustainable travel. In many cases the reverse is true, as transport investment does not effectively support development strategies. But urban structure and mobility are inextricably interrelated. In this chapter the case is argued for much stronger links, as the actual achievement of integration has been limited because of the practical and institutional barriers.

In the UK, a key policy objective has been to reduce the need to travel, especially by car, and this has been central to planning guidance, principally PPG13.[1] The means to achieving this objective has been through improving the choice of travel by different modes of travel, by reducing trip lengths through land use planning decisions, and by limiting the use of the car through a range of demand management methods. Much of this thinking has predated the more recent discussions about sustainable transport, yet the thinking behind PPG13 is still central to the integration of planning and transport in current practice.

The scale of traffic growth in the UK (Figure 3.1) has followed the growth in the economy, and it is difficult to identify the impacts of any policy initiatives. Traffic levels, congestion and carbon dioxide (CO_2) emissions continue to rise in many areas. In most places walking, cycling and bus use are increasing slowly at best, and it is only rail that has experienced a significant growth in usage. Car dependency still remains dominant, and the need to use the car is often 'built in' under current and planned development patterns. This could be interpreted as a failure of policy,

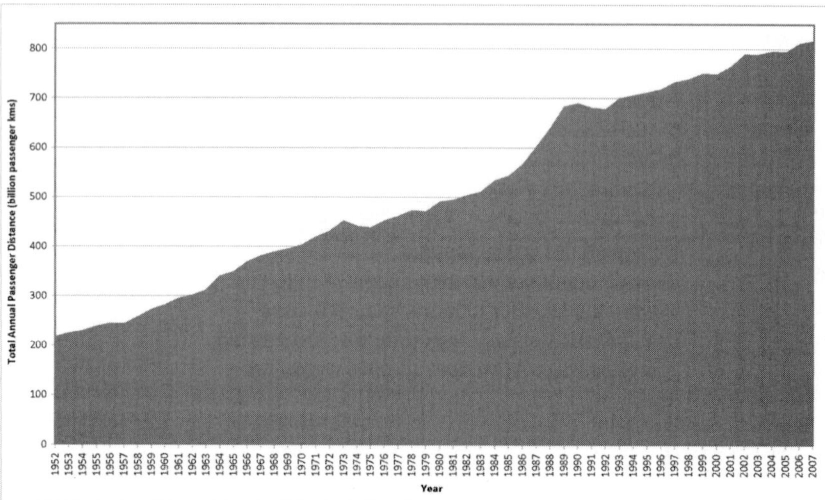

Figure 3.1 Growth in passenger distance travelled in the UK
Source: DfT, Transport Statistics Great Britain, 2008

but the counterfactual is always difficult to describe. The policy dilemma needs to reconcile the role of the car with that of the environment and quality of life in cities. Present policies are not achieving greater sustainability in travel behaviour, as people still want to use their cars wherever possible.

This chapter examines the relationships evident between urban structure and travel. It draws on a study, 'Planning for Sustainable Travel', carried out for the Commission for Integrated Transport (Hideman *et al.*, 2008–09). It seeks to illustrate the important role that spatial planning can play, particularly at the strategic scale, in enabling greater sustainability in travel patterns. It is recognized that integration of planning and transport means that the car needs to be accommodated in the city, and that the distances between activities need to provide the opportunity for choice of travel modes and destinations. Integration is seen as a mixture of modes of transport and not the replacement of one mode. This means that a greater emphasis needs to be placed on exploring the opportunities for reducing average per capita travel by car, so that a greater integration in spatial planning and travel can be achieved. This chapter is structured in five main parts, covering the literature, the drivers for an improved approach, some recent trends, opportunities for policy makers, and conclusions.

A central argument developed is that spatial planning, although important to the generation of travel, only has limited impact in promoting sustainability. A complementary transport planning framework, including traffic demand management measures and parking policies, is also essential. Thus the focus should be on developing a wide range of policy levers, and in combining the spatial planning and transport planning disciplines as an integrated and complementary effort explicitly focused on achieving greater sustainability in transport. Such a 'connected' approach means that the existing organizational and institutional structures may not be the most appropriate for implementation.

It should be noted that the spatial framework is only one of the underlying reasons for travel, and is often overridden by socio-economic factors, preferences and attitudinal characteristics, or choice of activity. Policy measures need to be combined with information, awareness raising, and engagement processes that communicate not only what is being proposed to the user but the rationale behind that implementation. Otherwise there are likely to be unintended consequences and rebound effects. Sustainability as an objective is also multidimensional (including economic, social and economic dimensions), which means that there are different and sometimes conflicting directions. The integration of spatial planning and transport is much more than just getting the right measures to work in the same direction (which may be hard enough), but also requires that the institutional and organizational structures are also complementary, and that the public are engaged and 'own' the approaches being taken.

2. The previous literature

The interrelationships between urban structure and travel are complex and have many dimensions. The location of activities – homes, workplaces, leisure, health, education and other facilities – act as the 'physical structuring framework' for travel. Authors have debated whether spatial planning variables are associated with energy consumption in travel, with much discussion over the existence of relationships (e.g. Newman and Kenworthy, 1989, 1999; Breheny, 1992; Banister *et al.*, 1994; Cervero, 1996; ECOTEC, 1993; Kitamura *et al.*, 1997; Headicar and Curtis, 1998; Boarnet and Crane, 1999; Stead, 2001; Schwanen and Mokhtarian, 2005; Handy *et al.*, 2005; Hickman, 2007; Hickman and Banister, 2007b; and many more). The analysis has become more sophisticated over time, with increased consideration of multivariate relationships and attitudinal and cultural contexts, including a recent focus on self-selection issues (the influence of attitudes on people's location choices and hence travel behaviour). Socio-economic, attitudinal and contextual characteristics all play important roles in the demand for travel, alongside the requirements of the trip. However, the empirical evidence demonstrates that there are significant correlations between spatial planning and travel, even after accounting for wider influences (Ewing and Cervero, 2001; Bohte *et al.*, 2009; Naess, 2009; Cao *et al.*, 2009; and others), but often the direction of the relationships and implied causalities are much less clear.

Figure 3.2 presents some of the factors influencing travel demand from the perspective of urban planning. The current debate concerns the nature and strength of relationships, and assesses how urban and spatial planning can be more effectively used alongside other interventions. This builds upon the sustainable mobility paradigm and the means by which sustainable travel can be integrated into the pattern of urban form (Banister, 2008).

The spatial planning characteristics identified as important to travel demand have been characterized as the '3 Ds', covering Density, Diversity and Design (Cervero and Kockelman, 1997), or as an extended version of '5 Ds', covering Density, Diversity, Design, Distance (to public transport) and Destination Accessibility (Cervero *et al.*, 2009). For example, by increasing density and accessibility to a range of local facilities, it is possible to reduce mobility and the need to travel. As well as

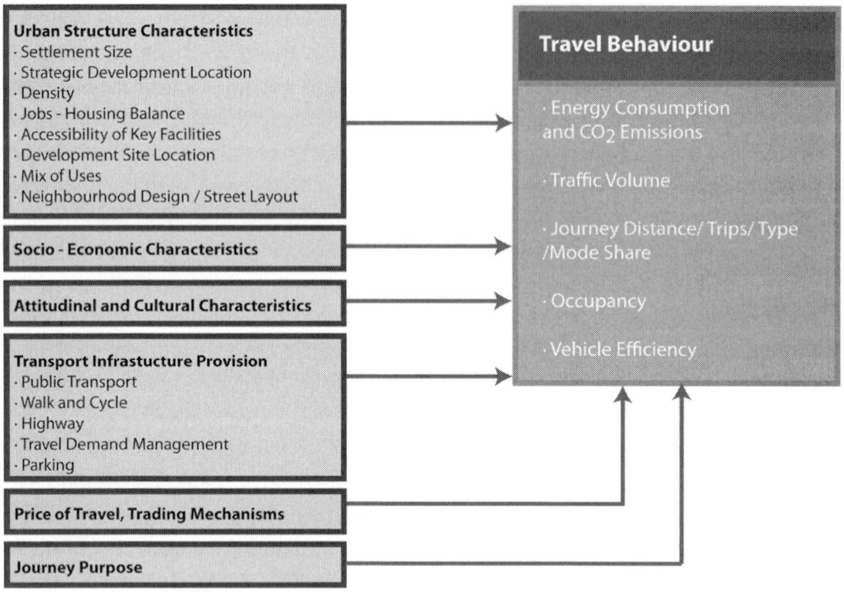

Figure 3.2 Urban structure as an enabler of sustainable travel

having environmental benefits and moving towards sustainable transport, it also promotes greater social inclusion and equity.

In this chapter, the 5 Ds have been extended to cover 11 themes that need to be addressed. At the strategic level, there is discussion concerning settlement size (Theme 1), strategic development location (Theme 2) and the strategic transport network (Theme 3). The 5 Ds cover the more local level of actions, including density (Theme 4); jobs-housing balance (Theme 5); accessibility (Theme 6); and design issues, here interpreted as development site location (Theme 7); mixed use (Theme 8); and neighbourhood design and street layout (Theme 9). The distance characteristic has been extended to include travel demand management (Theme 10)

Table 3.1 Strategic and local themes and the 5 Ds

Strategic themes	*Local themes*	*5 Ds (from Cervero et al., 2009)*
T1: Settlement size	T4: Density T5: Jobs-housing balance T6: Accessibility to key facilities	Density Destination accessibility
T2: Strategic development location	T7: Development site location T8: Mixed use T9: Neighbourhood design and street layout	Design
T3: Strategic transport network	T10: Travel demand management T11: Parking	Distance to public transport Diversity

and parking (Theme 11). These are the two key transport policy options over which planners have considerable control, and the operation of the public transport system is an important element here. The 11 themes map well onto the 5 Ds (Cervero *et al.*, 2009), as shown in Table 3.1. The exception is the diversity heading that seems to be more of a generic characteristic, as it covers all the themes. It is not the intention here to create 'clone settlements' that all seem to be the same, but to create a variety of sustainable settlements that allow for local distinctiveness, character and individuality (Simms *et al.*, 2005). So each particular situation requires a 'unique' solution. The 11 themes, covering three strategic areas and eight more local ones are described in greater detail in Section 5 of this chapter.

3. Drivers for an improved approach

There are four key factors that need to be addressed in integrating planning and transport:

1. The climate change imperative – many countries are beginning to adopt stringent CO_2 reduction strategies. The UK government, for example, has recently adopted an 80 per cent CO_2 emission reduction target by 2050 on 1990 levels (Climate Change Act, 2008). Personal motorized travel currently represents 24 per cent of total UK CO_2 emissions by source and is steadily increasing (BERR, 2008).
2. The long-run availability and price of oil (Strahan, 2007) – an oil dependent transport system is very vulnerable in the medium to long term to volatility and increased prices.
3. The projected growth in traffic and worsening congestion, especially on inter-urban roads (Headicar, 2009) – the UK government's Ten Year Plan acknowledged that increased highway investment would not accommodate traffic demand over the longer term, and the 2004 White Paper proposed a national system of road pricing as a central element in ameliorating the situation. The implementation of road pricing has proved extremely difficult.
4. Improving the accessibility and 'inclusion' among people without the use of a car – reflects the needs of the increasing number of elderly people and those living in less urbanized areas. Expected cuts in public spending during the post-recession era (e.g. for welfare services and subsidized transport) are likely to accentuate these problems.

The objective of 'reducing the need to travel' is already recognized in UK government guidance (PPG13) (DOE, 1994, 1995), but the practical response has been articulated largely in terms of the form and management of individual towns and small developments. Much greater attention needs to be given to the increasing proportion of car mileage which arises from medium and longer distance trips *between* or *wholly outside* urban areas. The growth in longer distance car trips means that more attention should be given to analysing strategic location options in terms of their ability to minimize unnecessary car trips, or in selecting places where the greatest potential exists to improve accessibility by all modes of transport.

Institutionally, this means that a greater strategic emphasis is required in policy analysis and implementation work in the planning and transport fields. This would complement the present micro-level focus on the design of internal layouts (inspired by the publication of *Manual for Streets* (DfT, 2007a)), 'Smarter Choice' behavioural measures (after Cairns *et al.*, 2004) and Transport Assessments of individual development proposals. A more comprehensive approach would aim to reduce the car share of trips between towns, and even the trip distribution, but this would require more fundamental changes in transport and planning policy to create the requisite opportunities and to extend the principles of traffic demand management into the inter-urban sphere (Headicar, 2009). There are a number of measures already employed to encourage modal shift on individual trips, but these need to be set against the high and rising levels of household car ownership. Car availability is the most important single factor contributing to levels of car use (Stead, 2001). It follows that if per capita car use is to be reduced, more attention needs to be given to the combination of development planning, travel planning and traffic demand management policies within a given urban area, which in turn would create an environment in which it is both feasible and attractive to reduce levels of car use. This then needs to be extended to the inter-urban trips where the opportunities for change may be less obvious.

The changes that occur within a single period of the statutory development planning system (15 years or so) are relatively small, when compared with the continuing influence of the established building stock and the patterns of activity and travel associated with it. However, when placed against the background of a longer-term vision, planning can begin to alter the inherited pattern of travel behaviour and establish a changed trajectory for the future. A much stronger forward-looking 'futures' aspect to policy making, building on the recent developments in transport scenario building, would be one way to progress this longer-term, more holistic perspective (Hickman and Banister, 2007a).

4. Some evidence

Two key sets of relationships between urban structure and travel can be identified using the latest National Travel Survey (NTS) (DfT, 2008).[2] Residential population density, settlement size and type, accessibility, and various socio-economic characteristics are all significantly related to travel distance and mode share.

Density – There is an inverse linear relationship between density and travel, where increased density is associated with reduced travel distance, particularly by car (Figure 3.3). The 15–30 and 30–50 persons per hectare (pph) density bands (which embrace most built-up areas) have similar per capita travel, making it difficult to identify the trends within the mid-density ranges.

Car drivers in Great Britain average 3,660 miles per annum (51 per cent mode share) in areas that have an average density of 2.5 pph. In London, a lower average distance travelled by car is evident at 1,876 miles per annum (35 per cent mode share, and a higher average density of 46 pph). The car driver and passenger mode share also reduces markedly, from 83 per cent at densities of 5–15 pph to 44 per cent at densities over 50 pph. Distance and mode share by public transport

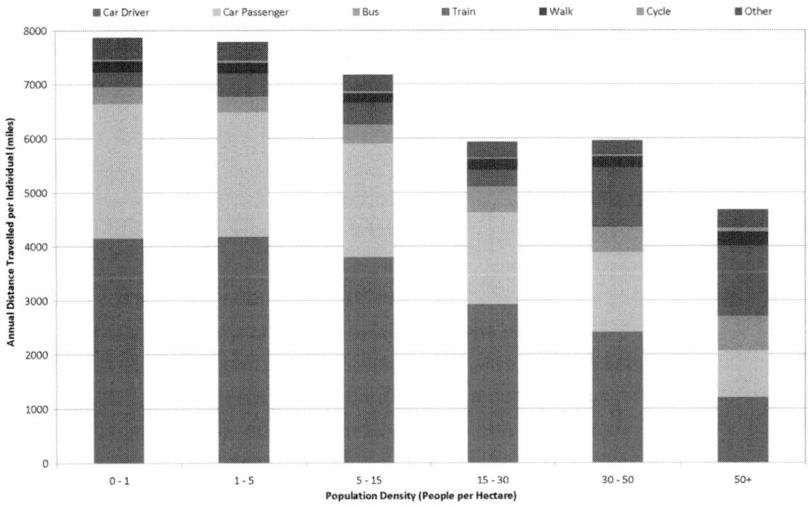

Figure 3.3
Density and travel in the UK
Source: DfT, 2008

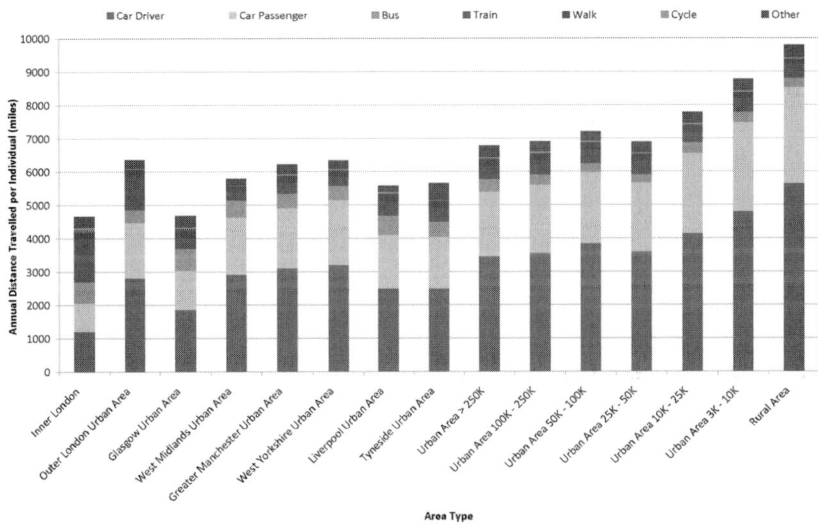

Figure 3.4
Settlement size and travel in the UK
Source: DfT, 2008

increases with density, particularly over 30 pph. Walking distance is similar over most densities but greater in the highest density category of over 50 pph.

Settlement size/area type – Again, there is an inverse linear relationship between increased average distance travelled and settlement size. The largest differential in Great Britain is between inner London (an average of 4,673 miles per person per annum) and rural areas (an average of 9,806 miles per annum). Outer London performs more like the other metropolitan areas in terms of average distance travelled. The highest distances travelled in non-rural areas are found in the smaller urban areas, particularly those with a population under 25,000 (Figure 3.4).

The car driver mode share also increases, so that residents of rural areas have a per capita car driver mileage over 50 per cent higher than the national average. The highest distances travelled in non-rural areas, both in total and as car driver, are found in the smaller urban areas, particularly with a population under 25,000.

5. Opportunities for policy makers

There are many opportunities for effective integration of spatial planning policies with transport decisions. Based on an extensive literature review and interviews with planners, and others working at various levels of government across the UK, the key principles are presented for encouraging more sustainable travel (Hickman *et al.*, 2009) and summarized here under 11 themes.

Theme 1: Settlement size

Settlement size refers to the total population or number of dwellings within a contiguous built-up area (Figure 3.4). Larger settlements provide a greater mix of employment, shops and other services (including specialized services). There is greater likelihood of residents finding jobs and utilizing facilities, or of services drawing their employees and customers, from within the same urban area, leading to the possibility of greater 'self-containment'. This tends to lessen average trip lengths and in particular reduces the need for inter-urban travel, which at present is heavily car based (Newman and Kenworthy, 1999; Naess and Sandberg, 1996; Banister, 1997; Hickman and Banister, 2007b).

Larger settlements are also associated with higher development densities and higher travel volumes on their main corridors. This allows higher levels of public transport service to be operated on a commercial basis, for more journeys to be made by walking and cycling, and this can be coupled with a managed approach to car parking. Policy makers could consider the advantages of locating the majority of new development in or adjacent to the largest urban areas within the region. New development would not replicate the existing distribution of population across urban settlements or reflect recent growth patterns, but it should be planned with an explicit concern over sustainable travel.

The concentration of development in or adjacent to the largest urban areas may not always be desirable, because of housing need in other areas, support for smaller communities and their services, restricted land availability, and environmental or infrastructure constraints. As a general rule, the expansion of larger settlements (at a 25,000 minimum population) is generally preferable to 'leapfrogging' development to smaller towns or 'spreading' development across a number of settlements. A review of Green Belt[3] (or other designations utilized to reduce urban dispersal) policy would be included here.

Theme 2: Strategic development location

Strategic development location refers to the selection of areas for major new development (residential, employment, leisure and retail) between regions and sub-regions, including the spatial distribution of housing and employment within Growth Areas and between Growth Points and other urban centres (Figure 3.5).

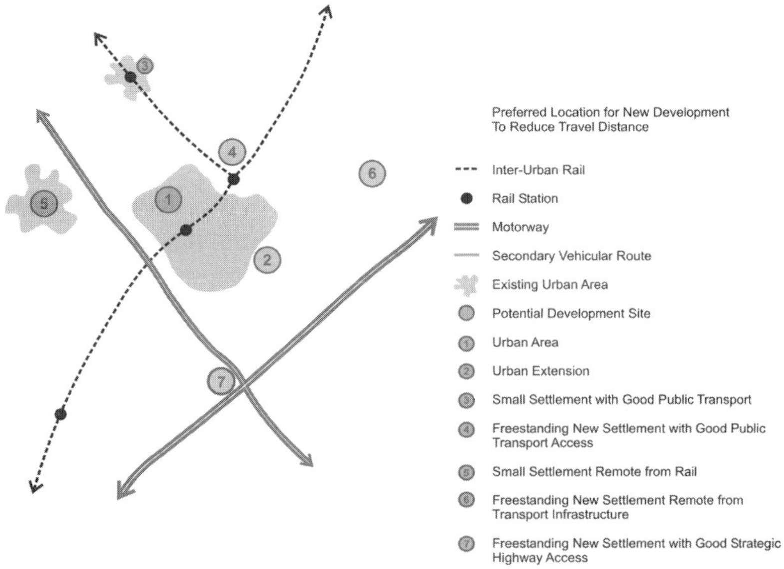

Figure 3.5 Strategic development location and travel

The allocation of planned development (e.g. between local planning authority areas) needs to identify locations where there is a jobs-housing balance and where highly accessible places can be identified (Headicar, 2003; Hickman *et al.*, 2009).

Strategic development location is also relevant for the allocation of development within the areas of individual local planning authorities, e.g. in their Local Development Framework Core Strategies. The aim should be to locate development where travel by car is likely to be low, both in terms of trip length and mode share, and where the use of non-car modes can be promoted, usually in locations where good public transport accessibility is available.

To achieve integrated regional and sub-regional planning, policy makers are advised to locate major employment, retail and leisure uses with a sub-regional catchment, in existing city and town centres, or at other locations which can be accessed conveniently by public transport. Development in non-central locations, close to junctions with motorways and similar dual-carriageway routes should all be avoided, unless they enjoy exceptional public transport accessibility (e.g. a rail 'parkway' station). The aim here is to discourage short and medium-distance travel by car on strategic highways, in particular where new housing results in a 'dormitory community'. Locations for additional housing should also have regard to the proportion of trips likely to be made within the home settlement (i.e. the degree of 'self-containment'), the average distance of trips to places outside the home settlement, and the likely proportion to be made by public transport. Where significant out-commuting is seen as inevitable, new housing should be located in places which already enjoy good public transport accessibility to the relevant external destinations.

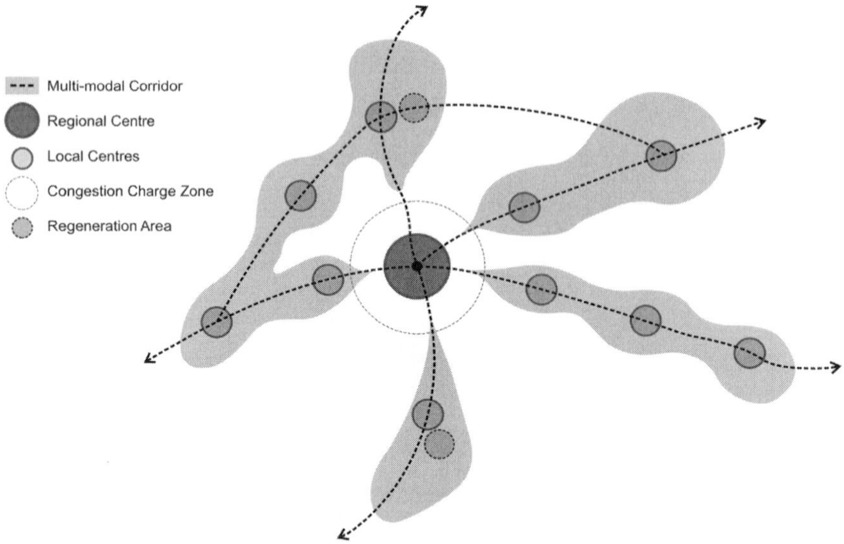

Figure 3.6 Strategic transport network

Theme 3: Strategic transport network

The strategic transport network refers to transport infrastructure that supports medium and long-distance travel, generally between towns and cities or along major corridors in urban areas (Figure 3.6). To increase the sustainability of longer-distance travel generated between settlements, the efficiency and reliability of the existing public transport network (rail and bus) should be improved and with the necessary supporting investment. Development patterns are then created to support public transport usage and discourage the use of the strategic highway network for car trips (Breheny and Rookwood, 1993; Hall and Ward, 1998; Headicar and Curtis, 1998; Hickman *et al.*, 2009).

In regional and sub-regional planning, policy makers are advised to develop sub-regional and city-regional governance structures (e.g. Multi-Area Agreements) that support an effective process for achieving integration in transport and urban planning. Key public transport linkages and networks between cities and towns and within larger conurbations (in collaboration with national government) need to be developed, and major development should be located near to nodes on this network where capacity exists or can be developed.

The efficiency of the strategic transport network can be improved by increasing integration between modes, for example at important urban and edge-of-town interchanges and park-and-ride sites. Public transport infrastructure investments should be prioritized so that they support desired development patterns, and more efficient use of available road capacity through travel demand management measures (Theme 10).

Theme 4: Density

Density refers to the intensity of use of land by different uses. In UK planning practice, housing density is generally measured in dwellings per net hectare (dph),

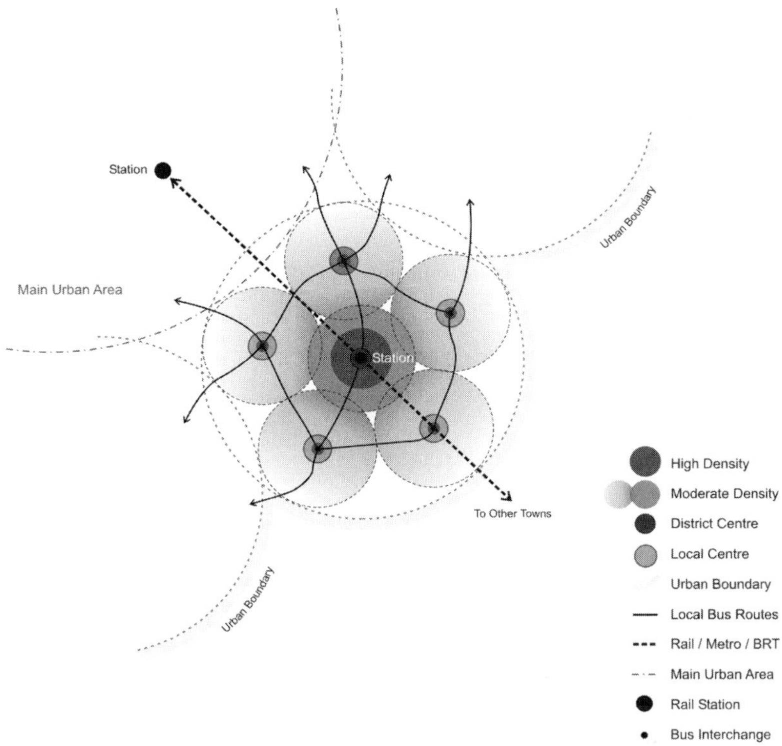

Figure 3.7 Density and travel
Source: Adapted from Rogers, 1997

where the area includes developable residential land.[4] Density is relevant to planning debates at a strategic level for the allocation of additional development land and for framing master plans for major development areas and for negotiations on development applications. Sustainable travel means raising the density of development, particularly around public transport nodes (Figure 3.7). This contributes to lower transport energy consumption and CO_2 emissions, greater scope for viable and attractive public transport services, and reduced car use in terms of both mode share and distance travelled.

There is continuing debate over appropriate density levels. PPS3 (DCLG, 2006) advises a 'working minimum of 30 dwellings per hectare'. Llewelyn Davies (1997, 2001), Rogers (1997) and others have called for the use of higher densities. Much higher densities can be achieved in many areas, up to 50–100 dwellings per hectare (dph), and even 100–200 dph plus around important public transport interchanges. Flexible density standards can be used according to each location, but minimum thresholds should be set at higher rather than lower levels. In regional, sub-regional and local planning, policy makers are advised to consider the interrelationships between public transport accessibility, parking and density in the early stages of strategic planning processes (Integrated Regional Strategies and LDF Core Strategies), including locations beyond town centres and inner urban areas so that suburban areas can be included.

Theme 5: Jobs-housing balance

Jobs-housing balance refers to the relationship of employment opportunities and workforce population within a geographic area, and it is measured in terms of the ratio of jobs per worker. For example, a jobs-housing balance of 1.25 means there are 5 jobs for every 4 workers. Qualitative matching between skills, aspirations and job type is critically important as well as numerical balance (Cervero, 1989, 1996; Ewing, 1995). Quality local employment opportunities will reduce overall commuting distance from residents and also reduce long-distance commuting into an area. Like most of the urban structure variables, it is a necessary but not sufficient condition for reducing travel distances. It is more important at the strategic travel to work area level, or in peripheral and remote urban areas where opportunities for cross-area commuting are less.

Policy makers are therefore advised to consider the different scales over which jobs-housing balance is best achieved. This can initially be set at regional, travel to work and urban area levels. Existing commute patterns, planned residential and employment locations and workforce characteristics should all be examined to ensure that there are limited mismatches which may encourage long commuting distances. Effective jobs-housing balances for urban areas are in the range 1.0–1.5 (Cervero, 1996). Increments of new growth should provide balance at the strategic level. Large employment generators should be located at the most accessible locations by public transport, walking and cycling.

Theme 6: Accessibility to key facilities

Accessibility refers to the ease of reaching destinations or activities.[5] Places that are highly accessible can be reached by many people quickly, whereas inaccessible places can only be reached by a few people in the same amount of time (Figure 3.8). However, there are important differences between the accessibility available by car and non-car modes. Typically, the gap between car and non-car accessibility is least in relation to urban centres and greatest in rural areas and areas of social deprivation (Ewing, 1995; Cervero and Kockelman, 1997).

Accessibility is conventionally perceived in physical travel terms, but social networking is becoming increasingly important. As yet there is little evidence of an aggregate substitution effect (with electronic 'virtual' interaction replacing physical travel), and virtual interaction seems to increase alongside physical travel (e.g. Castells, 1996; Hall and Pain, 2006; Choo *et al.*, 2005).

Key facilities serve a wider catchment than the immediate neighbourhood in which they are situated. For example, some employment centres, shopping centres, hospitals, educational institutions, leisure centres and cultural attractions all have regional impacts. The public transport accessibility of key facilities is therefore of particular importance because they are major travel generators and attractors, and because they determine the level of opportunities available across the social and geographical spectrum. Opportunities can be created for trips to be combined in one round trip to a single destination (i.e. a centre with a mix of uses). Integrated planning means that land use and transport policies can be pursued as a series of coordinated initiatives.

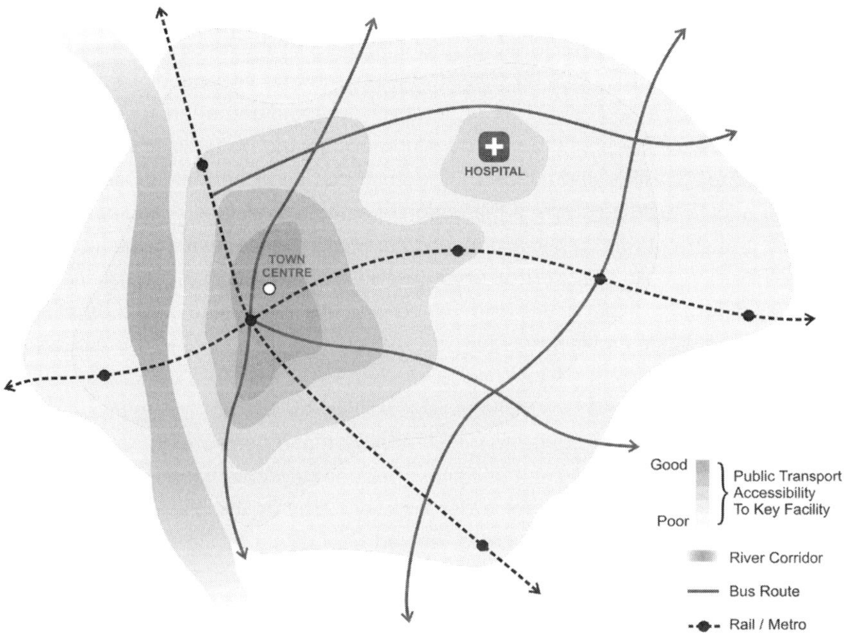

Figure 3.8 Accessibility and travel

Policy makers are therefore advised to locate key facilities within town, suburban and rural centres which relate to the catchment areas of the activities concerned in ways that will ensure high levels of accessibility to homes and workplaces by public transport, as well as by car. Priority needs to be given to established centres before consideration is given to other locations on the public transport network which offer similar levels of accessibility. Developments outside established centres need to include a mixed-use element to facilitate multipurpose trips, travel demand management measures, and controlled parking on site to complement other parking restrictions in the vicinity.

Theme 7: Development site location

Development site location refers to the selection of sites for new housing or other developments. It covers the type of decision that would generally be taken early in the Local Development Framework (LDF) process.[6] These decisions have a major impact on the volume of traffic, the proportion by non-car modes and the accessibility (opportunities) available to different social groups. Development site location is also often a catalyst for transport interventions, helping to make viable new or enhanced transport facilities or to remedy existing traffic or environmental problems. More generally, the utilization and support for existing public transport services and community facilities in the locality is also important. Services or facilities need to be incorporated within the development that will improve accessibility by sustainable modes, including the consideration of car-free or low-car provision housing.

Policy makers are advised to adopt a systematic process of identifying and assessing sites for development which includes noting existing accessibility by car and public transport to/from employment and other key facilities. Examples, for larger towns, include incorporating network links in the layout of development to enable existing urban bus services to be utilized and enhanced. Larger extensions may justify a dedicated bus or light rail service along a radial corridor with priority measures. For small towns, development should be focused on radial corridors in order to utilize and support inter-urban bus services that run along them. Consideration should be given to the option of selective release of land at the edge of larger settlements and in public transport corridors, taking into account the relative accessibility by public transport of alternative locations to jobs and major facilities, the likely difference in per capita car mileage, and the potential to 'swap' other locations for open space provision.

Theme 8: Mixed use

Mixed use refers to the degree to which different land uses are contained within a geographic area, generally at a building, street or neighbourhood level. It may be specified within a masterplan for a development area or the brief for a particular development site. In residential areas, 'mixed use' refers to the provision of local facilities which enable many day-to-day activities to be undertaken on foot. In business areas, 'mixed use development' provides similar local opportunities for employees during or at either end of the working day. Individuals can undertake multiple activities/stops in a single trip and non-motorized trips through a diverse urban environment are promoted (Frank and Pivo, 1994; Cervero and Radisch, 1996).

Policy makers are therefore advised to consider locating essential community facilities (e.g. grocers, local schools, banks) within walking distance of all homes in a neighbourhood in order to reduce travel distances and encourage social inclusion. This will require a certain density of housing in order to concentrate demand sufficiently for the shops and services to be economically viable. In addition, complementary uses should be identified, such as a day-care and fitness centres or bookstores and cafés, and support building types which facilitate co-location, so that individuals can reach more activities per trip. Where public transport is available, retail uses should be promoted that complement employment centres in order to increase public transport mode share.

Theme 9: Neighbourhood design and street layout

Neighbourhood design refers to the scale, form and function of buildings and open space (including streetscapes). Street layout refers to the pattern of local streets, for example as 'traditional' grid networks, cul-de-sacs or hybrid forms (Figure 3.9). Both can have an impact on generated travel patterns. Sustainability objectives move transport planning at this scale beyond a preoccupation with vehicular throughput, to include consideration of transport routes as 'places' as well as 'links' (DfT, 2007; Duany *et al.*, 1992; and for a useful working typology see Jones *et al.*, 2008).

Policy makers are advised to encourage walking, cycling and public transport use through permeable, well-connected, 'traditional' grid street networks. Design should avoid circuitous, 'surburban', cul-de-sac street networks with few access points and lengthy routes to nearby locations. In new developments, safe

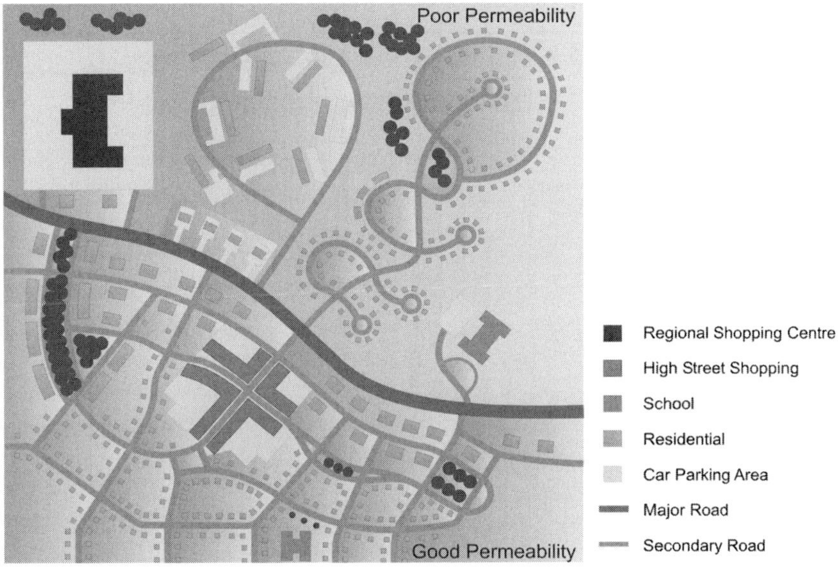

Figure 3.9 Street Layout and travel
Source: Adapted from Duany et al., *1992*
Note: The image contrasts poor permeability to the north of the road and better permeability to the south. The original diagram has been amended to provide linear high street style shopping to the south (originally mall style) and a more integral school location.

and high-quality walking and cycling environments should be provided throughout. In existing developments, consider retrofitting footpaths and adding cycle lanes to improve the travel experience of walkers and cyclists. Sustainable modes can be given priority in terms of journey length and time, and this is sometimes known as 'filtered permeability' (Melia, 2007). The integration between new development and adjacent built-up areas needs to be assured in terms of street network, public transport services, footpaths/cycle routes and design standards, especially where the links involved are critical to delivering good local accessibility. Many recent good practice design advice and resources are available, including the *Manual for Streets* (DfT, 2007) and the *Urban Design Compendium 2* (Roger Evans Associates, 2007). CABE (2007) summarizes a number of good practice streets and spaces in the UK.

Theme 10: Travel demand management

Travel demand management (TDM, and sometimes known as 'mobility management') covers a wide range of measures aimed at reducing car use and its adverse impacts (Litman, 2009). They complement the more traditional development components of spatial planning in promoting sustainable travel, and some of the organizational initiatives involve behavioural measures (or 'smarter choices' (Cairns *et al.*, 2004)). Possible TDM interventions can be listed under three main categories (Table 3.2), but there is not space here to discuss each of these separately (Ison and Rye, 2008).

Policy makers are advised to strengthen the land use component of spatial strategies at all levels, with the development of a rigorous TDM strategy which sets out to enhance the overall sustainability. These strategies should be implemented in an integrated manner in order to maximize their effectiveness, thereby securing best overall value for money and creating travel environments in

Table 3.2 TDM interventions

Organizational and operational	*Financial*	*Infrastructure*
• Travel plans (workplace, school, residential, area-wide). • Personalized travel planning. • Car pooling, car sharing and car clubs. • Company work hours, flexi-working, home working. • Home retailing and delivery. • Tele-activities and interaction. • Marketing/media campaigns. • Transport optimization, peak congestion avoidance. • Slower speeds and ecological driving styles.	• Pricing regimes, including (where applicable) road user cordon charging, area-licensing schemes, continuous charging. • Vehicle ownership taxes. • Public transport investment/subsidy. • Parking charges. • Bicycle investment/ subsidy.	• Improved public transport facilities, including (where applicable) National Rail, Underground, light and ultra light rapid transit, guided bus and bus, etc. • Demand responsive transport. • Park and ride. • Improved walking and cycling facilities. • Road space re-allocation and priority, traffic calming, access control and restrictions. • Streetscape design. • Parking.

which more sustainable choices can be made. The UK Travel Demonstration Towns pilot illustrates the type of integrated strategy that may be effective in reducing car-based travel (DfT, 2009). Such packages of interventions should be carried out across all major urban areas and new developments.

Theme 11: Parking

Parking refers to the amount of space planned for the storage of cars and other vehicles (on- and off-street) in new development and to the management of space in existing and new developments. It also includes provision for two-wheelers (powered) and bicycles. Because people do not necessarily park at or within their destination site, it is necessary to consider development related policies in the context of local parking conditions more generally and of policies for the provision and management of on-street space and publicly available car parks (English Partnerships, 2008).

Parking policy is a central element in TDM, and should therefore feature in spatial strategies at all levels. However it is much under-utilized as a TDM to encourage less car use in order to improve traffic and environmental conditions in an area and to contribute to broader transport and sustainable development objectives. Controlling parking through restriction of spaces and/or pricing typically complements a variety of measures designed to promote the use of non-car alternatives, and it can be linked to giving priority to low-emission vehicles. Both the amount of parking space and the form in which it is provided (i.e. within the curtilage of private developments, in allocated or unallocated off-street spaces, and in on-street bays) have implications for the wider issues of neighbourhood design and street layout. It can also be seen as an integral part of the design of 'car-free' and 'low-car' housing developments.

Policy makers are advised to apply separate considerations to the main categories of development (business, retail and leisure and residential). These should be interpreted in relation to local density and public transport accessibility conditions, and provide adequate access and temporary parking for delivery and service vehicles. Accommodation for future charging sites for electric or other alternatively fuelled vehicles should also be considered. It is important to coordinate policies across local authority boundaries, and within and between settlements, so that local action is consistent with regional and sub-regional TDM strategies. In the absence of a national strategy for comprehensive road pricing, parking policies at both home and non-home trip ends remain a key tool in managing the demand for travel.

6. Conclusions

Spatial planning can be used more effectively in helping to achieve sustainable travel in a more integrated way, particularly if a range of policy levers are utilized. The socio-demographic dimension to travel behaviour is central to decisions made by individuals and businesses, and the role of attitudes to certain types of travel and living circumstances is now better understood. However the pattern of settlement, the distribution of land uses and physical structure of urban areas all appear to remain important in setting the 'envelope for travel'. Urban planning (including strategic and local forward planning, masterplanning and development control) is one of the key factors influencing travel behaviour at all levels of policy intervention.

'Integrating land use and transport planning' is often put forward as a policy objective, but relatively little follows in practice at the strategic level (which most affects the volume and mode of travel). Traditional tenets of preventing urban sprawl remain in place, reinforced by more recent initiatives to resist car-dependent developments, utilizing previously developed (brownfield) sites and raising housing densities. But too often new development is spread between towns in an area, following the historical settlement pattern or more recent shifts in the residential population. Decisions made on the location of all types of new development have a key impact on the numbers of trips, the modes used and the distances travelled. Too often the lack of integration results in more, not less travel.

Planning and managing development in relation to a range of variables, including settlement size; development location; the transport network; density, jobs-housing balance; accessibility; mixed uses; site location; neighbourhood design and street layout; traffic demand management and parking (our suggested 11 themes), can help in moving towards greater sustainability in travel lifestyles. But the key question here is whether a clearer understanding of the relationships is sufficient to provide integration. There is also a need to change the culture of dependence on transport in all its forms, so that the total amount of travel is reduced. A key element here may be the concept of 'unnecessary mobility', where some trips can be reduced in volume relative to others which have more value. In a carbon and resource constrained world, more understanding is required of the types of trips that can be reduced. The various responsibilities for a changed approach also require some further thought. Integration means moving from cooperation and consultation to working more effectively together across disciplines, towards clearly defined objectives (e.g. sustainable transport), and this requires support from government at all levels, and from businesses and local communities. The problem is not only physical, but one that addresses social and cultural values, and motivations for travel.

Although the approach taken in this chapter is rather prescriptive in terms of what can be done within the existing decision frameworks, the messages are clear. Too often in the past the different traditions of transport planners and those of urban and regional planners have followed different and often conflicting paths. Now, more than ever, it is important to bring the professions and traditions together so that they can work more effectively in a coherent manner to achieve sustainable transport and land use systems. A very important part of this is in procedural and institutional practice, as multidisciplinary working is critical to the better integration of the planning and transport disciplines.

Recent history has demonstrated that the transition to achieving sustainable travel is very difficult, certainly beyond the limited successful experience in some urban areas. The debate over relationships between spatial planning and travel will continue to develop, with increasing sophistication. Often it is not the knowledge and research that is lacking, but the commitment to effective application of this knowledge in practice.

Acknowledgements

Many thanks to the Commission for Integrated Transport for funding the study on which this chapter is based, and also members of the CfIT Project Management Group and the wider CfIT Working Group for very useful comments throughout the project. Special thanks to the local authority practitioners and members interviewed as the basis for the case study material in the main study. Also the wider study team who have helped develop the work – from Halcrow, Oxford Brookes University and University of Oxford (Transport Studies Unit). The views expressed in this chapter are, of course, from the authors and do not necessarily reflect those of CfIT or any of the local authorities or practitioners interviewed.

Notes

1 In the UK there are a series of Planning Policy Guidance (PPG) notes and more recently Planning Policy Statements (PPS) that are intended to interpret policy legislation into action. PPG13 on transport was pioneering when it was first published in 1994, as it promoted the planning of cities and urban areas with an explicit aim of reducing the reliance on the use of the car.

2 The use of the NTS is limited in this chapter to bivariate considerations of density and settlement type and distance travelled. More detailed analysis, including multivariate analysis, is included in the background technical report to the main study (Hickman *et al.*, 2009).

3 The Green Belt is a strong land use policy that has been applied in the UK over many years, primarily to maintain open and accessible green space around cities, but it has also acted as a constraint on urban peripheral development.

4 Other metrics can be used. For example, gross density includes all land (i.e. including major roads, parks, service facilities, etc.) and is often measured in terms of dwellings per hectare or persons

per square kilometre. Gross density is useful for comparing densities across larger areas and for estimating potential public transport demand. Habitable room densities allow different house types to be accommodated. The more recent research is beginning to develop more effective measures of density and quality, e.g. number of useful facilities per area, such as bookstores or coffee shops.

5 Accessibility for persons with disabilities is also an important issue, and is covered in various guidance notes; a broader definition is taken here in terms of accessibility to destinations.

6 Local Development Frameworks are developed by local authorities in England and Wales setting out the development strategy for the local area. They were introduced by the Planning and Compulsory Purchase Act, 2004.

References

Banister, D. (1997) The theory behind the integration of land use and transport planning. London: Waterfront Conference.

—— (2008) The sustainable mobility paradigm. *Transport Policy*, 15: 73–80.

Banister, D. and Hickman, R. (2006) How to design a more sustainable and fairer built environment: transport and communications. *IEEE Proceedings of the Intelligent Transport System*, 153 (4): 276–291.

Banister, D., Watson, S. and Wood, C. (1994) The relationship between energy use in transport and urban form. Working Paper 12. The Bartlett School of Planning, University College London.

Boarnet, M.G. and Crane, R. (1999) *Travel by Design: The Influence of Urban Form on Travel*. New York: Oxford University Press.

Bohte, W., Maat, K. and van Wee, B. (2009) Measuring attitudes in research on residential self-selection and travel behaviour: a review of theories and empirical research. *Transport Reviews*, 29 (3): 325–357.

Breheny, M. (1992) The contradictions of the compact city: a review. In Breheny, M. (ed.) *Sustainable Development and Urban Form*. London: Pion, pp. 138–159.

Breheny, M. and Rookwood, R. (1993) Planning the sustainable city region. In Blowers, A. (ed.) *Planning for a Sustainable Environment*. London: Earthscan, pp. 150–189.

Cairns, S., Sloman, L., Newson, C., Anable, J., Kirkbride, A. and Goodwin, P. (2004) *Smarter Choices: Changing the Way We Travel*. London: DfT.

Cao, X., Mokhtarian, P. and Handy, S. (2009) Examining the impact of residential self selection on travel behaviour. A focus on empirical findings. *Transport Reviews*, 29 (3): 359–396.

Castells, M. (1996) *The Information Age: Economy, Society and Culture. Volume 1: The Rise of the Network Society*. Oxford: Blackwell.

Cervero, R. (1989) Jobs-housing balancing and regional mobility. *Journal of the American Planning Association*, 55 (2): 136–150.

—— (1996) Jobs-housing balancing revisited. *Journal of the American Planning Association*, 62 (4): 492–511.

Cervero, R. and Kockelman, K. (1997) Travel demand and the 3Ds: density, diversity and design. *Transportation Research Part D*, 2 (3): 199–219.

Cervero, R. and Radisch, C. (1996) Travel choices in pedestrian versus automobile oriented neighbourhoods. *Transport Policy*, 3: 127–141.

Cervero, R., Sarmiento, O., Jacoby, E., Gomez, L. and Neiman, A. (2009) Influences of built environments on walking and cycling: lessons from Bogotá. *International Journal of Sustainable Transportation*, 3 (4): 203–226.

Choo, S., Mokhtarian, P. and Salomon, I. (2005) Does telecommuting reduce vehicle-miles travelled? An aggregate time series analysis for the U.S. *Transportation*, 32 (1): 37–64.

Commission for Architecture and the Built Environment (CABE) (2007) *This Way to Better Streets*. London: CABE.

Department for Communities and Local Government (DCLG) (2006) *PPS3 Housing*. London: Stationery Office.
Department for Transport (DfT) (2007) *Manual for Streets*. London: Stationery Office.
—— (DfT) (2008) National Travel Survey, 2002–06 data. London: DfT.
—— (DfT) (2008) Transport Statistics Great Britain. London: DfT.
—— (2009) Sustainable travel demonstration towns. Available online at: http://www.dft.gov.uk/pgr/sustainable/demonstrationtowns/sustainabletraveldemonstrati5772 (accessed October 2009).
Department of Business, Enterprise and Regulatory Reform (BERR) (2008, March) *Energy Trends. A National Statistics Publication*, London.
Department of the Environment (DOE) (1994) *PPG13: Transport*. London: HMSO.
—— (1995) *PPG13 A Guide to Better Practice. Reducing the Need to Travel through Land Use and Transport Planning*. London: HMSO.
Department of Environment, Transport and the Regions (DETR) (2001) *Planning Policy Guidance Note 13: Transport (Revised)*. PPG 13. London: HMSO.
Duany, A., Plater-Zyberk, E. and Speck, J. (1992) *Suburban Nation: The Rise of Sprawl and the Decline of the American Dream*. New York: North Point Press.
ECOTEC (1993) *Reducing Transport Emissions Through Planning*. Department of Environment. London: HMSO.
English Partnerships (2008) *Parking. What Works Where?* London: EP.
Ewing, R. (1995) *Best Development Practices: Doing the Right Thing and Making Money at the Same Time*. Chicago: Planners Press.
Ewing, R. and Cervero, R. (2001) Travel and the built environment: a synthesis. *Transportation Research Record*, 1780: 1–3515.
Frank, L.D. and Pivo, P. (1994) Impacts of mixed use and density on utilization of three modes of travel: single-occupant vehicle, transit, and walking. *Transportation Research Record*, 1466: 44–52.
Hall, P. and Pain, K. (2006) *The Polycentric Metropolis. Learning from Mega-City Regions in Europe*. London: Earthscan.
Hall, P. and Ward, C. (1998) *Sociable Cities. The Legacy of Ebenezer Howard*. Chichester: Wiley.
Handy, S., Cao, X. and Mokhtarian, P. (2005) Correlation or causality between the built environment and travel behaviour? Evidence from Northern California. *Transportation Research Part D*, 10 (6): 427–444.
Headicar, P. (2003) Land use planning and the management of transport demand. In Hine, J. and Preston, J. (eds) *Integrated Futures and Transport Choices*. Aldershot: Ashgate.
—— (2009) *Transport Policy and Planning in Great Britain*. Abingdon: Routledge.
Headicar, P. and Curtis, C. (1998) The location of new residential development: its influence on car-based travel. In Banister, D. (ed.) *Transport Policy and the Environment*. London: Spon.
Hickman, R. (2007) Reducing travel by design. A micro analysis of new housing location and the commute to work in Surrey. Unpublished PhD. London: UCL.
Hickman, R. and Banister, D. (2007a) Looking over the horizon: transport and reduced CO_2 emissions in the UK by 2030. *Transport Policy*, 14: 377–387.
—— (2007b) Transport and reduced energy consumption: what role can urban planning play? Transport Studies Unit (Working Paper, Ref. 1026). Oxford: Oxford University Centre for the Environment.
Hickman, R., Seaborn, C., Headicar, P., Ashiru, O., Saxena, S., Banister, D. and Pharoah, T. (2009) *Planning for Sustainable Travel*. Summary guide, web-based guidance and background technical report. London: Halcrow Group for CfIT. Details at http://www.plan4sustainabletravel.org
Ison, S. and Rye, T. (2008) (eds) *The Implementation and Effectiveness of Transport Demand Management Measures: An International Perspective*. Aldershot: Ashgate.
Jones, P., Boujenko, N. and Marshall, S. (2007) *Link and Place: A Guide to Street Planning and Design*. London: Landor.
Kitamura, R., Mokhtarian, P. and Laidet, L. (1997) A micro-analysis of land use and travel in five neighbourhoods in the San Francisco Bay area. *Transportation*, 24: 125–158.

Llewelyn Davies (1997) *Sustainable Residential Quality. New Approaches in Urban Living*. London: Llewelyn Davies.

—— (2001) *Better Places to Live: By Design. A Companion Guide to PPG3.* London: DTLR and CABE.

Litman, T. (2009) *Online TDM Encyclopedia*, Canada: Victoria Transport Policy Institute. Available online at: http://www.vtpi.org/tdm/ (accessed October 2009).

Melia, S. (2007) Eco-town mobility. *Town and Country Planning.* November. London: TCPA.

Naess, P. (2009) Residential self selection and appropriate control variables in land use: travel studies. *Transport Reviews*, 29 (3): 293–324.

Naess, P. and Sandberg, S.L. (1996) Workplace location, modal split and energy use for commute trips. *Urban Studies*, 33 (3): 557–580.

Newman, P.W.G. and Kenworthy, J.R. (1989) *Cities and Automobile Dependence. An International Sourcebook.* Aldershot: Gower.

—— (1999) *Sustainability and Cities: Overcoming Automobile Dependence.* California: Island Press.

Roger Evans Associates (2007) *Urban Design Compendium 2.* REA for English Partnerships. London.

Rogers, R. (1997) *Cities for a Small Planet.* London: Faber.

Schwanen, T. and Mokhtarian, P. (2005) What affects commute mode choice: neighbourhood physical structure or preferences towards neighbourhoods? *Journal of Transport Geography*, 13 (1): 83–99.

Simms, A., Kjell, P. and Potts, R. (2005) *Clone Town Britain: The Survey Results on the Bland State of the Nation.* London: New Economics Foundation. Available online at: http://www.neweconomics.org/publications/clone-town-brita n (accessed 24 February 2010).

Stead, D. (2001) Relationships between land use, socio-economic factors and travel patterns in Britain. *Environment and Planning B: Planning and Design*, 28: 499–528.

Strahan, D. (2007) *The Last Oil Shock.* London: John Murray.

Chapter 4

The need for integrated institutions and organizations in transport policy

The case of transport and climate change

Karen Anderton

1. Introduction

Institutional considerations should not be underestimated in the pursuit of integrating transport. As was noted in Chapter 2 of this volume, in order to achieve policy integration, the institutional basis for intersectoral or intergovernmental working is crucial. The primary issue under examination here is the interplay between organizations and institutions. In Chapter 17, we are presented with an example of how the disconnect between responsibilities and levels of government played a role in the failure of achieving integrated transport in a city in Israel; and in Chapter 18, we see how the complexities of existing organizational structures are problematic to better integrate public transport in France.

From these examples and from the evidence discussed in this chapter, it is clear that there is a need for integration at the institutional level to achieve certain policy goals. Here, the challenge posed by climate change is used to highlight this need and to demonstrate exactly why integration – of disparate organizations and the institutions they deliver – is fundamental to achieving integrated transport and in this case, reducing emissions.

A successful response to climate change requires joined-up thinking and collaboration between otherwise disparate entities, as can be seen when confronting transport-related emissions. While institutions are being introduced to tackle the climate impact of our mobility, conventional forms of decision making may be inhibiting the progress of these measures in reducing emissions. This chapter argues that in order to effectively address the immense challenge posed by transport-related greenhouse gas (GHG) emissions, the institutions which govern transport infrastructures, and those responsible for commanding climate change mitigation also need to be integrated.

As a case in point, this chapter will examine California's Senate Bill 375 (SB 375), passed in October 2008 – an institution designed to integrate land use and transportation funding in order to reduce GHG emissions. It will discuss some of the challenges and barriers, institutional in nature, which need to be overcome in order for the Bill to be successfully implemented and highlight how existing bureaucratic structures may prevent this positive intervention to address climate change delivering on its aims and objectives.

Before looking specifically at SB 375, the chapter will first clearly define 'institutions' and 'integration', as used here, and identify the types of barriers which stand to challenge policy implementation. Finally, lessons taken from California's experiences will be considered in the broader context and further research needs identified.

2. Institutions

The institutional factors of integration cover the following elements (see also Chapter 2 in this volume):

- The provision of capacities to cope with the emerging issues and duties
- The definition of a governmental organization to ensure integration
- The exchange of information and possibilities of interventions between different sectors
- The definition of leading and participating agencies and their respective duties and responsibilities.

(Drawn from Stead, this volume, adapted from Eggenberger and Partidário, 2000)

This is a useful typology and will be returned to later, but it is important to note that the scope covered by the term 'institution' is itself contentious, and thus should be closely examined and considered in the context of this chapter.

March and Olsen (1989) define an institution as an 'established law, custom, usage, practice, organization or other element in the political or social life of a people'. North's (1990) later definition, however, views institutions as 'the socially devised constraints that shape human interaction'. Importantly, the distinction made by North omits reference to organizations in the scope of the term. Given that this is the later and most commonly held-to reference definition, North's understanding will be used in this chapter: institutions are the structures through which organizations operate. While not covered in this definition, organizations are nonetheless a central consideration, as in this context.

Rietveld and Stough's (2004–2006) collective work has articulated and refined the examination of institutional dimensions of sustainable transport. Given that the parameters set around achieving 'sustainable transport' have many synonymous characteristics to those specifically needed to reduce GHG emissions from the transport sector, utilizing Rietveld and Stough's understanding of the distinctions surrounding the relationships between institutions and organizations is extremely useful to set the contextual constraints here. They endorse North's idea that institutions are distinct from organizations and elaborate as to their relationship: 'organizations are groups of actors that share a common interest or goal; institutions structure and define the relationships between actors and organizations'. In this chapter *how* organizations interact, when the institution *is* their common goal is the element under scrutiny.

According to Foxon (2002), institutions include formal constraints, such as legislation, economic rules and contracts, and informal constraints, such as social conventions and codes of behaviour. However, in drawing on the work of Williamson (1994) in their understanding of four distinct types of institution: formal; informal; governance; and resource allocation (Table 4.1), Rietveld and Stough (2005) assert that this framework provides means for examining policy arenas as well as relationships between different arenas and thus is a way to identify and understand the forces that are guiding action and behaviour in specific transport contexts.

Williamson's typology is useful here to elaborate on the complexities surrounding the current mismatch between transport and environmental institutions. Simply articulated, transport institutions have typically been more formal, while climate change has generally been dealt with through informal and governance institutions. As Hansen (2006) explains, they respectively represent both the 'old' and 'new' spheres of politics. While the environmental, particularly climate change, sector is synonymous with diverse forms of policy development (informal/governance/resource allocation institutions), transport has much more traditional and well-established systems of control (formal institutions). This idea shall be explored in more depth in the case study later in this chapter (p. 64).

National governments have traditionally favoured non-binding bilateral or multilateral agreements (informal/governance institutions) usually deliberated on an international stage to deal with climate change. Formal climate change institutions, such as binding legislation, are a relatively recent occurrence and something that has been pioneered by lower echelons of government (state/municipal). Indeed, only very recently have mandatory measures been adopted at the national level,

Table 4.1 Typology of institutions

Types of institution	*Examples*
Informal	Values, norms, customs, practices, traditions
Formal	Codified statutes, conventional provisions, laws, regulations and high-level administrative orders
Governance	Rules (minor laws, administrative orders, regulations and policy directives)
Resource allocation	Diverse actions and behaviour patterns of multiple actors in the decision environment

Source: Adapted from Williamson, 1994; as cited in Rietveld and Stough, 2005

as seen by the introduction of the UK Climate Change Bill, which became law in November 2008. This was however, preceded by numerous examples of earlier compulsory climate change institutions enacted by sub-national governments, such as the South Australian Climate Change and Greenhouse Emissions Reduction Act 2007, and Assembly Bill (AB 32) – California's 2006 Global Warming Solutions Act, a pre-cursor to SB 375 and a number of subsequent pieces of legislation related to its delivery. So we can see a shift towards more formalized approaches to climate change, but these responses have ultimately been pioneered independently of the top-down system of government. Indeed, several authors note the importance of local, sub-national governments in delivering a more sustainable transport system (Banister *et al.*, 2007; Rietveld and Stough, 2004).

This formalization helps to break down the barriers of institutional incompatibility that addressing transport-related climate change has previously faced. Transport organizations (departments and agencies) are now facing legislation – formal institutions to address climate change, rather than just calls for action to be taken – they are now obligated to respond.

This is a positive step, but to implement them, traditional areas of responsibility and jurisdiction may need to be rethought and redefined, and new organizations may ultimately be needed to address these new challenges (Sachs, 2009). Moreover, it requires that in the short term, the existing organizations responsible for implementing these now mandatory measures acknowledge that they may also need to evolve and integrate climate change into their operations in more than just a superficial manner. This is the codification of climate change. Yet in order to appreciate what this means in practice, it is imperative to define 'integration' in this context.

3. Integration

Literally, 'integration' can be described as 'an act or instance of combining into an integral whole' (MLA, 2009). Practically, what this means for governance, at least in this sense, is the 'de-silo-ing' of departments/agencies responsible for specific issues in order to make policy development and implementation more collaborative and inclusive. Importantly, it may also incorporate sectors outside of the public domain and their interests/role in delivery of institutions. The following definition of integrated transport was coined in 2003, relating to urban transport:

> The organizational process through which the planning and delivery of elements of the transport system are brought together, across modes, sectors, operators and institutions, with the aim of increasing economic and social benefits.
>
> (NEA *et al.*, 2003)

While this is a useful definition to conceptualize the organizational parameters around which integration occurs, it refers solely to economic and social benefits as the key drivers for integrating transport. However, in order for transport-related climate change institutions to work, increasing environmental benefits also need

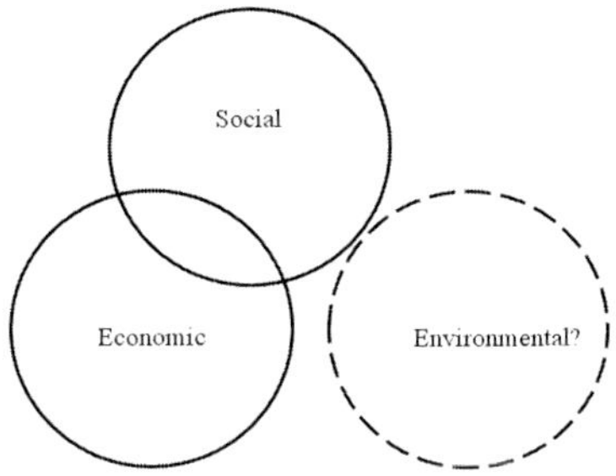

Figure 4.1 Aims of integrated transport *Source: After NEA, OGM and TSU, 2003*

to be seen as a driver for integration and changes to organizational processes (Figure 4.1).

Barriers to integration

Rietveld and Stough (2005) stated that getting beyond the decision to adopt, deploy or use was the primary bottleneck. However, it is important to examine the many other factors that stand in the way of institutional integration, once the decision has been made.

Fundamental to implementing effective institutions is the degree to which the systems of governance and the organizations responsible for implementation are suitable for the task in hand. Although there will always be unforeseen problems in the development and implementation of institutions, which need to be overcome (Bardach, 1977), there is always scope to ensure that the responsibility lies with organizations that are appropriate and have sufficient capacity to undertake the task in hand. May and Crass (2007) acknowledged the importance of such challenges in terms of changing behaviour in passenger transport use and subsequently developed a comprehensive list of institutional barriers (Table 4.2)

This list is by no means all-encompassing, since it does not speak to the complexity of addressing multiple potentially conflicting policy agendas nor does it acknowledge that there may be a need to support legislation in and of itself. Nonetheless, May and Crass provide a useful starting point to identify potential shortcomings in how institutions are implemented and offer logical insight into overcoming these barriers.

By taking into consideration the conceptual understanding of institutions, integration, and the typology of institutional barriers identified here, the following case study will go some way to highlighting that the organizational structures through which institutions are developed may be a significant factor in their success.

4. Case study: California's SB 375

Background – Why climate change?

Global GHG emissions totalled almost 27 billion tonnes in 2004, equating to a 28 per cent rise on 1990 levels (IEA, 2006). In 2007, the Intergovernmental Panel on Climate Change (IPCC) released its Fourth Assessment Report which suggested that in order to limit changes in global temperature to 2–2.4 °C above the pre-industrial average,

Table 4.2 The ECMT barriers to implementation

Institutional barriers	*Overcoming those barriers requires*
Split or duplicated responsibility	**More coordination between the tiers of government, and between agencies at each level**
Process	**Consistency in planning over the long term**
Identifying objectives, specifying problems; selecting possible solutions, appraisal, implementation	**A problem-led approach to developing solutions and strategies**
Political and public acceptability	Political champions and more positive involvement of the public and media
Information and skills	More effective use of data, models and appraisal methods
Financial	Financial support for strategies, without inducing policy bias
Legislative and regulatory	Legislation and regulations to support these requirements

Source: May and Crass, 2007

Note: The bold sections are particularly relevant here.

GHG emissions need to peak before 2015, with 50–85 per cent reductions on 2000 levels by 2050 (IPCC, 2007). Current debates are now beginning to speak of the likelihood of a 4 °C rise before the end of the century and our options to cope with this scenario in the run-up to the December 2009 Copenhagen UN Climate Change Conference (Hadley Centre, 2009).

In 2004, the transport sector produced 6 billion tonnes of carbon dioxide (CO_2), 23 per cent of the world's total energy-related emissions, with road transport accounting for 74 per cent of this figure, a significant share of which coming from passenger vehicles. The transport sector's growth rate is the highest among the end-user sectors (IPCC, 2007). According to the Organisation for Economic Co-operation and Development (OECD), global CO_2 emissions from transport are expected to double by 2050 (OECD, 2008), with projections suggesting that road passenger vehicles will remain the primary source of transport-related GHG emissions (WBCSD, 2004).

Historically, conventional transport agencies and the institutions they have overseen were responsible for governing issues very different to those of decarbonizing the infrastructure, reducing vehicle miles travelled (VMT) and encouraging alternatives to the fossil fuel-powered car. Yet these are exactly the challenges they face today and addressing these issues will question the way things are done on every level and across all scales.

Indeed, it could be said that climate change in the broader context is questioning established notions of government. As a fluid, overarching problem, which transcends boundaries, climate change will ultimately test the limitations of rigid hierarchical governments and allow us to examine the potential strengths of more inclusive, dynamic forms of governance.

Yet whether it is possible to integrate management of such disparate interests as transport infrastructure (which is currently not managed solely from one level of government) on the one hand and GHG reduction on the other is little investigated or understood. How the organizations responsible for these distinct areas will work to implement the institutions designed to address climate change is an issue, for the most part, still in discussion. Indeed, although much research has been carried out to assess differences in policy outcomes, relatively little effort has gone into examining the processes of policy implementation, and the factors contributing to success or failure (Hössinger *et al.*, 2003). California has passed legislation to address climate change and must now focus on how the institutions it has created will be delivered.

Background – California

California is a state synonymous with the automobile. It also has a history of leadership on environmental issues dating back nearly 40 years, to when the state established standards more stringent than those set by the federal government in the 1970 Clean Air Act (CAA). This was mainly due to the immense impact that automobiles were having on the state's air quality and since then California has pushed the boundaries on environmental action.

Unique in the sense that California's Air Resources Board (ARB) preceded the CAA, the state was afforded the right to set standards above the national average, by requesting special dispensation (a waiver) from the federal Environmental Protection Agency (EPA) to make their standards legal. No other state is permitted by federal law to establish an ARB or equivalent, but may meet California's standards over those set by the federal government.

Over the past decade and with the introduction of the Clean Cars Act – Pavley Bill (AB 1498) in 2002, California has consistently demonstrated a commitment to reduce GHG emissions. In 2006, Governor Schwarzenegger signed AB 32, which requires a reduction in state-wide emissions to 50 per cent below 1990 levels by 2030. The Governor also signed an Executive Order (S-3–05) stating that emissions should be reduced 80 per cent by 2050, and while not mandatory, many of the measures being developed at the current time are done so with consideration given to meeting this longer-term goal. These actions demonstrate that in California, there has been a transition of climate change institutions from governance/informal to formal status, based on Williamson's definition. Moreover, the assertion that environmental benefits should be a factor in the aims of 'integrated transport' is justified here, since SB 375 has been designed with GHG emission reduction as its central aim, through integrating land use and transport policy.

Given that 40 per cent of California's emissions comes from transportation, 74 per cent of which comes from passenger vehicles, prioritizing emission reduction from this sector is logical. The mandated reductions laid out in AB 32's 'Scoping Plan' (ARB, 2009b) – the roadmap to achieving the targets legislated for in the Bill – rely on the implementation of a number of diverse supplementary

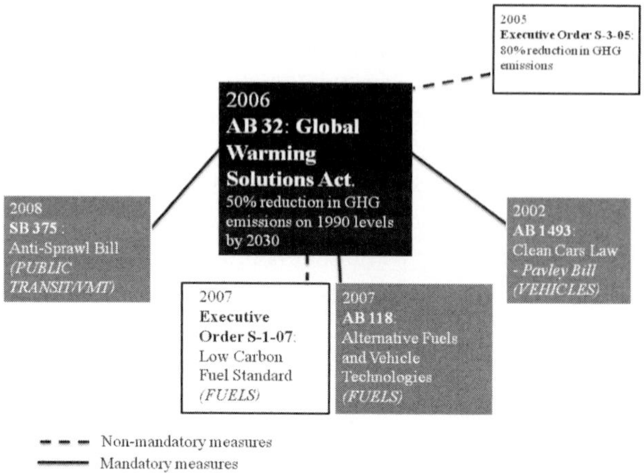

Figure 4.2 California's transport-related climate change institutions: three-legged approach

institutions, each of which deals with a specific source of emissions. For dealing with transport emissions, the institutions developed have collectively been coined the 'three-legged stool' as there are three distinct areas from which transport emissions can be reduced: a) vehicles; b) fuels; and c) land use/VMT (Figure 4.2). In October 2008, the Governor signed SB 375, designed to tackle the third leg – land use/VMT.

Through the Scoping Plan, ARB recognises that these are distinct but related components to a bigger picture and that addressing each through separate institutions is an effective means of achieving reductions from the sector as a whole. By examining the implementation process of these institutions in California, it is possible to see how the set-up into which they are being introduced may serve as a potential barrier. The following briefly outlines each 'leg of the stool' in turn, before taking an in-depth look at SB 375.

a) Vehicles

The Pavley Bill directs ARB to adopt regulations to achieve the maximum feasible and cost effective reduction of GHG from motor vehicles beginning with model year 2009. The law requires a 30 per cent reduction in GHG from tailpipes by 2016 through measures implemented by auto manufacturers to improve the fuel efficiency of their vehicles. It was signed by Governor Gray Davis in 2002, adopted in 2004, and has subsequently been adopted by 14 other states. However in 2007, the Bill was not granted an EPA waiver under the Bush administration and has therefore been subject to much delay and legal wrangling. It is the only Bill not granted a waiver in the 40 years of California's unique relationship with the CAA. President Obama took initial steps to grant the waiver after three days in office and it was formally approved in June 2009.

Additionally in 1990, ARB initiated a Zero Emission Vehicle (ZEV) mandate, designed to command car companies to make a certain amount of their vehicles emission-free. The programme never really gained momentum, and in 2008 the mandate was simplified and the percentages of vehicles required lowered as the mandate was regarded as too complex and no longer in line with California's GHG emission reduction goals.

Vehicle-focused institutions under AB 32 are expected to deliver 31.7 million metric tonnes of CO_2e reductions towards the 2020 target.

b) Fuels

Both Executive Order S-1–07 – the Low Carbon Fuel Standard (LCFS) and AB 118 – Alternative Fuels and Vehicle Technologies are focused on achieving reductions through providing investment in alternatives to fossil fuel energy sources to power the state's cars.

LCFS was issued in January 2007 and while not a binding legal requirement, it calls for a carbon intensity reduction of at least 10 per cent in the state's transportation fuels by 2020. It instructed California's Environmental Protection Agency Commission (CalEPA) to coordinate activities between the University of California, the California Energy Commission (CEC) and other state agencies to develop and propose a draft compliance schedule to meet the 2020 target. Furthermore, it directed ARB to consider initiating regulatory proceedings to establish and implement the LCFS. In response, ARB identified the LCFS as an early action item for regulation to be adopted and implemented by 2010 (ARB, 2009). In April 2009, Draft Resolution 09–31 regarding a LCFS regulation was subject to a hearing by the ARB board (ibid.).

AB 118 directs CEC to develop and implement the Alternative and Renewable Fuel and Vehicle Technology Program (CEC, 2007). The Bill was designed to promote alternative, lower carbon fuels and authorized CEC to spend up to US$120 million per year for over seven years (from 2008–15) to develop, demonstrate, and deploy innovative technologies to transform California's fuel and vehicle types (ARB, 2008).

Fuel-based institutions under AB 32 are expected to deliver 15 million metric tonnes of CO_2e reductions towards the 2020 target.

c) Land use/VMT

SB 375 – the Anti-Sprawl Bill – is designed specifically to deliver a proportion of the transport-related emissions reductions in the Scoping Plan by reducing VMT, increasing transit and promoting higher density growth.

SB 375 is expected to deliver 5 million metric tonnes of CO_2e reductions towards AB 32's 2020 reduction targets. Because of the long-term nature of land use planning, this 'leg' has the potential to deliver much larger reductions towards 2035, 2050 and beyond.

Through the current and planned implementation process of the Bill we begin to see how the intricacies of existing governance structures may impact on delivery. First, however, it is important to clearly define why VMT is central to emission reduction and why the link between land use planning, transportation and climate change mitigation has been made.

Why VMT?

According to Metz (2008), at the end of the twentieth century, the compact urban cities of the US had been replaced by suburban sprawl, made possible by the ubiquitous automobile. The result is that home and job location are increasingly uncoupled, radial travel from suburb to city centre is less common, trip patterns are more complex and journeys (VMT) longer.

VMT by California residents increased by more than 3 per cent a year between 1975 and 2004, and over the same period the state's population grew by less than 2 per cent (CEC, 2008). Californians are driving further and making more trips than ever before and if these trends continue, any reductions in emissions achieved through the measures outlined above would be negated by state-wide VMT growth. When this is coupled with population increase projections, the imperative to address how people get around becomes clear.

SB 375 is a Bill designed to make mobility different in California, to encourage people to use other modes of transport for shorter journeys and to generally drive less. It aims to do so by reshaping how neighbourhoods look and by providing better alternatives. By taking areas such as land use planning, urban transit provision and high density development into consideration, SB 375 is trying to thwart the 'car monoculture' (Sperling and Gordon, 2009) and sprawling suburbia for which California is famed.

SB 375: Introduction

The California Governor's Office describes the Bill as the 'nation's first law to control greenhouse gas emissions by curbing sprawl. SB 375 provides emissions-reducing goals for which regions can plan, integrates disjointed planning activities, and provides incentives for local governments and developers to follow new conscientiously-planned growth patterns. SB 375 enhances ARB's ability to reach AB 32 goals' (Office of the Governor, 2008).

It is an ambitious institution, in that it directly links funding for transport projects with land use planning and GHG emissions reduction. This connection is rarely formally acknowledged, much less legislated for, and demonstrates the state's commitment to reduce emissions. It is the type of institution, which if successful, could integrate the diverse policy areas of transport, planning and climate change and address them through a holistic regulatory measure. But it needs to be accompanied by organizational integration.

While SB 375 is a fairly unprecedented attempt to limit the GHG emissions associated with sprawl and unabated growth, California is working to implement the institution without significantly modifying the organizations through which it will be delivered. Whether the current system has the capacity to deliver SB 375 is an issue which can only be speculated on, given its infancy. But as far as it can, this overview offers a detailed glance into the planned implementation process and an understanding of how it is expected to be rolled out. In doing so, it highlights some of the factors which may become issues as the process unfolds.

SB 375: Mechanisms

Currently, major infrastructure funding comes through the federal Department of Transport (DoT) to the California Transport Commission (CTC), which has responsibility to allocate these funds based on applications: Regional Transportation Plans (RTPs) submitted by California's regional authorities; the Metropolitan Planning Organizations (MPOs), of which there are 18. Each MPO is required to submit a five-year RTP to CTC (Figure 4.3).

SB 375 aims to address GHG emissions from VMT by tasking the MPOs to incorporate 'ambitious and achievable' emission reduction targets, identified and allocated by ARB, and measures to reach them in their future RTPs. This element of the RTP is the Sustainable Community Strategy (SCS). While the inclusion of an SCS is not mandatory, there are incentives included in the Bill specifically to encourage the MPOs to produce the strategy. Indeed if regions develop integrated land use, housing and transportation plans that meet the SB 375 targets, new projects in these regions can be relieved of certain review requirements of the California Environmental Quality Act (CEQA).[1] Transportation projects that are inconsistent with the SCS would not be funded. Those MPOs that do not produce a SCS may submit an Alternative Planning Strategy (APS) separate to the RTP, with details as to how it could meet the target, alongside a justification for the lack of an SCS.

It should be noted that DoT funding is not the exclusive means of revenue and that increasingly private finance projects offer another stream of income to the state and local authorities for transport projects. This is important if the primary driver behind buy-in to SB 375 is access to funding.

SB 375: Responsibility

As mandated by the Governor, ARB has the authority to devise and distribute the GHG reduction targets to all of the MPOs. It set up the Regional Targets Advisory Committee (RTAC), consisting of state and regional representatives alongside leading thinkers, tasked with providing recommendations to ARB, by September 2009, as to how the targets should be set. The RTAC delivered this document to ARB on schedule and ARB is currently considering the recommendations made and will set the targets by September 2010. Alongside its recommendations, the RTAC delivered a series of outstanding questions to ARB about the implementation of the bill and

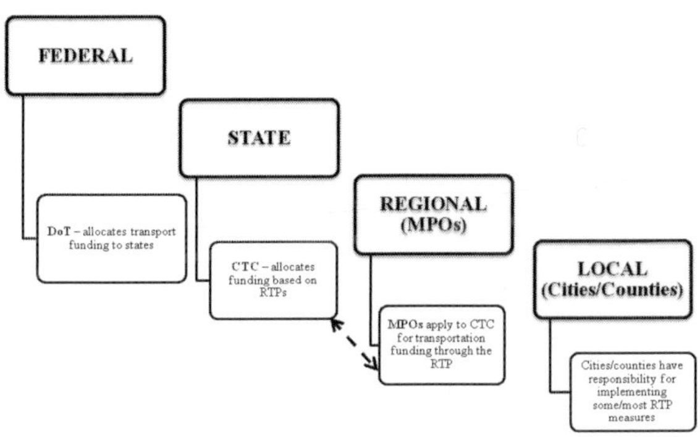

Figure 4.3 California transportation funding before SB 375

recommended that it convene one last time in 2010 to review the results of the MPOs' scenario planning activities. It is not envisioned that this group will continue to guide ARB once targets have been set.

As previously mentioned, ARB was established to manage California's air quality and has a history dealing with vehicle fuel standards. Yet it has newly been accorded management of all climate change affairs, centrally coordinating AB 32 and all supplementary institutions. None of these areas has previously been ARB's focus. Moreover, it is not a regulatory authority and therefore cannot command others to act. ARB's capacity to deliver SB 375 effectively, in addition to all of the other responsibilities recently added to its remit, may be held to question.

In delivering SB 375, authority to allocate transportation funding will continue to reside with CTC. ARB will assess the credentials of any SCS it receives and report back to CTC who will allocate funding to the MPO accordingly. While communication between the CTC and the ARB exists, it is a relatively recent development within the Californian bureaucracy, and the extent to which a truly collaborative relationship between the organizations will be established remains to be seen – but this too may be an important hurdle to overcome.

And while the Governor tasked ARB to deliver SB 375, responsibility for implementing the Bill and reducing emissions is ultimately held by both the MPOs and their respective local governments who are responsible for land use policy (Figure 4.4). If these local authorities decide not to prioritize GHG reduction and take the APS route (so long as they can justify this decision), while they may forego funding under the RTP and certain dispensation from CEQA requirements, there is little that the state government can do to enforce the targets set by the ARB.

While there are indications that some of the more powerful MPOs have the capacity and will to develop an SCS, in that many of them have already developed voluntary 'blueprints' around the issues, there are other MPOs and smaller cities and counties that may lack adequate resources to devise and deliver a comprehensive SCS. This is an issue that both the MPOs themselves and ARB/CTC need to acknowledge. It seems that there are some fundamental tensions in

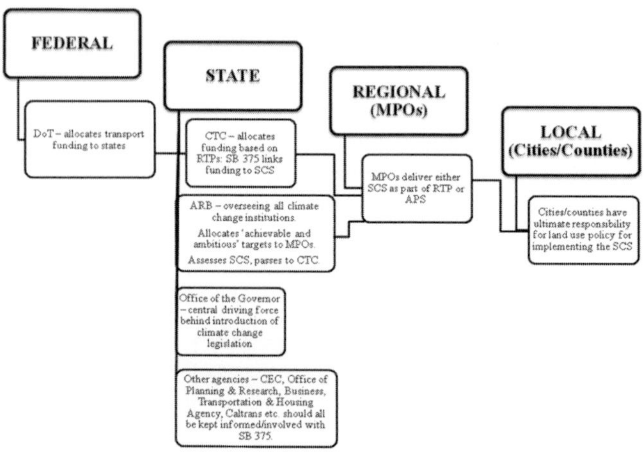

Figure 4.4
SB 375 – responsibilities and roles

terms of jurisdiction and authority within the operation of the Bill which may prevent successful implementation.

SB 375: Implementation

At the current time ARB is developing its methodology based on recommendations from the RTAC for allocating GHG reductions targets to the MPOs.

But other fundamental issues, outside the scope of target setting, need to be addressed surrounding SB 375, which have more to do with the organizational set-up through which the Bill will be delivered. We return to the list outlined earlier in the chapter around the institutional elements of integration outlined by Stead (Chapter 2 in this volume) to posit these questions.

- **The provision of capacities to cope with the emerging issues and duties**
  Is integrating an entirely new institution (SB 375) into existing structures of government through existing mechanisms (transportation funding) the best way to manage this complex issue?
- **The definition of a governmental organization to ensure integration**
  Is a non-regulatory authority (ARB) the best organization to oversee the delivery of SB 375?
- **The exchange of information and possibilities of interventions between different sectors**
  Can the separate authorities for transport, planning and air quality management realistically come together to deliver emission reductions through reducing VMT?
- **The definition of leading and participating agencies and their respective duties and responsibilities**
  Does SB 375, a land use Bill enforced by the state government, essentially undermine the authority of California's cities and counties with whom jurisdiction over this issue lies?

(Following Stead, this volume, adapted from Eggenberger and Partidário, 2000)

While this chapter does not attempt to offer comprehensive answers to these questions, it does highlight that some of the major features of SB 375 are incompatible with the system into which it is being integrated. I return to May and Crass (2007) to underline the areas which remain problematic for California and flag the barriers that need to be overcome (Table 4.3).

That California has clearly outlined how SB 375 will be delivered is not disputed, yet the choices it has made to govern the institution could be held to question. California's response attempts to integrate a group of separate organizations to deliver the Bill from various levels and disparate agencies of government.

But giving a lead agency which has no regulatory authority overarching responsibility for an issue which is cross-cutting in terms of scope, and involves stakeholders, which have previously had limited or no engagement, will make SB 375 challenging to implement. Moreover, this is coupled with split or vaguely defined

Table 4.3 Overcoming institutional barriers to SB 375

Institutional barriers	*Overcoming those barriers requires*	*Attempts to overcome these barriers – SB 375*	*Suggested responses*
Split or duplicated responsibility	More coordination between the tiers of government, and between agencies at each level.	Currently unknown how coordination between tiers of government and capacity of organizations to deliver will be arranged once targets have been set.	ARB central authority. Main 'connector' in the network between state agencies/levels of government. Creation of MPO forum would allow collaboration/knowledge sharing on SCS creation/implementation. State/MPO engagement with local authorities.
Process	Consistency in planning over the long term.	Initial implementation process managed by ARB by order of Governor. Unclear what changes will occur after targets are set. No major changes to levels of government responsibilities are expected. The Bill allows reassessment of targets – this may stall progress towards reductions.	Clearly define roles and responsibilities during target-setting phase of the Bill. Utilize periodic benchmarks and targets towards to 2020/2035 dates. Acknowledge that short-term election cycles may impact delivery of SB 375. Awareness of role developers may play in may influencing MPO/local government decisions over planning/development agendas with regard to SB 375.
Identifying objectives, specifying problems Selecting possible solutions, appraisal, implementation	A problem-led approach to developing solutions and strategies.	SB 375 makes climate change central to existing transportation funding process. SB 375 establishes clear timeline to deliver targets but existing timeline for RTP submission remains the same.	Consider organizational parameters as well as institutional framework. Acknowledge in RTAC recommendations that capacity of MPOs/cities/counties to implement SB 375 varies both in terms of expertise and resources.

Institutional barriers	*Overcoming those barriers requires*	*Attempts to overcome these barriers – SB 375*	*Suggested responses*
Political and public acceptability	Political champions and more positive involvement of the public and media.	Climate change agenda driven by Governor. RTAC makes target-setting process representative and equitable. All meeting/committees agendas, notes and outcomes available online. Public consultations regular. Free access to listservs for all.	Ensure climate change leadership continues after the next state elections. Transparency is crucial to keeping all levels/agencies of government engaged in policy developments – should continue after RTAC recommendations submitted. Media has role to play in informing public of Bill's implications and importance of reducing emissions.
Information and skills	More effective use of data, models and appraisal methods.	Experts in various fields relevant to SB 375 consulted by RTAC committee. The modelling capabilities of MPOs assessed by RTAC and feeding into target-setting process. MPOs working to normalize model assessment and incorporate information on state-wide data collection. Too early to assess levels/methods of appraisal/ implementation occurring at this time.	Keep abreast of newer modelling methods/technologies to ensure data collection/analysis is as robust as possible. Promote skill/information sharing between MPOs. Ensure data/monitoring of progress and reporting/verification of measures implemented and resulting reductions consistent and up-to-date across MPOs.
Financial	Financial support for strategies, without inducing policy bias.	Funding assessed based on SCS submitted to ARB/ CTC in RTP. In current economic crisis, changes in federal funding structures may impact delivery on SB 375 – remains uncertain.	Role/implications of private investment for SB 375 is unclear and little investigated. Private money could detract from conventional transport funding and thus fall outside RTP (and thus scope of SCS). Local governments interact with private investors – they don't need to wait for state-approved funding to be signed off and processed through MPO. May be significant consideration.
Legislative and regulatory	Legislation and regulations to support these requirements.	SB 375 passed to support emissions reductions mandated in AB 32. Provision of SCS not mandatory. Issues over whether state can regulate land use.	ARB is non-regulatory, has no means to ensure SB 375 will be met. This arrangement may need to be reconsidered if MPOs are not adequately incentivized to prepare an SCS.

Source: After May and Crass, 2007

responsibilities and overlapping jurisdictions among the other organizations involved, so while being 'integrative', this may not be the most effective organizational set-up into which to introduce the Bill.

5. Conclusion

SB 375

AB 32 goes a long way to harmonizing activity on climate change in California and should be commended. Moreover, the passage of SB 375 delivers an innovative institution which acknowledges the need to link disparate, although related, issues in response to climate change. It takes land use, planning, housing and transportation and highlights the fact that integrated approaches that break down existing boundaries are needed to reduce emissions. This in itself is progress and highlights a shift in mindset. Emissions reduction is no longer seen as peripheral policy activity to be managed through informal institutions.

The Scoping Plan approximates that SB 375 will contribute 5 million metric tonnes CO_2e reductions towards the 2020 target, yet many believe that it has potential to contribute reductions several orders of magnitude higher than this. And given that planning and development processes are relatively long term, projections to 2050 indicate that this leg of the stool could be much more significant to climate change mitigation than current thinking reflects (ClimatePlan, 2008; Hanak *et al.*, 2008).

However, land use is a policy which has traditionally been an exclusively local issue and while the local governments of California approved the passage of SB 375, reluctance to surrender authority to the state government may prove to be the biggest barrier to the Bill's success. While there has been some indication that the MPOs are willing to take this step, through the voluntary preparation of blueprint strategies, ultimately they are not mandated to produce an SCS.

Moreover, if local authorities are not adequately incentivized to implement them, or if MPOs are not sufficiently supported in the process of developing these strategies, emissions will not be reduced. Its discretionary nature is a worrying shortfall of this landmark institution.

ARB, CTC and all of the MPOs need to collaborate, messages need to be clear, the central purpose of the Bill prioritized. By setting the bar for other states, sub-national and national governments alike in developing 'formal', binding climate change legislation, California now needs to demonstrate that it can be appropriately integrated across the bureaucratic board to ensure successful implementation and in doing so it will continue to push the boundaries.

ARB is unique in the US and it may be that the challenges California faces will also be unique to its administration. Moreover, the state, along with its counterparts, is currently at the behest of impending national legislation, so the long timeframe to which it is working, together with changes in the direction of federal climate policy and the current budgetary crisis affecting the state, may leave California's climate change institutions facing an uncertain future. Yet the state is likely to continue to lead, and may play a role in guiding this federal policy.

SB 375 has drawn attention to the fact that in order to reduce emissions, we need to reassess the way we live, travel and fundamentally grow. Its development and implementation so far has seen unprecedented levels of collaboration and discussion, across areas of expertise, jurisdiction and authority. But whether it has gone far enough and whether the disparate stakeholders involved truly appreciate how dependent its success is on their ability to adjust to their intertwined roles remains to be seen.

What an alternate structure in the Californian context should look like is an area where further research is needed and may well stem from this preliminary overview. Similarly, the degree to which the situation in California is unique, or whether similar problems are being experienced elsewhere, is also ripe for investigation.

Integrating transport and climate change

While climate change policy has previously had a more informal institutional nature, we are seeing a 'formalizing' of the issue as the institutions designed to address it become more regulatory. While legislation is a powerful tool, simply mandating action on climate change is not enough without organizations set up around it to support its implementation, especially if overarching authority for these laws resides with non-regulatory, lone organizations. It may be that while we need more formalized institutions to prioritize climate change on political agendas, we actually need more fluid informal/governance-based organizational structures.

Environmental benefits are clearly a measure through which we can aim to achieve better integration. However, climate change is a multifaceted issue and simply identifying its importance on the policy agenda is not a sufficient response. It fundamentally challenges the way things are done. Can effective policy implementation actually be achieved without making the necessary changes to the governmental structures that deliver these institutions? Is integration actually occurring if we don't look at both aspects combined?

By maintaining conventional areas of jurisdiction, funding patterns and operation, climate change cannot be addressed. We need holistic approaches and organizations geared up to integrate institutions into their practices across the board.

While not advocating overarching, all-powerful climate change agencies with jurisdiction over all related issues through central emissions-reducing mandates, this chapter emphasizes the barriers to implementing multisector, cross-cutting institutions to deal with climate change. Climate change cuts across geographical, political and cultural boundaries. It cuts across issues and scales of government and requires a response that may not necessarily fit within the compartmentalized constructs of conventional political wisdom. This needs to be acknowledged during policy development.

Institutional and organizational integration

It is not just climate change for which this is an issue. More generally speaking, delivering new institutions through existing organizations is standard practice. Adding to existing measures or introducing supplementary features to legislation is appealing because it minimizes the need for another layer of bureaucracy in an already convoluted system. But as more new, different legislation comes into being and emerging issues become more prominent, at some point the existing structures of government are in danger of buckling under their ever-increasing remits.

This chapter serves to highlight the complexity of existing systems into which new institutions are introduced and the need for existing relationships to evolve in order to deal with these institutional developments. As we see in Chapter 17 in this volume, even with the highest degree of planning, organizational issues and responsibilities hinder implementation, and closer integration between those involved in particular transport projects is needed. And in Chapter 18, the importance of defining perimeters of competence, smoothing out regulatory differences and allowing one regulatory authority central control are defined as fundamental to public transport management. These are logical recommendations to address the organizational barriers to the implementation of any institution.

But perhaps most importantly, there is a need to pay more attention to implementing policies that have been set, rather than designing the 'best' policies and focusing solely on their outcomes, as Hossinger (2003) rightly surmised. Getting the right organizations involved in delivering these institutions is a critical element in this respect. Addressing this area of policy integration may actually alleviate some of the other challenges to integration discussed in this book.

Note

1 The California Environmental Quality Act (CEQA) is a law passed in 1970. The law requires development projects to submit formal documentation as to likely environmental impacts.

References

Banister, David, John Pucher and Martin Lee-Gosselin (2007) Making Sustainable Transport Politically and Publically Acceptable: Lessons from the EU, USA and Canada. In Piet Rietveld and Roger Stough (eds) *Institutions and Sustainable Transport: Regulatory Reform in Advanced Economics*, Cheltenham: Edward Elgar.

Bardach, Eugene (1977) *The Implementation Game: What Happens after a Bill Becomes a Law*, London: The MIT Press.

—— (1998) *Getting Agencies to Work Together: The Practice and Theory of Managerial Craftsmanship*, Washington, DC: The Brookings Institute.

California Air Resources Board (ARB) (2009a) *Low Carbon Fuel Standard Program*. Available online at: http://www.arb.ca.gov/fuels/lcfs/lcfs.htm (accessed 24 October 2009).

—— (2009b) *AB 32 Scoping Plan*. Available online at: http://www.arb.ca.gov/cc/scopingplan/scopingplan.htm (accessed 24 October 2009).

—— (2008) *Air Quality Improvement Program (AB 118)*. Available online at:http://www.arb.ca.gov/msprog/aqip/aqip.htm (accessed 24 October 2009).

California Energy Commission (CEC) (2007) *Development of Regulations for the Alternative and Renewable Fuel and Vehicle Technology Program*. Available online at: http://www.energy.ca.gov/ab118/index.html (accessed 24 October 2009).

—— (2008) California Leadership on Land Use and Climate Change. Presentation delivered by Panama Bartholomy, Advisor to the Chairman, California Energy Commission at New Partners for Smart

Growth, Washington, DC. Available online at: http://www.energy.ca.gov/2008publications/CEC-999-2008-014/CEC-999-2008-014.PDF (accessed 24 October 2009).

Climate Plan (2008) *Why 5 MMT Isn't Enough: The 2020 Target for Regional Transportation-Related Greenhouse Gases Should be 11–14 MMT.* Available online at: http://www.climateplanca.org/Why%205MMT%20Isn't%20Enough.pdf (accessed 24 October 2009).

De Borger, Bruno and Stef Proost (2007) Transport Pricing when Several Governments Compete for Transport Tax Revenue. In Piet Rietveld and Roger Stough (eds) *Institutions and Sustainable Transport: Regulatory Reform in Advanced Economics*, Cheltenham: Edward Elgar.

Eggenberger, M. and Partidário, M. (2000) Development of a framework to assist the integration of environmental, social and economic issues in spatial planning. *Impact Assessment and Project Appraisal*, 18(3): 201–207.

Foxon, Timothy J. (2002) Technological and Institutional 'Lock-in' as a Barrier to Sustainable Innovation. ICCEPT Working Paper, November. Available online at: http://www.iccept.ic.ac.uk/public.html (accessed 24 October 2009).

Hadley Centre, Met Office (2009) *Four Degrees and Beyond.* Available online at: http://www.metoffice.gov.uk/climatechange/news/latest/four-degrees.html (accessed 24 October 2009).

Hanak, Ellen, Louise Bedsworth, Sara Swanbeck and Joanna Malaczynski. (2008) *Climate Policy at the Local Level: A Survey of California's Cities and Counties*, California: Public Policy Institute of California.

Hansen, Carsten Jahn (2006) Urban Transport, the Environment and Deliberative Governance: The Role of Interdependence and Trust. *Journal of Environmental Policy and Planning*, 8(2): 159–179.

Hössinger, Reinhard, Andreas Witte and André Wolf (2003) Barriers and Factors of Success in Urban Transport Planning. 18th Annual Polis Conference, Cologne. Available online at: http://www.isb.rwthaachen.de/guidemaps/downloads/Conference_papers/Barriers_and_factors_of_success.pdf (accessed 24 October 2009).

Hull, A.D. and R.C. Tricker (2005) Assessing barriers to sustainable UK urban transport solutions. *Proceedings-Institution of Civil Engineers Engineering Sustainability*, 158(3): 171–180.

Intergovernmental Panel on Climate Change (ICPP) (2007) *Fourth Assessment Report, Working Group 3 Mitigation, Transport and its Infrastructure.* Available online at: http://www.ipcc.ch/pdf/assessment-report/ar4/wg3/ar4-wg3-nectar5.pdf (accessed 24 October 2009).

International Energy Agency, *IEA exhorts Ministers to "bridge the gap" between current policies and the new measures that will empower a more secure, environmentally acceptable energy system*, 14 November 2006, http://www.iea.org/textbase/press/pressdetail.asp?PRESS_REL_ID=189

March, J.G. and J.P. Olsen (1989) *Rediscovering Institutions: The Organizational Basis of Politics*, New York: The Free Press.

May, Tony and Mary Crass (2007) Sustainability in Transport: Implications for Policy Makers. *Transportation Research Record: Journal of the Transportation Research Board*, 2017: 1–9. Available online at: http://www.internationaltransportforum.org/Topics/Workshops/WS2MaySlides.pdf (accessed 24 October 2009).

Metz, David (2008) *The Limits to Travel: How Far Will You Go?*, London: Earthscan.

Modern Language Association (MLA) Integration. *Merriam-Webster's Dictionary of Law*, Merriam-Webster, Inc. Available online at: http://dictionary.classic.reference.com/browse/integration (accessed 28 May 2009).

NEA, OGM and TSU (2003) *Integration and Regulatory Structures in Public Transport: Final Report.* Brussels: DGTREN.

North, Douglass C. (1990) *Institutions, Institutional Change and Economic Performance*, Cambridge: Cambridge University Press.

Office of the Governor (2008) *Senate Bill 375: Redesigning Communities to Reduce Greenhouse Gases.* Available online at: http://gov.ca.gov/fact-sheet/10707/ (accessed 24 October 2009).

Organisation for Economic Co-operation and Development (OECD) (2008) *Climate Change: Meeting the Challenge to 2050*, Policy Brief, February. Available online at: http://www.oecd.org/dataoecd/6/21/39762914.pdf (accessed 24 October 2009).

Pavley, Fran (2002) *Climate Change and California*, NRDC Presentation. Available online at: http://www.nelson.wisc.edu/outreach/special/pavley.pdf (accessed 24 October 2009).

Potter S. and Skinner M.J. (2000) On Transport Integration: a Contribution to Better Understanding. *Futures*, 32: 275–287.

Rietveld, Piet and Roger Stough (2004) Institutions, Regulations and Sustainable Transport: A Cross-national Perspective. *Transport Reviews*, 24(6): 707–719.

—— (eds) (2005) *Barriers to Sustainable Transport: Institutions, Regulation and Sustainability*, Oxford: Taylor & Francis.

—— (2006) Institutions, Regulations and Sustainable Transport: A Review. *European Journal of Transport and Infrastructure Research*, 6(1): 99–112.

Sachs, Jeffrey (2009) Repairing Economic Governance. Public lecture delivered at James Martin 21st Century School, University of Oxford, 20 October. Available online at: http://www.21school.ox.ac.uk/resources/podcasts.cfm (accessed 24 October 2009).

Sperling, Daniel and Deborah Gordon (2009) *Two Billion Cars: Driving Toward Sustainability*. Oxford: University Press US, 2009.

Williamson, O.E. (1994) *Institutions and Economic Organization – The Governance Perspective*. Washington, DC: World Bank.

World Business Council for Sustainable Development (WBCSD) (2004) *Mobility 2030: Meeting the Challenges to Sustainability*. The Sustainable Mobility Project, Overview. Available online at: http://www.wbcsd.org/web/publications/mobility/overview.pdf (accessed 24 October 2009).

Chapter 5

Integrated transport policy in freight transport

Julian Allen, Michael Browne and Allan Woodburn

1. Introduction

This chapter is concerned with integration issues in relation to freight transport. Integration is an important topic in freight transport as enhanced integration has the potential to facilitate more sustainable freight transport operations through improving the efficiency of freight operations while at the same time reducing the negative environmental and social impacts. From an economic perspective improved integration in freight transport can result in reduced delivery times, improved delivery reliability, enhanced vehicle loading and utilization, reduced logistics costs, and less traffic congestion. From a social and environmental perspective improved integration in freight transport can lead to reduced vehicle trips and vehicle kilometres, less fossil fuel consumption, reduced air pollution, greenhouse gas and noise emissions, and fewer traffic casualties.

Many integration issues arise in freight transport and policy makers have an important role to play in overcoming problems concerned with integration. These include improving the integration of physical transport networks (and conflicts that occur on these networks between freight transport and other users); legal and social issues; standardization of handling units; issues concerning transport modes and vehicles (including using each mode to capitalize on its strengths and interoperability between modes); integration of the supply chain through physical distribution and logistics management; and joint working between the public and private sector. Section 2 of this chapter addresses each of these integration issues in turn. Section 3 considers the various types of policy interventions that can be made to improve the integration of freight transport. It also presents examples of major policy initiatives addressing freight transport integration at various geographical scales, including international level actions to improve the integration of physical transport networks, national level actions to enhance the integration of physical distribution networks, and regional/local level action to enhance integration between the public and private sector in relation to freight. Section 4 provides some closing thoughts and conclusions.

2. Aspects of integration in freight transport

The different aspects of integration that can be considered in relation to freight transport are more numerous than for passenger transport. This is due to several features of freight transport that distinguish it from passenger transport. First, there are many different companies involved in production and supply of a particular product (including shippers, carriers, receivers, etc.) This results in complex commercial relationships and decision-making processes. To understand why goods are transported by a particular mode, using a particular route, at a particular time, and at a particular frequency often requires gathering insight and knowledge from several different companies involved in the process. By comparison, to gain such an understanding of a passenger journey requires information from just one individual: the passenger.

Second, when aiming to obtain an understanding of passenger transport activity, the unit of analysis is normally the passenger transport vehicle (e.g. buses, cars, trains, etc.). For freight transport there are two possible objects of study: either the vehicle (e.g. the heavy goods vehicle, the van, the rail wagon, etc.) or the goods. Ogden (1977, p. 106) pointed out the need for care in distinguishing between the flow of goods and the vehicle movements that carry the flow:

> Goods vehicle movements are an important area of study as many of the costs and problems of urban freight such as delay and traffic congestion are related to vehicles. The study of goods and their flow is also of importance as the (urban) freight system is fundamentally concerned with commodity flow not vehicle movements.

In the course of a journey from origin to destination, goods may pass through numerous facilities (processing, manufacturing, stockholding and transhipment facilities) and may be carried by several different vehicles and operators over long distances using more than one mode. Freight journeys are therefore inherently more complicated than passenger journeys. This makes decisions about how best to model and forecast freight flows extremely challenging.

Third, the effort that goes into logistics management (i.e. the management of the flow of the goods rather than the transport vehicle or mode) and the associated information flow and data capture that supports this do not have a corollary in passenger transport. This therefore represents an additional network to passenger transport.

Fourth, for the public and private sectors to work effectively together to promote and achieve efficient, sustainable freight transport activity requires the joint working of numerous organizations that may have very different perspectives and rationales. These organizations include: manufacturers, retailers, wholesalers, freight transport operators, service providers, local authority transport planners and policy makers. This is different to, and typically more complex than, the relationship between public sector policy makers and passengers (i.e. the general public).

The complications of integration issues are greater for international freight activities than for national movements due to customs clearance, security and other regulatory issues that such flows raise – often resulting in additional stops, checks, delays and administration, all of which increase transport costs.

Therefore, aspects of freight transport integration range from physical transport networks, to the physical flow of product from point of production to consumption (and the relationships between companies that facilitate this), to logistics management, to relations between the public and private sectors. Bringing about greater integration in freight transport can therefore mean many different things, and is likely to be a relatively complex process requiring the active support and involvement of many organizations. Table 5.1 outlines these different aspects of integration that can be considered in relation to freight transport, together with a summary of the integration issue that it raises.

2.1 Transport networks

The transport network refers to the physical transport infrastructure associated with a particular mode. For instance, this could be a road network or a rail network. The network comprises links and nodes, nodes being the terminals and interchange points. In the case of freight transport, nodes include facilities such as distribution centres, rail freight terminals, seaports, wharves and airports.

These nodes are the fixed points within the distribution system at which goods need to be handled and/or stored. The types of important functions that these facilities can serve in the distribution system include:

- storage of goods;
- consolidation of loads;
- break-bulk operations;
- transfer of goods between modes.

Goods may be stored for a number of reasons: it may be simply to provide a buffer point between supply and demand or to cover for seasonal fluctuations. Stockholding can also help to guard against unforeseen circumstances and disruptions within the distribution system.

Consolidation is necessary when goods are collected in relatively small quantities from a number of different locations. By taking these goods to a central point it is possible to combine these goods into bulk loads and thereby improve the efficiency of the secondary stage of the transport operation (Beuthe and Kreutzberger, 2001; Hesse and Rodrigue, 2004).

Efficiency in the transport of goods can be achieved by moving the products in bulk loads over the main trunking stage of the distribution system. It is then possible to break the load into smaller consignments at a static facility for local delivery to the customer.

Table 5.1 Different aspects of freight transport integration

Topic	*Explanation*	*Freight transport integration issue*
Transport networks		
(i) Integration of the physical network	Physical transport infrastructure associated with a particular mode (e.g. road network, rail network, etc.).	Extent to which each of these transport networks is integrated in itself (e.g. connectivity of the road network) and extent to which there is integration between these transport networks (e.g. connectivity between road and rail networks).
(ii) Conflict between freight and other users on the physical network	Different types of vehicles/users on a physical transport network (e.g. cars, van, heavy goods vehicles, buses, motorcyclists and cyclists all using the same road network).	Extent to which freight vehicles impede/are impeded by other users of the network. This can be simplified as a potential conflict between passenger and freight transport.
Handling units	The units in which goods are placed to assist handling and transportation (through time savings and cost reduction).	Extent to which standardized handling units are used in different modes/vehicles/physical distribution networks.
Transport modes/vehicles		
(i) Modal split	Whether each freight transport mode is being used in such a way to capitalize on its strengths.	Extent to which each mode is used to the best of its strengths and in combination ('co-modality').
(ii) Interoperability	Relates to the interoperability of a vehicle on more than one physical transport network (e.g. road and rail, or road and sea).	Whether the ability exists to use the same vehicle on more than one transport network.
Physical distribution and logistics management networks	The physical transportation of a product between manufacturing and stockholding points from point of extraction/production to point of consumption (and reverse flows of waste/recycled product). This can involve the use of more than one transport mode and physical transport network in the total journey from point of extraction to consumption, or even on a single stage between two facilities.	Extent to which products flow smoothly and reliably along physical distribution networks from point of extraction/production to point of consumption. Where more than one mode is used this is dependent on the efficiency of intermodal transport networks (e.g. road-rail, sea-rail, etc.). Extent to which different supply chain parties work successfully together to manage the flow of the product and associated information.

(continued)

Topic	Explanation	Freight transport integration issue
Physical distribution and logistics management networks (*cont.*)	The management of the flow of the product from point of extraction/production to point of consumption (and reverse flows). This involves management and coordination between different supply chain parties (e.g. carriers, shippers, receivers, forwarders, etc.) and management of the information associated with these product flows.	
Public and private sector joint working	Freight transport involves the interaction of public and private sector organizations. Private sector companies comprise the supply chain parties involved in the management of the logistics network. Public sector organizations are responsible for the supply of the physical transport network and the regulations and other policy interventions associated with the use of this network by the private sector.	Extent to which the public and private sectors are able to work together to ensure that freight transport is conducted as efficiently as possible (maximizing economic benefits while minimizing negative social and environmental impacts).

Some nodes are unimodal whereas others offer interchange options between two or more transport networks (e.g. road-rail, road-inland waterway, sea-rail, etc.). In addition to offering the possibility to transfer goods or handling units between vehicles or modes, many freight terminals offer warehousing, inventory management and other value-added logistics services. Generally, the larger the size of the terminal site and the greater the throughput of product, the greater the likelihood that it will offer connections to more than one transport network. For example, a national or regional distribution centre is more likely to be rail-connected than a local fulfilment centre. Virtually all freight terminals are connected to the road network. Therefore by definition, seaports, airports and rail freight terminals are all at least bimodal.

Improvements to transport networks by reducing bottlenecks, filling in gaps and upgrading interchange facilities enhance the speed and reliability of freight transport movements and make it possible to transport goods over longer distances at lower costs. These developments may encourage the sourcing of raw materials, parts and finished products over ever-increasing distances. However, ultimately company decisions about sourcing strategies are determined by the trade-off between production and transport/logistics costs. The quality of the transport network is one important determinant of transport costs. Improvements to transport networks can help facilitate increases in the market area over which companies produce and sell their goods. In this sense, improving the integration of transport networks can help facilitate economic growth and prosperity, but at the same time can result in less environmentally sustainable outcomes as goods are sourced and sold over ever-growing distances. As trade patterns become increasingly global, a growing proportion of goods flows pass through major seaports, airports, rail terminals and international distribution centres.

Traditionally transport networks have been planned and developed on a national scale. However, as trade flows and the movement of people become increasingly international, it has been necessary for supra-national bodies and national governments to work together to enhance transport networks and improve the reliability of international goods movement.

Border crossings and customs clearance points may cause significant journey time-delays, poor journey time-reliability and additional costs for international freight movements. Landlocked countries can face particular difficulties in relation to border-crossing delays and costs.

In the case of rail networks, incompatibility between different national systems poses considerable challenges for cross-border rail movements. These network differences can include signalling, track and loading gauges, power sources, and driver requirements – all of which can potentially disrupt or prevent international rail freight movements.

Conflict between users of physical transport networks

Users of physical transport networks compete for limited resources in terms of both space and time on the network. Delays and unreliability problems in freight transport services are regularly caused by traffic congestion resulting from demand for road space at a given time exceeding supply. Congestion can occur either on the transport links that connect nodes (i.e. roads, railways, etc.) or at nodes (such as airports, rail terminals, seaports, and road-based distribution centres). Congestion that occurs on links is the result of the demand by all users exceeding the capacity of the link at a given time. Freight vehicles usually comprise a relatively small proportion of total vehicle traffic on transport networks (for example, in the case of roads, light and heavy vehicles typically account for approximately 10 per cent and 5 per cent of total vehicle movements respectively), however this can be far higher around major freight nodes and on key freight routes.

At nodes dedicated to freight transport (such as container ports or distribution centres) congestion is the result of freight industry demand exceeding capacity. However, in the case of shared nodes (e.g. airports) congestion is complicated by the interaction between passenger and freight activity competing for space where there is a lack of adequate capacity. There is also conflict between passenger and freight services on railways. Most railway lines are used for both passenger and freight trains, and priority is usually given to passenger trains when allocating train paths. Similarly, inland waterway barges carrying freight do not usually receive any priority over non-freight vessels in harbours (Beuthe, 2007).

Road-based distribution often requires collections and deliveries to take place on the public road network. As a result this reduces the space available on the road network for other activities. Conversely, other kerbside activity such as car parking and bus stops, together with reservation of the kerbside lane for buses (i.e. bus lanes), cause difficulties for freight transport operations needing to stop to load and unload at these locations. Much of this kerbside competition for road space takes place in busy, urban areas. The Freight Transport Association has noted that, 'While industry has achieved significant success in improving vehicle productivity and utilization, urban congestion imposes major constraints on further improvements' (Freight Transport Association, 1996). As well as competing for space and time on the road network, the interaction of passenger and freight vehicles also results in safety-related issues including collisions that can involve casualties.

2.2 Standardization of handling units

Traditionally many types of goods were handled in an uncontained form. Even where packaging or rudimentary containers were used this tended to be non-standard and non-uniform. For instance, the ancient Egyptians used straw baskets for ease of loading freight on ships. Later, the introduction of wooden crates (for shipping), barrels (for liquids), and bags (for grain) helped to increase the ease and speed of vehicle loading. However it was still extremely difficult and time-consuming to load and unload transport vehicles, and to transfer goods between vehicles or modes. Loading and unloading required enormous quantities of manpower; over time this was supplemented by machinery to assist lifting and winching, as well as forklift trucks, hand trucks and conveyors (Mercogliano, 2006).

However, until the introduction of the shipping container, containers used were not standardized in terms of dimensions, markings, corner fittings, etc. The shipping container was introduced in the 1950s and had become standardized through the International Organization for Standardization (ISO) conventions by the end of the 1960s. It rapidly became used in land freight transport as well. These containers offered the following advantages, all of which have made it faster and cheaper to transport goods:

- The standard dimensions facilitate easy transfer by machine.
- The standard dimensions facilitate better utilization of the vehicle's/vessel's loading space.
- The automated handling systems have made large reductions in manpower possible (for instance, a 40,000 ton container ship requires 750 person-hours for unloading, whereas traditional techniques would take 24,000 person-hours). (Slack, 1998)

The progressive introduction of enhanced container and material handling systems and equipment has facilitated the growth of larger freight vehicles and vessels, as well as the possibility to decrease the transfer time per loading unit. These materials-handling developments facilitated the growth of intermodal transport and enhanced global opportunities for trade since containers could be transferred between ships, lorries and trains, along international supply chains. As container throughput increased, this had the effect of further reducing the unit cost of transport. It has been argued that, although it is not possible to attribute changes in the world economy to one cause, the reduction in freight costs as a result of the introduction of container shipping in recent decades has played a key role in increasing the integration of the global economy (Levinson, 2006).

As well as the spread of containerization and the related handling equipment, other developments in handling systems for road transport that have led to bundling and unitization such as the use of pallets, roll cages and packaging trays systems, have also had a similar effect on the load that can be carried and the time taken to load and unload the vehicle.

2.3 Transport modes/vehicles – modal split and interoperability

The importance of freight transport for economic development has been noted above. To fully illustrate the global scale of freight transport activity is beyond the scope of this chapter. However, to give an impression of the freight volumes moved according to mode, summary data are provided in Tables 5.2 and 5.3. It is important to note that these data reflect the conventional approach of considering freight activity on a strictly modal basis. This unfortunately hinders the debate about the importance of integration between the modes and the importance of intermodal freight movements.

There are many ways in which the different modes can work together to provide a more integrated freight transport system. At the broadest level, the European Union aims to achieve integration in freight transport through the concept of co-modality, by which they mean 'the efficient use of different modes on their own and in combination', leading to 'an optimal and sustainable utilization of resources' (European Commission, 2006). For many freight flows, this is likely to relate to just one single mode of transport, most likely road, but this may still require integration to provide an efficient and sustainable door-to-door service. In other cases, particularly over long distances, an efficient and sustainable outcome may better be achieved through the use of more than one mode of transport. Several different terms are used to describe such movements (Beuthe, 2007):

- 'Multimodal' transport – transport of goods by at least two different modes regardless of the method by which it is organized.
- 'Intermodal' transport – the movement of goods in the same handling unit or vehicle, which uses more than one mode of transport without any handling of the goods themselves in changing modes (i.e. either the handling unit – such as a container – or the vehicle are transported on different transport networks).
- 'Combined' transport – a form of intermodal transport in which the major part of the journey (the trunk movement) is by rail, inland waterways or sea, and the initial or final leg (or both) are carried out by road over as short a distance as possible (the term combined transport is most commonly used in relation to road-rail transport).

Table 5.2 Comparative goods transport performance. EU-27, USA, Japan, China and Russia, 2006 (billion tonne-km and %)

	EU-27		*USA*		*Japan*		*China*		*Russia*	
Road	1,888	46%	1,890	30%	347	60%	975	11%	201	4%
Rail	435	10%	2,705	43%	23	4%	2,195	25%	1,951	41%
Oil pipeline	135	3%	854	14%	–	0%	166	2%	2,499	53%
Inland waterways	138	3%	486	8%	–	0%	1,291	15%	58	1%
Sea (domestic/ intra-EU-27)	1,545	37%	332	5%	208	36%	4,258	48%	48	1%
Total	**4,140**	**100%**	**6,266**	**100%**	**578**	**100%**	**8,886**	**100%**	**4,757**	**100%**

Source: Eurostat, 2009a

Note: The data concerning different geographical entities are not fully comparable.

Table 5.3 Goods transport performance by mode: EU-27, 1995–2006 (billion tonne-km)

	Road	*Rail*	*Oil pipelines*	*Inland waterways*	*Sea*	*Air*	*Total*
Mode share	46%	10%	3%	3%	37%	0%	100%
2006	1,888	435	135	138	1,545	3.0	4,143
2005	1,800	413	136	138	1,530	2.9	4,020
2004	1,747	413	131	136	1,488	2.8	3,918
2003	1,625	391	130	123	1,445	2.6	3,717
2000	1,519	401	126	133	1,348	2.7	3,529
1995	1,289	386	115	121	1,150	2.0	3,062
AAGR 1995–2006	3.5%	1.1%	1.5%	1.2%	2.7%	3.8%	2.8%

Source: Eurostat, 2009a

Note: AAGR – average annual growth rate.

The majority of rail, inland waterway, and coastal and sea shipping, is at least multimodal, if not intermodal. This is because these modes tend not to serve the origins and destinations of product flows, so road transport is usually required for initial collection and ultimate delivery. For many flows, particularly of manufactured goods, intermodal units are used to allow easier and more efficient transfer between modes. Air freight can also be considered a form of multimodal transport as it typically involves road legs at the start and end of the journey, sometimes using intermodal units but often with more standard road freight feeder operations (with goods being transported via road-based distribution centres).

An essential feature of combined transport is that the trunk movement should be quicker or cheaper (or both) compared to the equivalent journey by road. However, the performance of the combined transport system can be influenced by the efficiency of the unit load transfer operation. For conventional rail wagon movements, the goods themselves are handled during loading and unloading at the rail terminal. Combined transport overcomes this constraint by only transferring unit loads (e.g. containers) between transport modes. As a result, several systems have evolved in order to suit the needs of the combined transport market.

The main systems used in combined road-rail systems fall into two main categories: unitized and trailer movements (Figure 5.1):

- Unitized freight – generally carried in 20-, 40- or 45-foot-long ISO standard containers capable of stacking. Although double stacking by rail is not possible in Europe due to infrastructure constraints (e.g. the height of bridges and tunnels and the catenary overhead equipment). Specialized containers exist to carry liquids and temperature-controlled goods. A popular alternative in mainland Europe has been the demountable lorry body, also known as the swap body. Swap bodies are similar to ISO containers, but generally are more lightly built and therefore not stackable and are transferred using different lifting points to

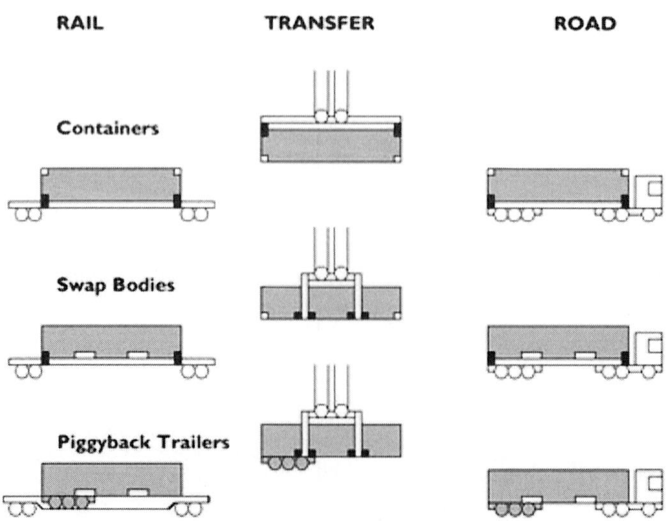

Figure 5.1 Combined transport techniques
Source: Network Rail website (www.networkrail.co.uk)

a container as can be seen in Figure 5.1. Their main benefit is that dimensions are optimized for road operations, often leading to better loading efficiencies. Swap bodies often have curtain sides whereas containers are typically rigid metal 'boxes'.

- The trailer train (or 'piggyback') – complete road semi-trailers or entire vehicles are carried on either ordinary flat wagons or in specially designed wagons that allow the trailer wheels to be below the deck of the wagon platform. This type of operation has implications for the overall size and weight of trains, and 'piggyback' operation has decreased in importance in Europe. It is most widely used in Europe on Alpine corridors.

As noted at the start of this section, comprehensive and reliable data relating to intermodal transport activity are fairly limited, since statistical sources tend to focus on activity using specific modes and intermodal flows are counted separately for the different legs of the journey. According to Eurostat (2009b), there is no standardized information about intermodal transport activity across the European Union relating to:

- previous or next mode of transport for intermodal units (i.e. containers, swap bodies and semi-trailers), which would allow flow visibility along the transport chain rather than on an individual leg-by-leg basis;
- transhipment terminal performance (e.g. waiting times);
- criteria related to mode choice, allowing informed decision making about ways in which rail, inland waterway and sea transport can be made more attractive to complement road haulage.

Despite these shortcomings, Eurostat periodically publishes a short statistics document focusing on the unitization of freight transport in the European Union (see, for example, Eurostat, 2008). Data availability issues mean that an incomplete picture is provided, even for basic measures such as mode share. Critically, statistics are published by specific mode and are not integrated across the different modes. Further, data for inland waterways are not available at all, and a number of countries are unable to provide unitized transport statistics for road and rail. Where comprehensive national statistics do exist, the methodologies often differ, which makes international comparison difficult. Initiatives are under development to try to improve the statistical basis for intermodal freight movements. In Germany, for example, an approach has been trialled which combines existing mode-based transport statistics for shipping, rail, inland waterway and road, supported by expert interviews in ports and leading to more detailed information about mode share and flow origins and destinations.

2.4 Physical distribution and logistics management networks

The physical distribution network refers to the flow of goods along supply chains as well as the management of the flow by freight transport operators and forwarders. The physical distribution network for a product can involve a wide range of activities involved in the movement of the goods (including vehicle loading and unloading, transhipment, consolidation and break-bulk) and a wide range of transport modes using transport networks. It can also comprise many commercial organizations (including manufacturers, wholesalers and retailers together with transport companies) that make use of transport networks. Integrating these activities so that they function efficiently and reliably in such a way as to meet consumer demand is a major challenge.

There are many different distribution network structures for the organization of product flow from point of despatch to point of delivery. The most important of these distribution network structures are shown in Figure 5.2. Some of these structures involve direct flows from the point of despatch to point of delivery, while others involve intermediate handling. The structures depicted in Figure 5.2 therefore vary in terms of the number of vehicles, modes and transport networks used in the journey. Each journey that a product makes in the physical distribution network can involve a different distribution network structure.

It should be noted that the efficiency of the different distribution network structures depicted in Figure 5.2 is strongly influenced by the (optimal) location of the terminals/hubs (see, for example, Limbourg and Jourquin, 2009).

Growing importance of logistics management

Logistics is now a widely used and understood term throughout the business world, and refers essentially to the management of supply chains in commerce and industry. Precise definitions vary, but the common thread is a concern for the movement and storage of goods, together with the associated information flows, from the beginning to the end of the product supply chain (i.e. the point of extraction/production to the point of consumption), as well as reverse flows of return product, waste and recycled material. So, for a manufacturing company, logistics management could include:

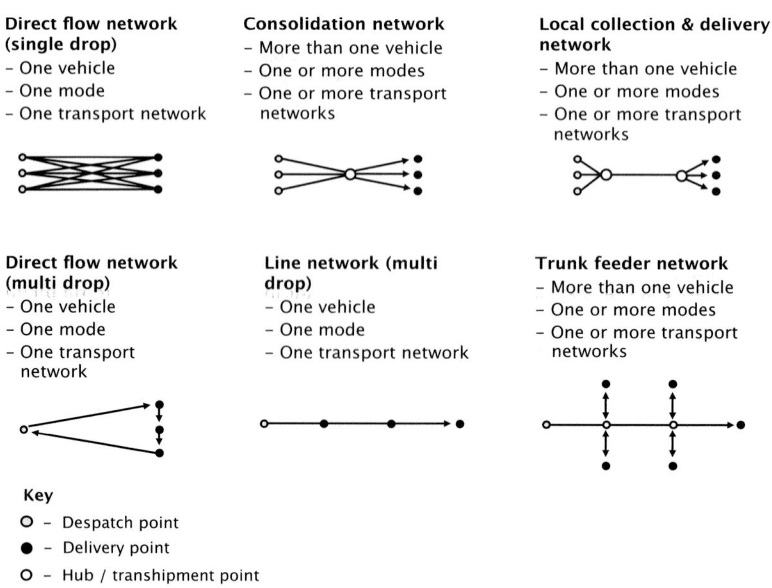

Figure 5.2 Possible distribution network structures for a freight transport journey, taking into account vehicles, modes, and transport networks

- the procurement and sourcing of raw materials or components;
- inwards transport;
- materials handling, storage, inventory management, packaging and labelling and the link to production processes;
- order and related information management;
- final distribution of finished products to customers.

Figure 5.3 shows a conceptual framework for a logistics management network and also illustrates the connection between the transport movements discussed above and the information management aspects.

Many commercial organizations will be involved in the logistics management for a product as it passes along its supply chain, and they will need to work closely together to ensure it meets all the requirements of the consumer. No single party is usually responsible for determining how the supply chain is organized and exactly how and when transport operations take place. Instead these decisions are made jointly between the supply chain parties, taking into account existing constraints within each of their businesses.

Therefore, logistics planning requires the active involvement of many organizations. Integrating the activities, motivations, and ideas of this wide range of organizations in order to provide products to consumers in accordance with their quality, price and availability expectations is the major challenge for logistics management.

Logistics management trends are increasingly leading companies to:

- coordinate all activities in the supply chain;
- compress the time from product development to market introduction;

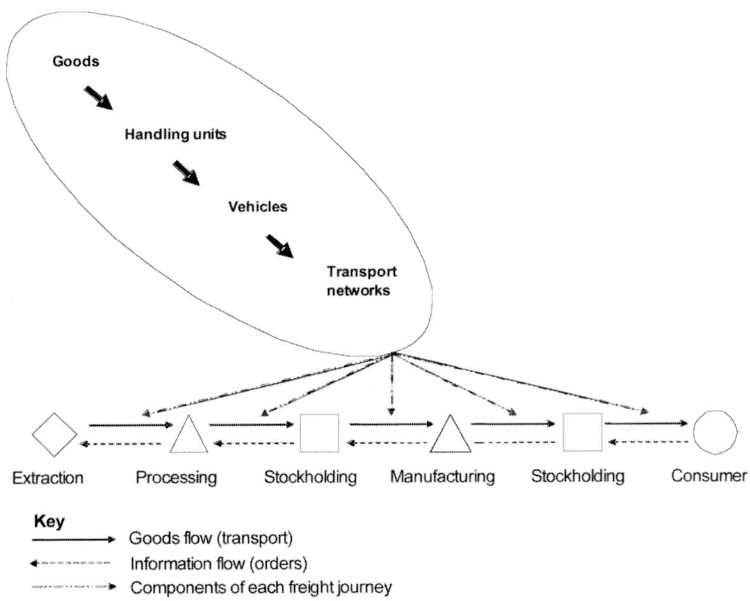

Figure 5.3 Components in the physical distribution and logistics management network for a product

- shift from forecast-driven to demand-driven supply chains;
- reduce the actual order to delivery time (order lead times);
- increase speed and reliability of movement between all parties in the chain (including inter-company moves and the return cycle);
- increase the geographical area over which products are produced and sold.

The various actors in supply chains are also becoming more demanding – for example final consumers are demanding better services, wider choices and lower costs together with product innovation. Major retailers expect order-cycle time reductions and lower costs (sometimes in order to buffer pressures on their own margins). Manufacturers have achieved significant production lead time reductions and now expect parts and finished goods to be shipped from and to any point in the world in a flexible and responsive way.

Closer political integration (such as has taken place in the EU) has a major effect on the level of trade between countries and this has important implications for logistics demand. The prospect of a Europe free from internal borders spurred many companies to review and then reconfigure their logistics systems (Cooper *et al.*, 1994). This, together with the logistics management trends listed above, has resulted in a rationalization of both production and stockholding sites among some large companies operating across Europe, thereby increasing the demand for national and, especially, international transport services. At the same time there has been a trend to reduce stock levels by managing production much more carefully and implementing just-in-time (JIT) production and ordering techniques. Both these developments lead to an increase in the consumption of transport services within the supply chain. This can occur as a result of either increasing trip length (as is the case with the concentration of production and storage) or greater frequency of deliveries (as occurs in a JIT system). At the same time there has been an increase in the degree of outsourcing of freight transport and other logistics activities as the needs have become ever more complex and this results in a reduction in in-house freight transport activities. In addition, there is greater pressure on transport services to achieve high levels of reliability.

2.5 Public and private sector roles in freight transport

It is important to distinguish between the two different groups who are capable of implementing changes to the freight system in order to improve integration and the rationale for their doing so, namely: i) governmental organizations (at international, national, regional and local levels); and ii) companies.

Logistics activities are primarily performed by private companies. Companies (freight transport companies and other supply chain parties) tend to implement initiatives that will improve the integration of their supply chain operations because they will derive some internal benefit from this change in behaviour. This would usually occur because the company can achieve internal economic advantages from improvements in integration. Instances of company-led initiatives to improve integration include closer working and coordination with supply chain partners to enhance efficiency, the use of improved, standardized materials-handling technology, and load unitization techniques. Changes implemented by governmental organizations occur through the introduction of policy measures that force or encourage companies to change their actions.

Governments at all levels do not have a very good track record in involving freight transport actors in the decision-making process in recent decades. Instead participation of such groups in policy making has often been kept to a limited consultation exercise at best. However, this has begun to change in the last few years as interest in freight transport has grown among policy makers, and some have decided that a more inclusive approach is likely to result in more efficient and environmentally sustainable outcomes.

Efforts to establish close working relationships between the public and private sector to address freight transport problems and to discuss and review existing policies and operating practices can have important impacts on the efficiency of freight transport operations and in reducing negative externalities. Examples of this type of joint working between the public and private sector include the introduction of Freight Quality Partnerships in the UK in the last ten years (see Section 3 for more details).

3. Policy interventions and integrated freight transport

Policy makers can make a wide range of interventions in an attempt to improve the integration of freight transport operations. These can be grouped into several categories including:

- transport infrastructure provision;
- infrastructure management;
- pricing;
- behavioural change;
- information provision;
- modal shift.

These policy interventions can take a number of different forms including: i) technological approaches (that generally attempt to change the supply of transport by making better use of existing resources); ii) economic and fiscal approaches (that usually try to influence the demand for transport by making transport more expensive, or encourage a particular mode or fuel type through financial incentives or subsidies); and iii) regulatory approaches that either restrict the way in which infrastructure or vehicles are used or put in place qualitative and quantitative controls to help prevent poor standards and operating practices. The geographical scale at which freight policy measures are typically applied varies by country and by measure – this can be international, national, regional, urban or site-specific.

Examples of policy interventions concerning integration at three different levels – international, national and local – are provided below.

International: European rail network for competitive freight

European transport policy favours the development of rail freight as a means of meeting sustainability commitments, including the reduction of greenhouse gas emissions and a reduced reliance on fossil fuel resources. Rail's share of the European freight market has declined considerably over several decades due to an inability to compete with more responsive and flexible road haulage services, although a number of member states have reported recent increases in rail freight activity. Currently, half of all rail freight in the European Union is international, reflecting rail's general strength in moving freight over long distances. However, the European Commission is concerned that a lack of integration between national rail networks is hindering the development of new high-quality, reliable rail freight services across the continent. As a consequence, several legislative packages have been implemented to liberalize the rail freight market and further legislation is proposed. This includes the creation of a European rail network for competitive freight, based on a number of international corridors and in line with existing initiatives such as the Trans European Transport Network (TEN-T) programme; the introduction of the European Rail Traffic Management System (ERTMS); and the various railway packages aimed at market liberalization and enhanced interoperability (European Commission, 2008).

In particular, by giving priority to certain types of freight traffic, the proposed international corridors will encourage:

- the integration of national rail infrastructure resulting from closer cooperation between infrastructure operators;
- an improved response to rail freight operators' requirements;
- more effective management of infrastructure used by both passenger and freight trains so that the latter are no longer systematically disadvantaged;
- better connections between rail and other transport modes, aiding the development of co-modality.

The main focus of the initiative is on time-sensitive flows, such as consumer goods in unitized loads, where a high level of performance is needed to provide a realistic alternative to road haulage. Integration, both between member states and among those involved in rail freight network and service provision, is crucial in order to achieve a high degree of efficiency and punctuality. For an analysis of policies aimed at integration to facilitate intermodal freight transport, see Chapter 13 in this volume.

National: UK Department for Transport – Delivering a Sustainable Transport System (DaSTS)

The UK Government published *Delivering a Sustainable Transport System* (DaSTS) in November 2008 (DfT, 2008a). DaSTS emphasizes the government view that investing in national transport networks is essential to stimulating the country's economic growth and supporting the nation's position as a leading world economy, but that this needs to take place, 'in a way which is consistent with reducing greenhouse gas emissions overall' (DfT, 2008a).

This will be achieved through a new National Networks Strategy Group, which will focus 'on measures to ensure we make best use of investment in key strategic national corridors, and develop these for the longer term. The development of solutions for International Networks will also be led by the Department. Regions and local authorities will undertake similar work for the remaining networks' (DfT, 2008b). It is intended that predictable end-to-end journey time for freight movements will be a priority of this National Networks' focus.

As part of this process, the Government has identified 14 Strategic National Corridors (SNCs) which 'represent the connectivity that is vital to the national economy' and 'cover a large number of strategic type journeys' (DfT, 2008c). The selection of these corridors has been based on the most important strategic destinations in the country (defined as the largest urban areas, the busiest airports and the most heavily used seaports). Some SNCs will make use of road, rail and air infrastructure, while others 'may only have significant strategic movement on one or two modes' (DfT, 2008c).

For international networks, the Government intends to carry out a series of gateway and freight studies to generate future policy options. The development of options to address the challenges on International Networks will build on the same end-to-end journey approach as for National Networks.

The Government has acknowledged that traditionally in thinking about freight transport it 'has focused on mode, rather than commodity – how many HGVs or trains there were rather than on the logistical detail of what freight moved where'. However, in the new approach contained in DaSTS they will be looking more closely at the composition of freight traffic on these key routes and seeking to understand more fully the factors generating freight demand on them (DfT, 2008d).

Also many government activities address freight and non-freight transport issues simultaneously, especially investment issues concerned with key strategic routes. In this sense the Government acknowledges the importance of integrating freight into wider decision-making processes, especially in terms of taking account of benefits to freight and non-freight in investment decision making (DfT, 2008d).

Local: Freight Quality Partnerships

The Central London Freight Quality Partnership (CLFQP) covers the London boroughs of Camden, City of London, Islington, Kensington and Chelsea, Lambeth, Southwark, and Westminster (CLFQP, 2007a). Central London was home to 1.4 million residents and contained almost 1.4 million jobs in 2008 (GLA Economics, 2009; GLA, 2009; Mayor of London, 2008). While geographically compact, its unique freight operating environment is served by operators from all over the UK and Europe.

CLFQP is a public/private partnership set up in 2006 to develop a common understanding of freight transport issues, and to encourage and create innovative solutions for delivery and movement of goods specifically within central London. The partnership encourages a sustainable approach to the movement of goods to achieve a balance between business needs and the environmental and social needs of the central London community (CLFQP, 2007b).

Key stakeholders on the CLFQP steering committee include: freight transport, distribution and service companies and their industry associations; local businesses and employers; and local authorities and Transport for London. There is also a working group looking at loading/unloading and parking issues. Approximately 200 different organizations have been involved in CLFQP work and activities to date. CLFQP is very active in responding to consultations at all levels of government. These include the Oxford Street, Regent Street and Bond Street consultation by the City of Westminster; the Mayor of London's Low Emission Zone; the DfT Operational Parking Guidance consultation and the Parliamentary Transport Select Committee investigations into Freight Transport; and Taxes and Charges on Road Users (at both of which the CLFQP provided verbal evidence).

The partnership is also working to improve communication and understanding between vehicle operators and parking enforcement authorities. It also represents operators' views to ensure that central London can continue to function efficiently in terms of goods and service provision during the 2012 Olympics.

Delivery and Servicing Plans (DSPs) are currently being developed and it is expected that they will bring efficient, sustainable and environmentally friendly delivery practices. As part of DSPs, businesses voluntarily agree to implement measures to reduce the impacts of goods vehicles serving their premises. Other recent work includes development of local freight forums to improve communication and address freight transport problems at a very local level; the implementation of new loading bays in several local high streets; and improvements in road signage for goods vehicle drivers.

When freight issues are addressed in the FQP context, members take joint responsibility and commitment in discussing the problems and potential solutions, recognizing that working in partnership is necessary for delivering successful freight transport schemes. The London FQPs website (www.londonsfqps.co.uk) informs members of the latest CLFQP developments.

4. Conclusions

Freight transport plays an important role in providing the goods and services required to ensure economic vitality and quality of life. However, in doing so these transport operations impose negative social and environmental impacts, including fossil fuel consumption, air pollution, noise, accidents, and traffic congestion. This relationship between the economic, social and environmental impacts (both positive and negative) lies at the centre of the interaction between freight transport and sustainable development.

Enhancing the integration of various aspects of freight transport (as discussed in Sections 2.1–5) is important in facilitating more sustainable freight transport operations through improving the efficiency of freight operations, while at the same time reducing the negative environmental and social impacts. It is this relationship between the integration and sustainability of freight transport that makes integration such an important topic and objective. However, as mentioned in Section 2.1, while improved integration of physical transport networks can help facilitate economic growth and prosperity, this can also lead to less environmentally sustainable outcomes, as goods are sourced and sold over ever-growing distances.

Efforts to improve integration in transport networks are usually achieved through major infrastructure projects to remove bottlenecks and fill in gaps to improve network connectivity and transfer. Given that the provision of transport infrastructure is usually a public sector responsibility, the vast majority of efforts in this field will be led by the policy makers using public funding. The recent global economic crisis has led to reductions in the demand for freight movement and as a result some of the pressure on infrastructure and freight services has been reduced. Future developments will always be subject to uncertainty, however, economic and social aspirations in many countries suggest that it will continue to be essential to plan for enhanced integration between freight modes in order to ensure that future growth can be accommodated in a way that minimizes negative environmental impacts.

Conflicts that arise as a result of use of shared networks by passenger and freight transport are also a topic for policy makers responsible for space allocation decisions concerning the use of the network and safety policies. These conflicts are usually addressed through regulations concerning the type/sizes/weights of vehicles that are permitted to use all or part of the network, and time-based regulations concerning access to the network (as well as space allocation and timing regulations concerning the use of the kerbside in the case of roads). Safety issues arising from the use of this shared space, especially concerning vulnerable users such as pedestrians and cyclists in the case of road transport, all fall within the remit of public policy makers.

Work on standardizing handling units has helped to greatly improve the speed and cost of freight transport and has been a major facilitator of growing international trade. Much of this effort (by companies, trade bodies and policy makers) has, as would be expected, taken place at an international scale and significant progress has been achieved in relation to containers.

The issue of modal split (i.e. ensuring each mode is used to the best of its strengths to improve efficiency and reduce negative externalities in relation to freight transport) is usually addressed by policy makers through pricing policies and subsidies to encourage mode shift towards less environmentally damaging modes (e.g. from road to rail). Success in improving modal split (in favour of more sustainable modes) has been relatively modest to date but the European Union is exploring policies to integrate external costs into the charging regimes and to harmonize fuel taxation between countries. It is widely expected that these policies will favour rail and waterborne modes, with their rates becoming more competitive relative to road haulage charges (European Commission, 2001; see also Chapter 13 in this volume for further discussion and analysis).

Policy makers' efforts to improve public and private sector joint working on freight transport issues have only begun relatively recently. Most progress on this issue has been made at the local (especially urban) scale, as reflected by Freight Quality Partnerships in the UK.

Integration issues in physical distribution and logistics management networks are the responsibility of companies rather than policy makers. Improving integration in this respect requires that processes and activities in the supply chain are better coordinated, and that supply chain parties work more closely together.

Companies will strive to bring about these integration improvements as they seek greater efficiency and higher productivity in their operations. However, policy makers are showing interest in better understanding supply chains and the flow of products along them in order to appreciate problems (and potentially solutions) from a logistics perspective rather than focusing only on freight vehicles and modes.

References

Beuthe, M. (2007) 'Intermodal freight transport in Europe', in T. Leinbach and C. Capineri (eds) *Globalized Freight Transport: Intermodailty, E-commerce, Logistics and Sustainability*, Cheltenham: Edward Elgar.

Beuthe, M. and Kreutzberger, E. (2001) 'Consolidation and trans-shipment', in A. Brewer, D. Henscher and K. Button (eds) *Handbook of Logistics and Supply Chain Management*, Oxford: Elsevier.

Central London Freight Quality Partnership (CLFQP) (2007a) *CLFQP Brochure*. Available online at: http://www.londonsfqps.co.uk (accessed 10 June 2009).

—— (2007b) *CLFQP Annual Activity Report 2007/8*. Available online at: http://www.londonsfqps.co.uk (accessed 10 June 2009).

Cooper, J., Browne, M. and Peters, M. (1994) *European Logistics: Markets, Management and Strategy*, Oxford: Blackwell.

Department for Transport (DfT) (2008a) *Delivering a Sustainable Transport System*, London: DfT.

—— (2008b) *National Networks*, Statement to Parliament by the Secretary of State for Transport, 29 October 2008. London: DfT. Available online at: http://www.dft.gov.uk/press/speechesstatements/statements/wmsnationalnetworks (accessed 5 June 2009).

—— (2008c) *Delivering a Sustainable Transport System: Consultation on Planning for 2014 and Beyond*, London: DfT.

—— (2008d) *Delivering a Sustainable Transport System: The Logistics Perspective*, London: DfT.

European Commission (2001) *White Paper – European Transport Policy for 2010: Time to Decide*, Brussels: European Commission.

—— (2006) *Keep Europe Moving: Sustainable Mobility for Our Continent. Mid-term Review of the European Commission's 2001 Transport White Paper*, COM(2006) 314 final, Brussels: European Commission.

—— (2008) *Proposal for a Regulation of the European Parliament and of the Council Concerning a European Rail Network for Competitive Freight*, COM(2008) 852 final, Brussels: European Commission.

Eurostat (2008) *Unitization of Freight Transport in Europe 2005*, 20/2008, Luxembourg: Eurostat.

—— (2009a) *Panorama of Transport: 2009 Edition*, Luxembourg: Eurostat.

—— (2009b) *Statistical Initiatives: Eurostat's Intermodal and Maritime Statistics Task Forces 2008*, 3rd Session of the UNECE Group of Experts on Hinterland Connections to Seaports, 23 January, Geneva: UNECE.

Freight Transport Association (1996) *Lorries in Urban Areas – Delivering the Goods and Serving the Community*, Freight Matters 5/96, Tunbridge Wells: Freight Transport Association.

GLA (2008) *ONS Mid-year Population Estimates: 2008, GLA Demography Update*, London: Greater London Authority.

GLA Economics (2009) *London's Economic Outlook: Autumn 2009*, London: Greater London Authority.

Hesse, M. and Rodrigue, J-P. (2004) 'The transport geography of logistics and freight distribution', *Journal of Transport Geography* 12, 3: 171–184.

Levinson, M. (2006) 'Container shipping and the economy: Stimulating trade and transformations worldwide', in *The Intermodal Container Era: History, Security, and Trends*, Transportation Research Board of the National Academies, TR News Number 246, September–October: 10–12.

Limbourg, S. and Jourquin, B. (2009) 'Optimal rail-road container terminal locations on the European network', *Transportation Research Part E: Logistics and Transportation Review* 45, 4: 551–563.

Mayor of London (2008) *The London Plan: Spatial Development Strategy for Greater London – Consolidated with Alterations Since 2004*, London: Greater London Authority.

Mercogliano, S. (2006) 'The container revolution', *Sea History* 114: 8–11.

Network Rail (2010) Intermodal services, http://www.networkrail.co.uk/aspx/2210.aspx

Ogden, K. (1977) 'Modelling urban freight generation', *Traffic Engineering and Control* 18, 3: 106–109.

Slack, B. (1998) 'Intermodal transportation', in B. Hoyle and R. Knowles (eds) *Modern Transport Geography*, Chichester: Wiley.

—— (2001) 'Intermodal transportation', in A. Brewer, D. Henscher and K. Button (eds) *Handbook of Logistics and Supply Chain Management*, Oxford: Elsevier.

Chapter 6

The value of reliability and its relevance in transport networks

Luca Zamparini and Aura Reggiani

1. Introduction

The efficient and effective development of a transport network, at the local, national and international scales, requires the possibility for passengers and for firms to experience only a limited degree of heterogeneity between planned and experienced travel times. In order to accomplish this goal, the necessary infrastructural investments need not only to take care of the prospect to reduce travel times but, and maybe more important, also to increase the level of reliability of the transport system. This aim has also been part of the proposition at the basis of the White Paper (European Commission, 2001) that, at the beginning of this decade, stated the necessity to eliminate (reduce) the bottlenecks that are present in the European transport network. Moreover, the same document put changing the relative shares of using various modes at the core of the European Union middle- to long-term strategy in the transport sector. In this context, it should be kept in mind that any transport activity that includes a mode different than road is, in the vast majority of the cases, a multimodal trip. It is therefore necessary that the various legs of the trip are tightly integrated, from both the geographical and transport system viewpoints. As a consequence, nodal points become extremely relevant for the achievement of an integrated transport[1] system (Gorter *et al.*, 2000), and the possibility to minimize travel times and thus maximize the utility of economic actors depends on the reliability of scheduled travel times. In the case of public transport chains and networks, reliability must be based on a customer-oriented approach and on arrival and departure times, as this may influence the choice of a unimodal or multimodal transport alternative (Bruinsma *et al.*, 1998).

Reliability is also a very important attribute when freight transport firms' plans are considered. A recent study by Muilerman *et al.* (2005) has ascertained that firms that adopt time-based strategies refer to on-time reliability (regarded as a trade-off with costs) of their logistics networks as the most important factor driving the evolution of their strategies and the shift from unimodal to multimodal freight transport options. Logisticians may be more interested in increasing on-time reliability, even if this implies a more costly logistics structure. Moreover, traditional logistics literature has emphasized the relationship between transport-related uncertainty and inventory management costs (see, among others, Closs *et al.*, 2003 for rail transport and Saldanha *et al.*, 2006 in the case of maritime transport. See also Chapter 5 in this volume on integrated transport policy in freight transport).

Given the importance of reliability for both private economic actors and public administrations, several analytical specifications and empirical evaluations of the value of reliability (VOR) have been proposed for both passenger and freight transport. The VOR can be defined as the willingness to pay in order to reduce the randomness in experienced travel times on a determined route. This is certainly relevant nowadays in the presence of integrated transport systems/networks, where reliability (and its related value) can be considered as a user-oriented quality measure (Immers *et al.*, 2004). The costs connected to a low level of reliability are manifold. They can be related, for example, to late arrivals, to the necessity to spend more time than expected in travel, to the delays of subsequent logistics activities, and to the possibility of losing a connected leg of a multimodal trip. All these play an important role in the preference to use the private car over public transport.

The present chapter offers a survey of the transport economic literature concerning VOR and a discussion of its relevance in integrated transport networks. First, it considers the main micro-economic models that have been proposed in order to provide an identification of the VOR. Second, it presents an international overview of the VOR estimations, aiming to ascertain the similarities and differences among transport means and among countries, in both passenger and freight transport. Third, it proposes a discussion that pertains to the relevance of VOR in multimodal transport chains in both passenger and freight transport. Finally, the paper offers some concluding remarks and highlights future directions of research.

2. Theoretical survey of value of reliability models

2.1 Preface

Reliability can be considered as the consistency of travel times on a certain origin-destination among different economic actors or as the possible repetitions of the same travel by an economic actor at different hours of the day (i.e. a truck driver making several different short-distance shipments in a day) or on different days (i.e. a commuter travelling from home to work every morning). In passenger and public transport, reliability (or, more properly, the lack of) can be defined as the sum of residual and random variations in travel time once the economic actor has taken into account the variations during the day, week and season, and the other foreseen causes (i.e. road maintenance works) of heterogeneity in travel time. From a statistical viewpoint, reliability is closely related to the variance (or standard deviation) in travel times. Hence, the higher the reliability, the lower the variance. It is possible to consider, for example, the morning and afternoon trips of a commuter on a certain route to measure the travel time reliability. The lower the variability of his travel times, the higher the reliability.

From an economic viewpoint, the lack of reliability imposes an extra cost on a traveller or on a freight transport firm planning and implementing a predetermined journey. This cost can be related, as proposed in seminal contributions on this topic by Gaver (1968) and Knight (1974), to the necessity to depart earlier than in the case of reliable transport in order to have a safety buffer or margin. However, when transport networks or logistics chains are considered, unreliable travel times can impose economic costs on other legs of the network or to other phases of the logistics chain (i.e. the necessary increase of 'safety' stocks in warehouses or points of sale). Reliability has thus been defined, by part of the economic literature related to freight transport, as the percentage of consignments that arrive at destination within scheduled time.

The following sub-sections illustrate the main theoretical models that have been proposed in order to provide an identification of the VOR. The first will analyse the main models that have included reliability among the attributes that characterize passenger transport activities; the second focuses its attention on freight transport; and, lastly, the third proposes some remarks on the similarities and heterogeneities among the models presented in the previous sub-sections.

2.2 The value of reliability in passenger transport models

The paper by Small (1982), based on the time allocation model by Becker (1965), can be considered the seminal work on VOR in passenger transport. In this article, the analysis was devoted to the importance of scheduling choices with respect to leisure time, working time and 'consumption time' (the time spent on an activity that provides no direct utility but is a necessary complement to consumption of goods, leisure or working time). In addition, the author proposed a utility function (W) that was influenced not only by travel time but also by early arrivals and late arrivals. The utility function:

$$W = R(s) + \mu TIM + \beta_1 SDE + \beta_2 SDL + \beta_3 DIL \qquad (1)$$

where R(s) is the possibility of reporting errors rounded off to 10 or 15 minutes,[2] TIM (Travel Time in Minutes) is travel time expressed in minutes, SDE (Scheduling Deviation in case of Early Arrivals) is the early arrival variable, SDL[3] (Scheduling Deviation in case of Late Arrivals) is the late arrival variable, and DIL (Lateness Dummy) is an extra cost due to late arrival. The coefficients, which were estimated through an experiment conducted in the San Francisco Bay Area, showed that all coefficients are negative, as expected, and that $\beta_3 < \beta_2 < \mu < \beta_1$. This implies that the disutility of arriving early is lower than the amount of travel time gained and that the largest disutility is due to late arrivals, especially in the case in which the work trip is related to a job with an inflexible schedule. So, workers prefer early arrivals to late ones and are thus eager to sacrifice a determined amount of their time before their job shift begins in order not to incur in the larger penalty due to a late arrival.

Noland and Small (1995) and Noland *et al.* (1998) base their model on the assumptions proposed by Gaver (1968) (according to whom, since utility is a linear function of travel time, schedule delay early and schedule delay late, the economic actor maximizes the expected utility with a particular scheduling choice on the basis of the distribution of travel times) and by Polak (1987) (who took into account the risk aversion of individuals and opted for a concave transformation of the utility function). They add a discrete lateness penalty, taking into account the fact that the level of congestion is not stable during the morning commute but is characterized by a time varying pattern, and derive the consequent expected costs to the commuter given her expected scheduling choice.

Noland *et al.* (1998) consider the abovementioned utility function by Small (equation 1) and clarify the elements that constitute the total commuting time (T) as the sum of T_f (free-flow travel time), T_x (recurrent congestion), and T_r (unpredictable congestion, due to accidents or unexpected maintenance work). The latter term is based on a probability distribution function. They thus derive the following scheduling cost function (C_S):

$$C_S(T_r, T_e) = \alpha(T_f + T_x + T_r) + \beta(1 - DL)[T_e - T_r] + \gamma DL(T_r - T_e) + \theta DL \qquad (2)$$

where T_e is the maximum early arrival time and DL is the lateness dummy. Given that T_r is based on a probability distribution function, Noland *et al.* (1998) define P_L as the probability to arrive late and derive the expected cost function (EC_S):

$$EC_S = \alpha E(T) + \beta E(SDE) + \gamma E(SDL) + \theta P_L \qquad (3)$$

Another cost imposed by unreliability is given by the decreased degree of ability to plan one's activities. Noland *et al.* (1998) consider in this respect a planning cost (C_P) that is due to the standard deviation (STDEV) of the extra amount of travel time that is due to unpredictable congestion (T_r). The total expected cost (EC) is consequently equal to:

$$EC = EC_S + C_P = \alpha E(T) + \beta E(SDE) + \gamma E(SDL) + \theta P_L + \sigma f(STDEV) \qquad (4)$$

This latter specification can be very important when transport networks are considered. The planning cost for the individuals could also be interpreted as the necessity to rearrange subsequent or previous legs of an overall transportation activity given the standard deviation of the unexpected congestion on a determined leg. Reductions of this variability may also imply the possibility of reconsidering and optimizing the other legs of the network.

A different specification to model the choices of travellers among different possible routes was proposed by Lam and Small (2001) and Small *et al.* (2005) by using the stochastic utility function (U_{in}) for the individual *i* and modal choice *n*, in the framework of the random utility theory conceived by McFadden (1974), as follows:

$$U_{in} = V_i(t_{in}, v_{in}, c_{in}, x_{in}) + \varepsilon_{in} \qquad (5)$$

where *t*, *v*, and *c* are respectively the travel time, the variability in travel time and the cost of travel. Moreover, *x* is a meta-variable that summarizes a series of socio-economic attributes and the specific characteristics of the trip. The choice of the economic actor with respect to two or more alternatives from an origin to a determined destination is based on the comparison between the obtained values for the indirect utility. This function gives an economic estimate both to the value of time[4] (VOT, mainly based on *t* and *c*) and to the VOR (mainly based on *v* and *c*) and enables a comparison of them. When a transport network is considered, it is probably the case that the VOR exceeds the VOT.

Bhat and Sardesai (2006), aiming to analyse the commute mode choice, have underlined the impact of reliability when work trips are coupled with non-work activities through stop-making. They also use the random utility (U_{qit}) maximizing model of choice, as follows:

$$U_{qit} = \beta'_q w_{qit} + \mu' z_{qit} + \phi_{qit} \tag{6}$$

where q represents an individual, i is the chosen transport alternative, t is the occasion when the alternative is selected, and w_{qit} defines the choice of an economic actor, according to a revealed preference or a stated preference method, with a corresponding coefficient vector (β'_q QUOTE) of idiosyncratic observed and unobserved attributes. In equation 6, the stochastic component is divided in two error terms ($\mu' z_{qit}$ and ϕ_{qit}), which are respectively related to uncorrelated and correlated errors across alternatives and individuals. The estimation of a log-likelihood function will obtain the coefficients of the transport-related attributes of w_{qit}. It is consequently possible to estimate the VOR.

Table 6.1 summarizes the main similarities and heterogeneities among the theoretical models presented in this sub-section.

Table 6.1 Main characteristics of theoretical models related to the VOR in passenger transport

Author(s)	*Utility/Cost function*	*Attributes*
Small (1982)	Utility	Reporting errors, travel time, schedule delay early, schedule delay late, lateness penalty
Noland *et al.* (1998)	Expected scheduling cost	Expected travel time, expected schedule delay early, expected schedule delay late, lateness penalty, standard deviation of extra amount of travel time
Lam and Small (2001)	Stochastic utility	Travel time, variability in travel time, cost of travel, socio-economic attributes and characteristics of the trip
Bhat and Sardesai (2006)	Stochastic utility	Idiosyncratic observed and unobserved attributes, error term divided among uncorrelated and correlated errors across alternatives and individuals

It is possible to note that three models consider various forms of utility, while the model by Noland *et al.* (1998) takes into account the expected scheduling cost. With respect to the considered variables, the first two models concentrate on travel time and on a degree of reliability based on schedule delay, both early and late. Moreover, they consider a lateness penalty. The latter two models contemplate reliability jointly with a series of socio-economic and idiosyncratic attributes. To our knowledge, all the authors that have estimated reliability have used a logit specification. No author – to our knowledge – used a nested logit specification. However, especially in the case of integrated transport, the use of nested logit would allow one to fulfil the IIA (Independence of Irrelevant Alternatives) hypothesis and to consider the sequential choice process related to the various legs of an origin to destination journey.

2.3 The value of reliability in freight transport models

As in the case related to the evaluation of VOT in freight transport (Zamparini and Reggiani, 2007b), a prerequisite for all models concerning the VOR is the identification of the economic agent whose utility, or profit function, has to be maximized (i.e. the shipper, the receiver, or some other economic agent along the supply chain). This is particularly important when logistics chains are considered, given that the lack of reliability may impact in a remarkable way on the time and monetary costs incurred by firms that take part in the later stages of the supply chain. It may thus be the case that improving the reliability of a transport activity is more relevant than reducing the travel time. Reliability is therefore a crucial factor in promoting intermodal freight transport (see Chapters 5 and 13 in this volume).

The reasons why the freight transport industry values reliability have been surveyed in a paper by Fowkes *et al.* (2004), who divided demand side considerations and supply side ones. On the demand side, it is worth mentioning 'just in time' practices, deadlines for arrivals at ports (in case of a low level of reliability, strict arrival times at ports imply a rescheduling to start journeys earlier in order to achieve the preferred or booked sailing), quick-response operations for deliveries both to individual retail outlets and to regional distribution centres, and hub-and-spoke operations in regional, national and international transport networks. A particular freight transport network that requires an excellent degree of reliability is that related to express parcel operations, due to the possible refunds to customers for late deliveries. On the supply side, reliability is important for the efficiency of logistics operations, especially in the case of two-way loading practices related to the same industrial process (i.e. the inbound transport of raw materials or intermediate goods is connected to the outbound delivery of finished goods), of round-the-clock operations, and for the consolidation of heterogeneous deliveries. Moreover, the lack of reliability may imply a burden for transport and logistics firms in terms of driving hours and of management and warehousing regimes.

One of the first economic models taking into account reliability as one of the important attributes of freight transport was proposed by Winston (1981) in the context of the demand for intercity freight transport. After discussing the importance of choosing the appropriate economic agent in all cases that are characterized by imperfect competition and, more importantly, by uncertainty, he discussed the modal choice of a receiver on the basis of his desired inventory lot, of his production plans, of the daily quantity received, and of his attitude towards risk. This choice was modelled by Winston as a maximization problem of a random utility function (U_i) as:

$$U_i(Z_i, S) = V(Z_i, S) + \varepsilon_i(Z_i, S) \quad (7)$$

where U_i is the utility related to the ith mode. It can be divided into an observed part equal to $V(Z_i, S)$ and a random component that corresponds to $\varepsilon_i(Z_i, S)$. Z_i is the modal attributes of the ith mode and S is a vector of the characteristics related to firm and commodity. By considering the expected value of Z_i, it is possible to express the random expected utility (EU) of a receiver as:

$$EU(Z_i, S^k) = V(Z_i, S^k) + \eta_i^k \quad (8)$$

The error component (η_i^k) accounts for all unobserved heterogeneities in modal, firm and commodity attributes, as well as the attitude towards risk. Among the observed parameters, Winston considers reliability, defines it as the relative variation in transit times, and measures it as the coefficient of variation (i.e. the ratio between standard deviation and average transit time).

The model proposed by Winston is very important when transport networks are considered as it implies that modal shift (in the case of Winston's case study from road to rail) is not only due to a reduction in travel time but also to an increase in the reliability of a determined transport mode, or to a combination of both attributes.

A different definition of reliability in freight transport (considered as the percentage of consignments arriving within scheduled time) was provided by Fowkes and Shinghal (2002). They presented in detail the Adaptive Stated Preference Model that had already been used in previous works since the late 1980s (Fowkes and Tweddle, 1988; Fowkes *et al.*, 1991).

The model proposed by Fowkes and Shinghal is based on a utility function whose value is determined by a weighted sum of the modal or trip option attributes. The probability of choosing a determined alternative is based on a binary logit model. The parameters of the models are estimated through a weighted least squares regression. It is then possible to obtain monetary estimations of the various transport attributes other than the cost of the trip (i.e. time, reliability and frequency of service) by means of ratios between each attribute parameter and the cost parameter.

The adaptive stated preference approach was also adopted by Bolis and Maggi (2003) for their analysis of logistics strategy and transport service choices. They model the shipper as an industrial firm that has transport services among the inputs of its production function. This production function is based on the following cost structure (C_i):

$$C_i = C_i(Q, p_0, p_{ti}, Z_i, N) \quad (9)$$

where C_i is the cost related to choosing the ith transport alternative, Q is the produced output, p_0 is the price of all factors of production but transport, p_{ti} is the cost of transport when the ith alternative is adopted, Z_i is a vector of the transport and logistics attributes of the ith alternative (price, time, reliability, mode, frequency and flexibility) and N accounts for the network size and structure of the firm. The probability of choosing a determined transport alternative is then estimated by a binary logit model.

A similar approach was followed by Witlox and van Daele (2005), who studied the modal choice of a shipper. They identified six important attributes of transport activities (cost, time, loss and damage, frequency, reliability and flexibility) and used them to evaluate a non-linear utility function for the shipper. All attributes were then related to partial utility functions. By considering all of them, they were able to obtain the following equation:

$$\Delta U = \alpha\Delta u_1 + \beta\Delta u_2 + \gamma\Delta u_3 + \acute{E} \quad (10)$$

where ΔU represents the change in general utility due to changes in the single attributes times their related weights ($\alpha\Delta u_1$, $\beta\Delta u_2$, $\gamma\Delta u_3$, and so on). The presence of cost among the attributes of transport activities allows the computation of the monetary values of the other attributes, and among them reliability, as the amount of money that the shipper is eager to give up in order to observe an increase in one of the other attributes and keep constant their general utility level. As in the case of VOT (Zamparini and Reggiani, 2007a), the identification of VOR emerges from a maximization problem related to micro-economic models of discrete choice theory.

Analogously to Table 6.1 above, Table 6.2 summarizes the main similarities and heterogeneities among the theoretical models proposed for VOR in freight transport.

Table 6.2 Main characteristics of theoretical models related to the VOR in freight transport

Author(s)	*Utility/Cost function*	*Attributes*
Winston (1981)	Random utility function	Modal attributes of the ith mode of transport, characteristics of the firm and of the transported commodity
Fowkes and Shinghal (2002)	Weighted utility function	Modal or trip option attributes (cost, time, reliability, and frequency of service)
Bolis and Maggi (2003)	Cost function	Produced output, prices of all factors of production but transport, cost of transport, transport and logistics attributes of the ith alternative, network size and structure of the firm
Witlox and van Daele (2005)	Non-linear utility function	Quality attributes of freight transport (cost, time, loss and damage, frequency, reliability, flexibility)

Three models are on the (random, weighted, or non-linear) utility function, while the model by Bolis and Maggi (2003) considers a cost function. The modal (or trip option) attributes are taken into account by all models among the explanatory variables. The models by Winston (1981) and by Bolis and Maggi (2003) also include the characteristics of the firm.

3. An overview of empirical estimations of the value of reliability

The various models examined in the preceding section were the theoretical basis used in order to obtain the empirical VOR estimates that will be presented and discussed in the following sub-sections. The data collection methods used in the literature for VOR are the revealed preference (RP) strategy and the stated preference (SP) strategy. Both methods have strengths and weaknesses. The main advantage of RP is that it is based on the actual choices of the individuals or firms. However, SP studies allows the consideration of a wider array of possible alternatives (for a full description of the strengths and weaknesses of RP and SP studies, see Zamparini and Reggiani, 2007a).

The following sub-sections outline a survey of the empirical estimates of the VOR in both passenger and freight transport. The main aim of this survey is to introduce, at the widest possible extent, an heterogeneous sample of studies with respect to both geographical location and time span, and the transport mode considered. The main limitation encountered in the empirical survey was related to the fact that it was not always possible to identify a monetary VOR. Therefore, among the reviewed economic literature related to reliability, only the studies presenting a VOR that could be quantified in monetary terms were considered and included in the tables presented and discussed below.

3.1 Empirical survey of value of reliability in passenger transport

Table 6.3 proposes a summary of the transport economic studies that have been published since 2001 and that have estimated a monetary VOR for passenger transport. Five countries are considered (United States, New Zealand, United Kingdom, the Netherlands and Spain) and the period in which the data of the various studies were collected spans from 1997 to 2005. Most of the studies evaluate the VOR in terms of the local currency per hour. However, the two studies that have considered the United Kingdom have proposed a monetary VOR per minute.

In order to compare the results obtained for the VOR in the various studies, they were standardized in terms of percentage of the hourly wage rate (or of hourly wage rate divided by 60 in the case of the two studies that considered the VOR per minute in the United Kingdom). The data related to the hourly wage rates were gathered by means of the Laborsta Database of the International Labour Office.[5] The hourly wage rate considered was the one related to the year of data collection.[6] For the studies whose surveys were conducted in two adjoining years, an average of the wage rate in those two years was computed.

All studies performed outside the United States used a SP data collection approach. Among the four studies conducted in the United States, two studies used a RP method and two studies had a mixed RP/SP strategy.

Table 6.3 Survey of empirical estimates of the VOR in passenger transport

Author(s)	*Year of data*	*Data collection*	*Mode/Country*	*Value – percentage of hourly wage rate in parentheses*
Lam and Small (2001)	1998	RP	Road/US	$22.72 per hour (177%)
Liu *et al.* (2004)	1999–2000	RP	Road/US	$20.63 per hour (153%)
Small *et al.* (2005)	1999–2000	RP and SP	Road/US	$24.31 per hour (180%)
Bhat and Sardesai (2006)	2003–04	RP and SP	Multiple modes/US	$7.18 per hour – flexible work schedule (46%) $14.23 per hour – inflexible work schedule (92%)
Hensher (2001)	1999	SP	Road/NZ	NZ$5.15 per hour (29%) NZ$15.10 per hour in the Auckland area (86%)
Bates *et al.* (2001)	1999*	SP	Rail/UK	56 p/min early arrivals (410%)
				114 p/min late arrivals (838%)
				127 p/min value of delay (933%)
Hollander (2006)	2004	SP	Bus/UK	5.2 p/min early arrivals (30%)
				14.4 p/min late arrivals (84%)
Rietveld *et al.* (2001)	1997	SP	Public Transport/ Netherlands	€0.29 to move from 50% to 0% probability of 2 minutes delay (62%) €2.18 to move from 50% to 0% probability of 15 minutes delay (62%)
Asensio and Matas (2007)	2005	SP	Road/E	€34.40 per hour late arrivals (301%)
				€51.10 fixed work schedule late arrivals (448%)
				€21.40 flexible work schedule late arrivals (188%)
				€7.00 per hour early arrivals (61%)

Note:
* Presumed year of data collection, as explained in the text.

It is possible to note the extreme heterogeneity in the estimated VOR. It varies from a minimum value of 29 per cent of the hourly wage rate for the case of New Zealand in 1999 (Hensher, 2001) to a maximum value of 448 per cent of the hourly wage rate in the case of late arrivals with a fixed work schedule in the study by Asensio and Matas (2007), which took into account the commuters of the Barcelona area in Spain in 2005. It is noteworthy that passengers attach a disutility also to early arrivals, although it has a remarkably lower monetary value than the disutility related to late arrivals (see Bates *et al.*, 2001; Hollander, 2006; Asensio and Matas, 2007). Moreover, in the cases in which the commuting activity is connected to a fixed work schedule, the VOR is significantly higher than the value obtained for a flexible work schedule (see Bhat and Sardesai, 2006, and Asensio and Matas, 2007).

By taking into account the geographic location of the studies, it appears that Barcelona (Asensio and Matas, 2007) and Los Angeles (Lam and Small, 2001; Liu *et al.*, 2004; Small *et al.*, 2005) are the two areas where the highest VORs, in percentage terms with respect to the hourly wage rate, are observed. This similarity in results might be explained by the fact that these cities are characterized by a relatively high degree of sprawl. For example, in the context of the United States, the study by Bhat and Sardesai (2006) in the Austin area displays VORs that are remarkably lower than the ones obtained in the Los Angeles area. The study performed in New Zealand (Hensher, 2001) shows a sensible difference between the overall VOR and the VOR obtained for the sub-sample related to the metropolitan Auckland area, which has a suburban character and is characterized by the long distances travelled.

When the transport mode is considered, it is not possible to elicit – from Table 6.3 – a clear pattern of the estimated VORs. Private car transport (road) VOR values have in general a higher value than bus and public transport, with the exception of the overall VOR estimate for private car transport in New Zealand. A possible explanation of this result might be related to the higher income, and consequent higher willingness to pay of car owners with respect to public transport users. Concerning Europe-based studies, surprisingly the lowest VORs are observed in the Netherlands. This is probably due to the remarkable frequency and complementarity of transport modes in dense transport and infrastructure networks.

It is interesting to note that the case study by Bates *et al.* (2001) was performed in London at Paddington Station, which is an important node of the metropolitan multimodal transport network, taking into account rail transport. The resulting VOR estimates, normalized by the likely hourly wage rate at the time of the experiment, are remarkably higher than all the other VORs of the sample considered in Table 6.1. Part of the result may be due to the fact that the estimates, unlike the majority of the other VORs that are calculated per hour, are computed per minute. However, a large part of the difference is probably due to the fact that Paddington Station is one of the main nodal points of the multimodal transport network of London. Consequently, even a few minutes of delay with respect to the preferred arrival time at the station may result in a remarkably larger delay at the final destination of the trip.

3.2 Empirical survey of value of reliability in freight transport

This sub-section presents and discusses a sample of case studies in freight transport that display monetary VOR estimates (Table 6.4), spanning from 1981 to 2004. The studies were conducted in eight different countries (United States, Sweden, Finland, United Kingdom, Switzerland, Italy, India and Australia) and considered three different modes of transport. Seventeen estimated VORs are related to road haulage, five studies pertain to rail transport and one is related to air transport. The two studies conducted in the United States are based on RP while all other studies have used SP data collection strategies.

The VOR observations are very heterogeneous and a comparison among the studies is difficult for a series of reasons. First, the definition of reliability differs among the studies. Some consider reliability as the standard deviation of arrival at destination or the spread of arrival times (Fowkes *et al.*, 2001). Other studies consider reliability as the percentage of consignments that arrive in due time and estimate the VOR as the amount of money that firms are willing to pay in order to have a one per cent or one per thousand increase of this indicator. Second, the unit of value for the latter cases is varied. Some studies take into account the entire shipment, others the tons or the pallets. Lastly, while most of the studies consider one single country, two studies (Kurri *et al.*, 2000, and Bolis and Maggi, 2003) analyse the transport activities between two different countries.

It is, however, possible to draw some interesting conclusions through the analysis of the results of the various studies. From the studies by Transek (1990 and 1992), it emerges that when the consignments are due in the same day the VOR is higher than one obtained in the case of next-day consignments, regardless of the transport mode. It thus appears that reliability scores a higher value when the transport activity is inserted in a tight logistics schedule.

In general, the VORs for road transport are not comparable with the ones obtained for rail transport, given that the unit of value is an entire shipment (Winston, 1981; Transek, 1990; INREGIA, 2001), and the transported quantities in the two transport modes are of different orders of magnitude. However, the study by Kurri *et al.* (2000) considers the ton as the unit of measurement and obtains a VOR for road transport that is remarkably higher than the VOR for rail transport.

The VOR observations that emerge from the paper by Fowkes *et al.* (2004) allow us to consider the various statistical dimensions of reliability. It obtains the highest estimate for the value of delay time, followed by the value of spread time and the schedule delay. The study by Bolis and Maggi (2003) present a VOR related to international transport within two European alpine countries (Switzerland and Italy). Besides the estimated VOR, their study is important as it ascertained the necessary increase in reliability that firms would require in order to move from unimodal to multimodal (rail and road) transport. Lastly, Wigan *et al.* (2000) estimate VOR as the percentage of consignments in due time and compare the observations for three different road haulage activities. It emerges that intercapital freight transport (overnight transport, between two different cities, from one plant to another or from one plant to a warehouse) has a per pallet VOR that is higher than the one obtained for transport of pallets from one plant to a warehouse or to a different plant in the

Table 6.4 Survey of empirical estimates of the VOR in freight transport

Author(s)	*Year of data*	*Data*	*Mode/ Country*	*Value*	*Unit of value*
Winston (1981)	1975–77	RP	Road/US	US$404.00	Day, standard deviation
	1975–78		Rail/US	US$299–4,110	Day, standard deviation
Small *et al.* (1999)*	–	RP	Road/US	US$371.33	Hour and shipment
Wigan *et al.* (2000)	–	SP	Road/AUS	AUS$2.56 Intercapital	1% unit and pallet
	–		Road/AUS	AUS$1.25 Urban	1% unit and pallet
	–		Road/AUS	AUS$1.97 MMD**	1% unit and pallet
Transek (1990)*	1989–90	SP	Rail/S	SEK 60 same day	1% unit and shipment
	1989–90		Rail/S	SEK 40 next day	1% unit and shipment
	1989–90		Road/S	SEK 150 same day	1% unit and shipment
	1989–90		Road/S	SEK 30 next day	1% unit and shipment
Transek (1992)*	1991	SP	Road/S	SEK 280 same day	1% unit and shipment
	1991		Road/S	SEK 110 next day	1% unit and shipment
Kurri *et al.* (2000)*	1997	SP	Road/S-F	US$47.47	Hour and ton
	1998		Rail/S-F	US$0.50	Hour and ton
INREGIA (2001)*	1999	SP	Road/S	SEK 63	1 per th.nd and shipment
	1999		Rail/S	SEK 1142	1 per th.nd and shipment
	1999		Air/S	SEK 264	1 per th.nd and shipment
Fowkes *et al.* (2001)*	2000–01	SP	Road/UK	£61.5–167.6	Hour and spread
Fowkes *et al.* (2004)	2000–01	SP	Road/UK	107 p/min	Value of delay time
	2000–01		Road/UK	85 p/min	Value of spread time
	2000–01		Road/UK	66 p/min	Schedule delay
Bolis and Maggi (2003)	–	SP	Road/CH-IT	CHF 2.42	1% unit and ton

Source: Tavasszy and Bruzelius, 2005

Note:
* Tavasszy and Bruzelius studies
** MMD = Metropolitan Multidrop Deliveries.

same city. An intermediate VOR is obtained for an urban freight transport with many deliveries (Metropolitan Multidrop Deliveries).

4. The importance of the value of reliability in passenger and freight transport networks

Although no actual study displaying values of reliability in multimodal transport chains seems to have been performed, several elements emerging from the theoretical and empirical surveys of the previous sections seem to confirm the importance of the VOR for both passenger and freight transport in multimodal transport networks.

When passenger transport is considered, all possible causes influencing the travel time distribution, and consequently the reliability, in a transport network can be summarized as follows (van Lint *et al.*, 2008). The demand fluctuations may be due to seasonal effects, to the demand on parallel and connecting links and to the consequent spillback effects, to population or firm characteristics and to traffic information and user response. On the supply side, the fluctuations may be due to incidents, accidents, road maintenance works, road geometry and regulations, and traffic management and control. The interactions of supply and demand determine the volume of transport on the network and as a consequence the level of reliability. Improvements in the transport network may ameliorate reliability and, on the other hand, an increased reliability may determine the choice of mode and network among several alternatives, according to the theoretical models discussed in Section 2.

Some of the studies surveyed (Table 6.1) show that the VOR increases when the trip with a specific transport mode is part of an inflexible schedule. The importance of reliability for the evolution of transport networks has been clarified by Bates *et al.* (2001) who considered the case of fixed service intervals and theoretically compared the choice between two distinct services, the first one leading to the last possibility of arriving early and the second one to the first possibility of arriving late. Under the hypothesis of perfect reliability of both services, the chosen service will depend on the comparison between the disutility of arriving early and the disutility of arriving late. Bates *et al.* (2001) then introduce the notion of reliability, by assuming that services depart on schedule but travel time varies according to a determined probability distribution, and apply it to a rail journey with two legs. They assume that the train service in both legs is operated at a fixed interval and analyse all possible waiting times that are experienced by travellers at the interconnecting station. The factors influencing the waiting time are related to the random distribution of travelling time of the first train, the connection time and the headway of the second leg service. The arrival at the final station is then also dependent on the reliability distribution of the second train. If the minimization of general costs for travellers has to be taken into account, it is necessary to increase either the reliability or the frequency of the existing train services in a transport network.

In the case of freight transport, an interview-based survey by Muilerman *et al.* (2005) has ascertained that an improved on-time reliability, measured as the percentage of shipments on time, is the most important condition behind logistics changes and the move from unimodal to multimodal freight transport. Interviewed firms stated that their declared time-based logistics strategies are aimed at reducing the degree of uncertainty rather than increasing the speed of the logistics processes. Moreover, several studies (see, for example, Stank and Crum, 1997; Shinghal and Fowkes, 2002) have indicated that the VOR is higher when international transport activities are considered. The study by Stank and Crum (1997) has highlighted that the possibility of implementing sound 'just in time' distribution practices between the Mexican and United States border is tightly linked to the attainment of efficient transportation service performances. They have also stressed, in line with logistics-based literature, the need to increase inventory costs in response to unreliable transport networks. The study by Shinghal and Fowkes (2002) that provided an estimation of VOR for road transport in the Delhi-Mumbai corridor in India was based on a unit of measure of 8 hours per shipment and proposes a peculiar value of 3.35 per cent of overall costs. According to this estimate, firms would be willing to increase transport costs by 10 per cent in order to have a reduction of one day in the variability of arrivals. This value is mainly related to the fact that, in most cases, road transport in India is the first leg of a multimodal transport network. Firms are thus interested in arriving with a good degree of reliability to the node represented by the port of Mumbai.

Concerning the comparison between VOT and VOR in the case of passenger transport, VOT is normally lower in railways than in road transport (Zamparini and Reggiani, 2007a); on the contrary, VOR is higher in rail transport than in road transport when multimodal transport networks and nodes are considered. This is most likely related to the fact that reliability is the most important condition for having a seamless multimodal journey. Even small delays in one of the legs of the journey can cause remarkable delays at destination, given the presence of an integrated network. In the case of freight transport, the recent literature (see Bolis and Maggi, 2003, among others) has obtained econometric estimates that appear to confirm that the VOR is higher than the VOT.

For all the abovementioned reasons, public investments on transport networks should not only be aimed at reducing travel time but also at increasing the degree of reliability of the network. In this respect, cost-benefit analyses (CBA), multicriteria analyses (MCA) and other investment appraisal techniques (see Chapter 7 in this volume) should appropriately consider the relevance of this transport attribute. In the case of CBA, it would be important to estimate a proper monetary VOR for the users of the networks and to consider jointly passenger and freight transport. When MCA is considered, it is important to provide a proper weight to reliability; in this case, it might be important to consider both monetary VOR and the ratio between VOT and VOR.

5. Concluding remarks

This chapter aimed to outline the importance of VOR in transport networks by means of a theoretical and an empirical survey. The rich literature concerning VOR shows the increasing relevance of this transport attribute for different transport modes and at the various spatial scale levels.

From the analytical viewpoint, it appears that the random utility model is the most used approach. However, in the case of freight transport, there is a problem related to the identification of the decision maker (either the shipper or the sender or another economic actor along the supply chain).

Empirically, the examined studies show that no general VOR can be extrapolated for transferability and policy strategies. Spatial and socio-economic parameters play a determinant role in the estimated VORs. In the case of passenger transport, estimated VORs can be standardized and compared by normalizing them with respect to the hourly wage rate. The VORs in freight transport are far more heterogeneous both in terms of the identification of reliability (standard deviation of travel times or percentage of consignments arriving in due time) and of the unit of value (ton, pallet, consignment, and so on). Concerning the transport mode, no clear pattern emerges from the surveyed studies, even though inflexible activities schedules and the network topology structure seem to play a significant role in determining VOR. However, no empirical VOR estimate appears to have been carried out in the context of multimodal transport networks.

The necessity of extending the analytical and empirical surveys in order to better capture the relevance and values of VOR is clear. In this context, it would be worthwhile to extend the analysis to further transport attributes influencing VOR (i.e. frequency, safety, and so on). From the theoretical viewpoint, dynamic extensions of VOR formulations might be considered, by analysing, for example, different types of dynamic utility functions in an optimal control approach in the case of multimodal transport networks. From the empirical viewpoint, a clear research direction is the adoption of meta-analysis for both passenger and freight transport, in line with the ones performed for VOT (see Zamparini and Reggiani, 2007b, 2007c). The obtained results would certainly be useful for transferability, in the framework of evaluation and forecast analyses for integrated transport network policies.

Acknowledgements

The authors would like to thank David Banister and Moshe Givoni, who organized the NECTAR Cluster 1 Workshop 'Integrated Transport: from Policy to Practice' at OxTran (Oxford Transport Network, Oxford University) in September 2008, for their kind cooperation and useful observations. Thanks also go to the participants in this workshop and to the 1st Transatlantic NECTAR Conference, who contributed to the discussion of this paper. The remarks of Wout Dullaert and of two anonymous referees are also gratefully acknowledged.

Notes

1 The realization of an integrated transport has been considered essential by the British Commission for Integrated Transport (CfIT), established by the 1998 Integrated Transport White Paper (DETR, 1998) in the light of the following policy objectives: integration within and between different types of transport, integration with the environment, integration with land use planning, integration with policies of education, health and wealth creation.

2 Small (1982) specified R(s) as $R(s) = \beta_1 R15(s) + \beta_2 R10(s)$ where R15(s) is the possibility that individuals round off their travel times to the nearest 15 minutes and R10(s) is the possibility that they round it off to the nearest 10 minutes. β_1 and β_2 are the coefficients to be estimated.

3 Transport economic literature normally defines SDE as Schedule Delay Early and SDL as Schedule Delay Late.

4 The value of time can be defined as the monetary value (or the willingness to pay) attached to the possibility of saving a determined amount of travel time (Zamparini and Reggiani, 2007a).

5 The following were the values of hourly wage rates obtained: US 1998: US$12.78; US 1999: US$13.24; US 2000: US$13.76; US 2003: US$15.35; US 2004: US$15.67; the Netherlands 1997: €13.99; United Kingdom 1999: £8.19; United Kingdom 2004: £10.27; Spain 2005: €11.40; New Zealand 1999: NZ$17.62.

6 In the case of the study for which it was not possible to ascertain the year of data collection (Bates *et al.*, 2001), the standardization of the VOR was based on the hourly wage rate of 1999, as the authors in several passages of the paper state that the case study was 'recent'. However, the order of the VOR as percentages of the wage rate would be of comparable magnitude also by considering the hourly wage rates either in 1998 or in 2000.

References

Asensio J. and Matas A. (2007) *Commuters' Valuation of Travel Time Variability in Barcelona*, Barcelona, Document de Treball 07.01, Departament d'Economia Aplicada, Universitat Autonoma de Barcelona.

Bates J., Polak J., Jones P. and Cook A. (2001) 'The valuation of reliability for personal travel', *Transportation Research Part E* 37: 191–229.

Becker G. (1965) 'A theory of the allocation of time', *Economic Journal* 75: 493–517.

Bhat C. R. and Sardesai R. (2006) 'The impact of stop-making and travel time reliability on commute mode choice', *Transportation Research Part B* 40: 709–730.

Bolis S. and Maggi R. (2003) 'Logistics strategy and transport service choices: an adaptive stated preference experiment', *Growth and Change* 34, 4: 490–504.

Bruinsma F. R., Rietveld P. and van Vuuren D. J. (1998) *Unreliability in Public Transport Chains*, Amsterdam, Tinbergen Institute Discussion Paper, TI 98–130/3, Tinbergen Institute.

Closs D. J., Keller S. B. and Mollenkopf D. A. (2003) 'Chemical rail transport: the benefits of reliability', *Transportation Journal* 42, 3: 17–30.

DETR (Department for the Environment, Transport, and the Regions) (1998) *A New Deal for Transport: Better for Everyone. The Government's White Paper on the Future of Transport*, London, DETR. Available online at: http://www.dft.gov.uk/about/strategy/whitepapers/previous/anewdealfortransportbetterfo5695 (accessed 1 February 2010).

European Commission (2001) *White Paper – European Transport Policy for 2010: Time to Decide*, Luxembourg, Office for Official Publications of the European Communities.

Fowkes A. S., Firmin P. E., Tweddle G. and Whiteing A. E. (2004) 'How highly does the freight transport industry value journey time reliability – and for what reasons?', *International Journal of Logistics* 7, 1: 33–43.

Fowkes A. S., Firmin P. E., Whiteing A. E. and Tweddle G. (2001) 'Freight road user valuations of three different aspects of delay', paper presented at the European Transport Forum.

Fowkes A. S., Nash C. A. and Tweddle G. (1991) 'Investigating the market for intermodal freight technologies', *Transportation Research Part A* 25, 4: 161–172.

Fowkes A. S. and Shinghal N. (2002) *The Leeds Adaptive Stated Preference Methodology*, Working Paper 558, Leeds, Institute of Transport Studies, University of Leeds.

Fowkes A. S. and Tweddle G. (1988) 'A computer aided stated preference experiment for freight mode choice', *PTRC 16th Summer Annual Meeting 1988, Proceedings of Seminar D*, London, PTRC.

Gaver D. P. (1968) 'Headstart strategies for combating congestion', *Transportation Science* 2: 172–181.

Gorter C., Nijkamp P. and Vork R. (2000) 'Analysis of travellers' satisfaction with transport chains', *Transportation Planning and Technology* 23: 237–258.

Hensher D. A. (2001) 'The valuation of commuter travel time savings for car drivers: evaluating alternative model specification', *Transportation* 28: 101–118.

Hollander Y. (2006) 'Direct versus indirect models for the effects of unreliability', *Transportation Research Part A* 40: 699–711.

Immers B., Yperman I., Stada J. E. and Bleukx A. (2004) 'Reliability and robustness of transportation networks – problem survey and examples', Proceedings of the NECTAR Cluster 1 Meeting, Amsterdam.

INREGIA (2001) 'TidsvŠrden och transportkvalitet – INREGIA studie av tidsvŠrden och transportkvalitet fšr godstransporter 1999', underlagsrapport till SAMPLAN, 2001: 1, February.

Knight T. E. (1974) 'An approach to the evaluation of changes in travel unreliability: a "safety margin" hypothesis', *Transportation* 3: 393–408.

Kurri J., Sirkiä A. and Mikola J. (2000) 'Value of time in freight transport', *Transportation Research Record* 1725: 26–30.

Lam T. C. and Small K. A. (2001) 'The value of time and reliability: measurement from a value pricing experiment', *Transportation Research Part E* 37: 231–251.

van Lint J.W.C., van Zuylen H. J. and Tu H. (2008) 'Travel time unreliability on freeways: why measures based on variance tell only half the story', *Transportation Research Part A* 42: 258–277.

Liu H. X., Recker W. and Chen A. (2004) 'Uncovering the contribution of travel time reliability to dynamic route choice using real time loop data', *Transportation Research Part A* 38: 435–453.

McFadden D. (1974) 'Conditional logit analysis of qualitative choice behaviour', in P. Zarembka (ed.), Frontiers in Econometrics. New York: Academic Press, pp. 105–142.

Muilerman G. J., van der Hoorn T. and van der Heijden R. (2005) 'Determining the impacts of time-based logistics strategies in the Dutch food industry', *International Journal of Logistics: Research and Applications* 8, 3: 237–247.

Noland R. B. and Small K. A. (1995) 'Travel-time uncertainty, departure time choice, and the cost of morning commutes', *Transportation Research Record* 1483: 150–158.

Noland R. B., Small K. A., Koskenoja P. M. and Chu X. (1998) 'Simulating travel reliability', *Regional Science and Urban Economics* 28: 535–564.

Polak J. (1987) 'Travel time variability and departure time choice: a utility theoretic approach', Discussion Paper No. 15, Transport Studies Group, Polytechnic of Central London.

Rietveld P., Bruinsma F. R. and van Vuuren D. J. (2001) 'Coping with unreliability in public transport chains: a case study for Netherlands', *Transportation Research Part A* 35: 539–559.

Saldanha J. P., Russell D. M. and Tyworth J. E. (2006) 'A disaggregate analysis of ocean carriers' transit time performance', *Transportation Journal* 45, 2: 39–60.

Shinghal N. and Fowkes T. (2002) 'Freight mode choice and adaptive stated preferences', *Transportation Research Part E* 38: 367–378.

Small K. A. (1982) 'The scheduling of consumer activities: work trips', *The American Economic Review* 72, 3: 467–479.

Small K. A., Noland R. B., Chu X. and Lewis D. (1999) 'Valuation of travel-time savings and predictability in congested conditions for highway user-cost estimation', *Transportation Research Board NCHRP Report 431*.

Small K. A., Noland R. B. and Koskenoja P. (1995) 'Socio-economic attributes and impacts of travel reliability: a stated preference approach', University of California, Irvine, PATH Research Report.

Small K. A., Winston C. and Yan J. (2005) 'Uncovering the distribution of motorists' preferences for travel time and reliability', *Econometrica* 73, 4: 1367–1382.

Stank T. P. and Crum R. M. (1997) 'Just-in-time management and transportation service performance in a cross-border setting', *Transportation Journal* 36, 3: 31–42.

Tavasszy L. and Bruzelius N. (2005) 'The value of freight transport time: a logistics perspective – state of the art and research challenges', *European Conference of Ministers of Transport Round Table 127: Time and Transport*, Paris, OECD.

Transek (1990) 'Godskunders vŠrderingar', Banverket Rapport 9, 1990: 2.

—— (1992) 'Godskunders transportmedelsval', VV 1992: 25.

Wigan M., Rockliffe N., Thoresen T. and Tsolakis D. (2000) 'Valuing long-haul and metropolitan freight travel time and reliability', *Journal of Transportation and Statistics* 3, 3: 83–89.

Winston C. (1981) 'A disaggregate model of the demand for intercity freight transportation', *Econometrica* 49, 4: 981–1006.

Witlox F. and van Daele E. (2005) 'Determining the monetary value of quality attributes in freight transportation using a stated preference approach', *Transportation Planning and Technology* 28, 2: 77–92.

Zamparini L. and Reggiani A. (2007a) 'The value of travel time in passenger and freight transport: an overview', in van Geenhuizen M., Reggiani A. and Rietveld P. (eds) *Policy Analysis of Transport Networks*, Ashgate, Aldershot.

—— (2007b) 'Freight transport and the value of travel time savings: a meta-analysis of empirical studies', *Transport Reviews* 27, 5: 621–636.

—— (2007c) 'Meta-analysis and the value of travel time savings: a transatlantic perspective in passenger transport', *Networks and Spatial Economics* 7, 4: 377–396.

Chapter 7

Appraisal of integrated transport policies

Peter Bakker, Carl Koopmans and Peter Nijkamp

1. Need for assessment of integrated transport

In recent years broadly based and social science-oriented views have increasingly entered the discussion of transport behaviour and infrastructure, as transport policy is increasingly confronting important tensions between economic-technological potential and environmental-social constraints. Efficiency-equity dilemmas and efficiency-ecology dilemmas call for integrated policy perspectives (see Nijkamp and Blaas, 1994). A further network expansion of traditional infrastructure is generally incompatible with the need for a high quality of life. Environmental and safety considerations have become major factors in the social acceptance of our mobile society. Thus, new transport solutions and technologies will have to be implemented within the increasingly narrower limits imposed by society. The range of such solutions is even further limited by the simultaneous behaviour of all actors in our modern transport systems who generate congestion effects (including high accident rates).

Appraisal is an important component of decision making. While recognizing that decision making is not necessarily a rational process, appraisal information is still a prominent key factor. One might regard decision making as a complex interplay of information, interests and opinions of groups in society on the one hand, and democratic action and political skill on the other. Transport policies typically have a host of very different impacts (DfT, 2009). Examples are travel times, emissions to the environment, road accidents, wider economic effects, investment costs, and income distribution. In appraisal, it is important to find ways to make the impacts comparable, by computing *total impact* indicators for society as a whole. Often used methods are cost-benefit analysis (CBA), multicriteria analysis (MCA) and cost-effectiveness analysis (CEA) (Nijkamp *et al.*, 2003).

Appraisal of integrated transport policies is not straightforward, as these policies take different shapes. For instance, policies aimed at integration between transport modes (e.g. rail/bus or rail/bike) differ strongly from policies which integrate land use and transport. Another difference between policies is that between specific projects and full policy plans. A general characteristic of integrated policies, which we will use as a definition, is that they affect more than one relevant aspect or travel mode. This leads to serious challenges for appraisal compared with unimodal policies, as it is necessary to analyse not only separate aspects or travel modes but also the interaction between them. For instance, the integration of transport modes requires a proper analysis of transfers. And the appraisal of integrated land use/ transport policies needs to thoroughly address the land market. In this chapter, we focus on the merits and limitations of evaluation methods, especially in their application to integrated transport.

Section 2 below describes the current practice in the evaluation of integrated transport, looking first at estimating transport demand, and then at impact evaluation, in particular CBA and MCA. In Section 3, we present methods for estimating the demand for Park-and-Ride facilities and joint CBA of land use and transport. Finally, Section 4 concludes with a brief sketch of the achievements in appraisals so far, and possibilities for their further improvement in the future.

2. Current practice in evaluating transport plans

2.1 The importance and complexity of demand forecasting

The analysis of demand is an essential ingredient of the evaluation of any transport plan, integrated or not. A major issue of any transport plan evaluation will be to what extent the plan facilitates transport demand. In the past few decades, demand forecasting methods have been developed to be able to provide information about the effects of the new plans on future transport demand. But demand forecasting is less straightforward than one might think. If unimodal forecasting methods are already complex and only seemingly exact, one may wonder if it is worthwhile to combine these techniques for different aspects (e.g. land use and transport markets) or different transport modes. Before discussing how suitable these methods are for integrated transport policies, first it is useful to provide some insight into the principles of mainstream demand forecasting methods.

Mainstream travel demand forecasting methods are based on simplified models. Quinet and Vickerman (2004), for example, fit passenger traffic problems to the theory of consumer behaviour. Their main hypothesis assumes that each consumer has a utility function with a time and money constraint. They add a discrete choice process among transport modes. This type of model is the basis for mainstream passenger demand forecasting models (see, for example, Meyer and Millar, 1984). These models predict what will happen under hypothetical conditions: for instance, if a new motorway is built or a railway service is extended. First the study area is divided into hundreds of traffic zones. These zones are filled with socio-economic data estimated for the forecast year: land use, population (including distribution over categories of age and social participation), and economic activity. Generally, these models contain four main modules:

- Trip generation (trips per zone)

- Trip distribution (pairing origins and destinations)
- Modal split
- Traffic assignment (to the network)

Regarding the traveller's choice between public transport, car use and other modes, most models either do not pay much attention to individual socio-economics (trip interchange models), or do not pay sufficient attention to changes in level of transportation service (trip-end models) (see, for example, Black, 1995).

In many integrated transport studies, and certainly multimodal studies, public transport is an important element. Modelling supply is more complicated for public transport than for other modes. A first aspect is the involvement of the public transport operator. A premise in infrastructure appraisal studies is that the opening of new infrastructure will be followed by new public transport services. In markets where infrastructure administration is separated from transport operations, this is not as obvious as one might think. Services are sometimes started later or less frequently than is taken into account in the appraisal study, because of commercial or technical considerations of the operator. This is in contrast to the opening of a motorway that automatically gives travellers new possibilities for travel.

Also, for public transport it is not sufficient to model supply only with zones and links. In contrast to driving, cycling or walking, the length of the links only partially reflects travel friction. The supply of public transport is characterized by various quality aspects, such as:

- density of bus stops and railway stations in the service area, determining the time needed for access and egress of the system;
- frequency of the services, determining waiting times;
- equal intervals between services and the ease of remembering departure times;
- walking times during transfers;
- reliability, determining the need to incorporate time margins in trip planning;
- comfort in vehicles, stations and bus stops;
- possibilities to park bikes and cars near stations and bus stops;
- ease of use of the system: ticketing, reservations, timetable.

An often used approach is to dedicate each zone of the study area to one train station or bus stop. Sometimes artificial links are used, connecting zones without public transport to the transport system. Depending on the scale of the study, these artificial links may also be used for zones where in real life bus services are available, but not every single bus line is modelled in detail. Well-elaborated models use penalties and weighting factors reflecting the traveller's aversion to transfers and his relatively higher aversion to out-of-vehicle travel time.

Because public transport only accounts for a limited part of all travel (for example, 5 per cent of all trips and 11 per cent of all travelled kilometres in the Netherlands: see Bakker and Zwaneveld, 2009), it is possible that the public transport network is oversimplified in studies concerning all modes. On the other hand, the simplification of public transport supply ignores its 'true' nature: the fact that its attraction depends highly on the traveller's specific situation. Several studies show that the influence of public transport stops rapidly declines with increasing distances from the stop, especially on the 'not home-bound' side of trips. For instance, Debrezion *et al.* (2007) show that the attraction of public transport is very site-specific: real estate property values prove to be substantially influenced within a radius of 500 metres around a railway station. But, in the next 500 metres, this impact is reduced by two-thirds. The exact locations of bus stops and railway stations in relation to traveller's trip origin and destination address do matter. It is clear that, if the traffic zones in a model are larger, the effects of integrated transport plans that seek to optimize land use and public transport on a smaller geographical scale will be lost. And, regarding multimodal transport plans, their success may be more affected by the exact location of transfer facilities than models of this kind are able to predict.

In mainstream travel forecasting models, various other quality aspects of public transport are 'dissolved' in an average travel friction between zones. Some integrated transport plans aim to improve the ease of use of public transport and deliver substantial improvements from the traveller's point of view: for instance, integrated ticketing, integrated traveller information, better transfer facilities. These kinds of changes can only be modelled by means of small changes in the supposed general level of service. In general, the traditional four-step travel-demand estimation process 'is cumbersome, expensive and requires a large amount of data … The generation of trips is independent of the transportation supply characteristics and possible technical improvements, and the models are generally site-specific – that is they are not transferrable from one area to another' (Dickey *et al.*, 1983). Despite all those criticisms, this type of model is still the most used, primarily because it has been well tested and is completely operational.

In view of the high costs and long turnaround times of (integrated) travel demand forecasting studies, this process can be rather disappointing for policy makers. In some cases, a quick scan might be better: assess the impact of a single improvement on perceived travel times of travellers, and clarify the relative importance of the service in total public transport and other traffic volumes.

The question can be raised as to how the degree of integration can be improved in mainstream travel demand forecasting. Integrated policies that seek to optimize land use and transport are usually modelled satisfactorily, to the extent that land use is involved in transport by using spatial differences as input for trip generation. But, in the long run, the transport system will have a feedback influence on land use. The TIGRIS XL model, for instance, was developed as an extension of regular models in order to assess both the long-term impacts of transport policies on the spatial distribution of residents and jobs, and the effect of alternative land use policies on the transport system (Zondag and de Jong, 2005).

Integrated policies that aim to optimize the combined effect of various modes of transport are facilitated because various transport modes can be involved in the demand forecasting process. But is multimodality (the option of travellers to use different modes for one trip) considered sufficiently? Various models have been developed in order to better assess the effects of this kind of multimodal policy (particularly Park-and-Ride) (see, for instance, Cohn *et al.*, 1996; Fox, 2005; Li *et al.*, 2007; and Molin and van Gelder, 2008).

In particular, assessing integrated transport policy packages by considering important aspects such as marketing, integrated ticketing, or integrated traveller information appears to be a weak spot in regular demand forecasting methods. However, tailor-made studies have also been conducted (for instance, Chorus, 2007).

But is it worthwhile to use these new or extended models in regular transport planning and project assessment? By combining techniques, their complexity and uncertainty are multiplied. Extended models need additional data collection and computing time. New modules have to be calibrated with empirical data, and it may be doubtful whether these calibrations are valid for application in other regions. The contradiction is: the more completeness there is in combining the different aspects of integration, the less attention can be given to the separate aspects. The desire to be complete in any aspect creates a problem for the applicability of appraisal studies that often have to be undertaken under time and budget constraints. Maybe the challenge of integral studies is mainly to dare to use simple techniques in order to be able to predict, in a manageable way, effects on demand as an ingredient for evaluation studies.

2.2 Economic versus societal impact evaluation

This sub-section presents and compares different methods which can be used to evaluate integrated transport policies. First, the advantages and disadvantages of these methods are described. Then, the applicability of these methods to integrated transport policies is examined.

Rational evaluation of the total impact of a policy requires: a set of possible choices; relationships that determine the 'pay-off' (satisfaction, goal attainment) of each choice; and a preference-ordering among pay-offs (Simon, 1955). Assuming that the set of possible choices is given, we will look into the differences between methods in terms of the 'pay-off' criteria. Table 7.1 lists these characteristics for some popular appraisal methods.

Table 7.1 Decision criteria used in popular appraisal methods

Method	*Type*	*Criterion*	*Measurement through*
Cost-benefit analysis (CBA)	Economic	Welfare	Willingness-to-pay (WTP) for effects
Cost-effectiveness analysis (CEA)	Economic	Ratio of main effect to costs	Simple division
Multicriteria analysis (MCA)	Social	Weighted sum of effects	Political weights for effects
Balance sheet	Social	No integral criterion	No measurement; decision makers look at separate effects

Given the nature of each method, we may compare them in terms of advantages and disadvantages. Table 7.2 shows that CBA is firmly rooted in economic science and yields clear policy conclusions. On the other hand, CBA is often incomplete and does not connect well to the political process of decision making. MCA is a complement to – but in many respects also the opposite of – CBA: it is complete in terms of effects and combines research results with political input. However, this also opens the door for ambiguous weights or methods or even manipulation. Usually, MCA only ranks policies and does not show whether a specific policy is attractive or not. Cost-effectiveness analysis does not suffer from questionable weights, but it is often rather incomplete, as one only compares one (main) effect to the costs, while other benefits are ignored. Finally, the balance sheet method leaves the weighting of effects to decision makers, which, on the one hand, is very flexible, but, on the other, does not give much guidance from the research to decision makers.

Integrated transport policies have no single, central goal. Therefore, cost-effectiveness is less appropriate. Balance sheets are possible, but do not yield as much decision support as CBA and MCA. Therefore, CBA and MCA in most cases are the preferred methods. If (almost) all impacts can be monetized, CBA becomes more attractive. If, on the other hand, important impacts cannot be monetized, MCA, or a mix of MCA and CBA might be preferable.

Table 7.2 Advantages (+) and disadvantages (–) of appraisal methods

	Method			
Aspect	*CBA*	*MCA*	*CEA*	*Balance sheet*
Decision support	+ Discerns attractive policies from unattractive policies	+/– Usually ranks policies in terms of attractiveness	+/– Ranks policies in terms of attractiveness	– No attractiveness conclusion from the appraisal
General quality of weights	+ Based in economic science; analogy to utility theory	– Subjective weights or methods; risk of manipulation	+ Main effect and costs are weighted adequately	+/– No weights used
Completeness	– Some effects are hard to monetize (e.g. irreplaceable nature)	+ Can be applied to all effects	– Only the main effect and the costs are counted; other effects are ignored	+ All effects can be included
Connection with political process	– High-income people (and business interests) have high WTP; so they are overrepresented	+ Decision makers can apply their own weights (interests)	– No flexibility	+ Every decision maker can draw his/her own conclusions

2.3 Cost-benefit analysis: practice and extension

In this section, we 'zoom in' on three issues in CBA: (1) effects that are difficult to monetize; (2) distributional effects; and (3) the transparency of CBAs for decision makers.

CBA captures, in principle, all impacts on society with objective, market-based weights. This yields important policy information which is largely independent of the process of decision making. As such, it is an important benchmark for the quality of decisions, which may prevent the interaction among policy makers from resulting in 'negotiated nonsense' (van de Riet, 2003). However, some impacts are notoriously hard to monetize, and there are concerns about some of the methods employed (Pearce *et al.*, 2006). Examples include: the effects of transport projects on passenger comfort, on nature areas, and on 'beauty' aspects (e.g. the architecture of bridges). Lacking direct information from market prices, a 'good' CBA should use special techniques to estimate the willingness-to-pay of households and firms for such effects. However, such research is often difficult and expensive. Therefore, these quality aspects are often not monetized. For example, in the Netherlands, Annema *et al.* (2007) conclude that most transport CBAs do not monetize impacts of the project on landscape, nature and spatial quality. They do, however, give qualitative information on these impacts. This qualitative approach may result in a presentation bias in the final conclusions.

Also, market-based weights, based on willingness-to-pay, are higher for business travellers and high-income groups than for low-income travellers. Public transport is often seen as providing a minimum of accessibility to low-income groups, elderly people, handicapped users, etc. Using relatively low weights for the effects which accrue to these groups is considered to be awkward and unjust from a political point of view. This problem may be tackled by presenting not only total costs and benefits but also a Benefits Incidence Table, as practised in Japan (Hayashi and Morisugi, 2000). Table 7.3 contains an example of such an incidence table for both benefits and costs. The benefits and costs pertain to the introduction of an integrated system for electronic ticketing ('chipcards') and access gates in all modes of public transport in the Netherlands. Without going into detail about the specific characteristics of the project, we may note two interesting points:

- There are important economies of scale in the integrated system, as compared with each public transport mode taken separately or with each individual public transport operator on its own system.
- Some items do not constitute total benefits or costs, but only distributional effects, such as subsidies and chipcard sales. In fact, the price of the chipcards was chosen so as to create distributional effects which are generally positive for both operators and travellers. In this way, the incidence table has been used to optimize the project in political terms.

An important problem in the interaction between appraisal and policy making is the lack of transparency of CBAs, at least from the point of view of policy makers and the public. According to Pearce *et al.* (2006), CBA may be too complex for the busy

Table 7.3 Incidence table for electronic ticketing and access gates in public transport in the Netherlands (net present value, in million Euros)

	Total	*Travellers*	*Employers*	*Operators*	*Government*	*Wider economic effects*	*External effects (non-pecuniary)*
Costs of chipcard system, gates, etc.	–440 to –1000			–500 to –1060	60		
Time savings in buying tickets	500 to 620	490 to 610	10				
Reduction of violence	100 to 120	60 to 70	30 to 40				10
Reduction of non-paying travel	380 to 480			380 to 480			
Differentiated rates	240 to 480	20	–20	240 to 480			
Sales of chipcards	0	–310		+310			
Additional trips	50 to 80			50 to 80			
Efficiency in transport	130 to 160			70 to 80	–20	80 to 100	
Environmental effects	20						20
Subsidies	0			50	–50		
More efficient procurement	0			–70 to –320	70 to 320		
Total	420 to 1520	260 to 390	20 to 30	–30 to +660	60 to 310	80 to 100	30

Source: Koopmans, 2006

civil servant. Annema *et al.* (2007) perform a benchmark study of transport CBAs in the Netherlands, and conclude that poor transparency is one of three main points for improvement. The CBAs are not formulated to make the main results of the CBA clear to non-welfare economists. Economic jargon is used abundantly. Categories of costs and benefits of the project are often not explained, such as 'transport advantages', 'industrial site benefits', and 'exploitation surplus'. Pearce *et al.* (2006) note that 'Theoretical economists need a far better understanding of the pressures that affect actual decisions', but they do not add specific recommendations to this observation. In the Netherlands, the transparency issue was the reason to add a 'clear presentation of CBA results' guide (Koopmans, 2004) to the existing guidelines for CBA. The 'clear presentation' guide recommends:

- Present not only monetized values but also the concomitant physical effects (e.g. time savings in hours, CO_2 emissions in tonnes, number of road fatalities).
- If some effects are not monetized, present them as +?, -? or ?, and always include the non-monetized effects in the net benefits (the guide acknowledges that a net benefit of '340 to 460 million +?' is awkward, but considers this better policy information than the incomplete '340 to 460 million').
- Avoid jargon and technicalities in the summary.
- Link the CBA to policy objectives in the summary of the CBA.

Summing up, we may conclude that CBA has a great potential to be a factual counterweight in decision making. However, improvements are needed to fulfil this potential, such as more research into difficult-to-monetize effects, including incidence tables in CBAs and presenting CBA results in a clear and unbiased way.

2.4 Multicriteria analysis: practice and extension

Transportation has a great variety of both intended and unintended, desirable and undesirable, and local and supra-local effects. The 'undesirable' outcome of a highly mobile society (in terms of pollution, lack of safety and congestion) is – almost paradoxically – the result of rational and plausible actions of a great many individuals. Social science research has convincingly demonstrated that the neglect of social costs in individual decision making inevitably leads to a macro outcome that is far from optimal. This explains, for example, worsening quality-of-life conditions in major cities all over the world. Transport has become both a friend and an enemy, and has caused paradoxical feelings and views on its future. Such drastic changes are likely to exert a profound influence on the future spatial interaction pattern of our societies and will make it necessary for transportation planning to respond as efficiently as possible to new tendencies and new challenges. However, transportation planning is often marked by lack of resilience, so that flexible adjustments to new structural changes (e.g. deregulation, road pricing) often take place inadequately (see also Nijkamp *et al.*, 1992; Deakin *et al.*, 2008).

Policy has not addressed itself so far to such fundamental questions as the legitimation of minimal levels of accessibility, the effect of the geographical concentration of public facilities, the supply of public transport in different districts, and, the most important question of all, what the role of planning should be in society. One way out of this dilemma would be to increase the information-processing capacity of the planning system by installing computerized transport planning and management information systems. This would call for the design of tailor-made decision-support systems in transportation planning. Complementary to this 'high-tech' approach, there is – as mentioned above – a 'low-tech' alternative of making the methods used more transparent and comprehensible to planners, politicians and the public at large. This may also require the design of user-friendly decision-support systems and evaluation methods.

Planning concerns the integrated analysis of conflicting choice options. In general, the relative social ('public') value of the effects of planning projects (e.g. a highway project) is co-determined by political priorities at different institutional levels. Sometimes these values are – directly or indirectly – the result of prices generated by a market mechanism, but very often such values are more subjectively determined and based, for example, on the desiderata of individuals and groups in society (for instance, the value of a natural park or the visual beauty of an old theatre). Consequently, many conflicting views may emerge in evaluating alternative plans (e.g. different highway investment projects). In particular, modern approaches like MCA may serve as a meaningful evaluation vehicle for explicitly taking account of such conflicts regarding the foreseeable impacts of a plan. For example, everybody may agree that the implementation of a road project will destroy x hectares of a forest, but not everybody will attach the same value to these x hectares of the forest. MCA may then be helpful in taking into account such conflicting issues by considering priority schemes or weights as an ingredient in an evaluation analysis for investment projects. Of course, this will not always lead to a unique final solution, but the structure and consequences of conflicts among decision makers can be made more explicit, so that the range of politically feasible alternatives can also be analysed in greater detail.

Figure 7.1 shows the general structure of an evaluation problem in which MCA is used as the evaluation method. Starting at the bottom of the figure, the policy options are described in general terms. Then, criteria are defined. These are strongly related to preferences of actors in society. The next step is a more specific description of policy alternatives. The set of project effects to be researched is also influenced by preferences. Finally, the effects are combined in a MCA. The results might lead to direct decisions, but also to interactive procedures in which decision makers 'rethink' their preferences and weights.

Any evaluation technique for judging the desirability of public plans or projects should be logically and consistently connected to the nature of the decision problem concerned. Given the unique nature of many decision problems, there is no unambiguous method with universal validity, and hence each type of decision problem may require its own specific evaluation method.

Depending on the problems concerned, and on the precision of the data used, several sub-divisions of evaluation methods can be made:

- *Discrete* versus *continuous* evaluation problems.

- *Soft* versus *hard* evaluation problems. Soft problems include qualitative or ordinal information on the impacts of alternatives or on priorities/weights, whereas hard problems are based on quantitative (i.e. mainly cardinal) information.
- *Static* versus *dynamic* evaluation problems.
- *Multi-person* (or *multi-committee*) versus *single-person* (or *single-committee*) evaluation problems. In the case of multi-person or multi-committee problems, it is necessary to take into account the variation in preferences, while also considering the possibility of a multilevel decision structure.
- Evaluation problems based on the *generation of preferred alternative solutions* versus those based on the *selection of one ultimate alternative*. In the first case, the procedure aims to identify only non-dominated solutions (i.e. solutions for which the value of one policy objective cannot be improved without reducing the value of a competing objective) in the second case, the procedure aims to find one alternative which is considered as satisfactory after the articulation of preferences. An intermediate approach may be based on the identification of a set of dominating alternatives.
- *Single-step* versus *process* evaluation problems. The first category aims to find the most satisfactory solution as an unambiguous result at a certain point in time; the second category considers policy making as a process during which more information may be added, so that the ultimate solution is identified in a series of successive steps.

Multicriteria methods are appropriate to find a (complete or partial) ranking of choice alternatives that have to be judged on the basis of a broad (i.e. not exclusively

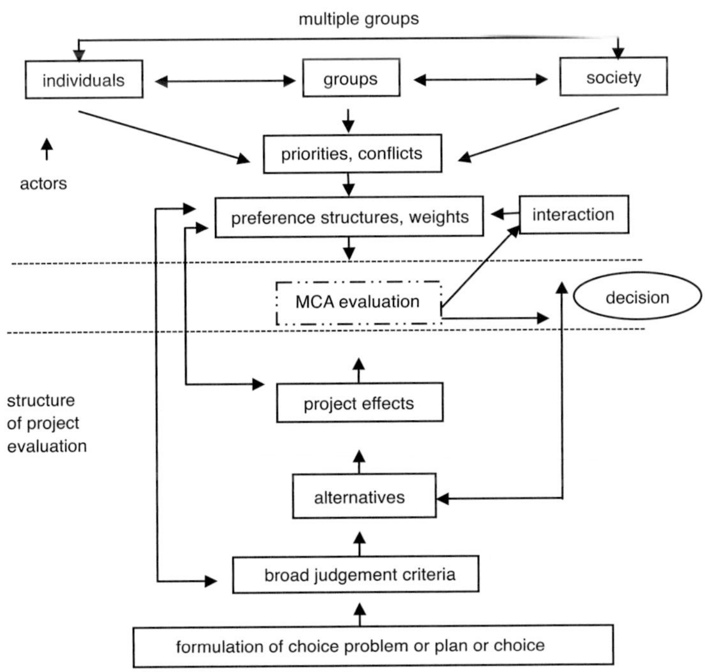

Figure 7.1 General structure of an evaluation problem

monetary) set of decision or choice criteria. In various cases, weighting procedures are used to arrive at an unambiguous solution, although the use of weights is not strictly necessary (see Nijkamp and Blaas, 1994).

As in all evaluation methods, the use of a plan effect (or impact score) matrix (or table) is a central step in multicriteria evaluation. This matrix contains, for all choice alternatives, the numerical estimates of outcomes of all relevant criteria, measured in their own appropriate dimensions (e.g. financial costs, reduction in traffic accidents, levels of air pollution).

Next, by confronting the a priori specified weights set for the judgement criteria with the plan-effect matrix, a ranking of alternatives may be obtained. There are, however, various procedures for confronting these two sets (depending on, among other things, the level of precision of the measurement of effects), and hence a wide variety of multicriteria evaluation methods has been designed in the recent past, ranging from extremely simple to fairly complicated. Table 7.4 offers a concise and simple illustration of a discrete MCA for plan evaluation.

Various classifications of multicriteria choice models may be made. In the literature, the following typology for these models has, inter alia, been proposed: *discrete* multicriteria models versus *continuous* multi-objective models, and *hard information* models versus *soft information* models.

Discrete choice models display only a finite number of distinct feasible choice possibilities (courses of action, strategies, solutions, alternative plans or projects, etc.), while continuous models may encompass an infinite number of choice possibilities (as is usually the case in programming models).

Hard information means information measured on a cardinal scale, while soft information means information based on a qualitative (ordinal or nominal) scale. Clearly, one may also distinguish mixed information, in which the information is partly cardinal, partly qualitative. Consequently, the typology in Table 7.5 may be used.

MCA may be seen as an important decision-support method for planning

Table 7.4 Illustrative MCA table for alternative routes

Alternatives	*Costs*	*Ecological damage*	*Accessibility*
Route 1			
Route 2			
Route 3			
Weights	W1	W2	W3

Table 7.5 A typology of multicriteria choice models

	Cardinal information	*Qualitative information*	*Mixed information*
Discrete multiple-criteria evaluation models	I	III	V
Continuous multiple-objective programming models	II	IV	VI

under uncertainty. Especially in the case of goal conflicts, it may serve to rationalize complex decision problems, by providing both a tool for communication between all actors involved and a rigorous analytical technique for examining (implicitly or explicitly) the implications of policy trade-offs. Flexibility in the design and use of such methods is necessary to ensure a tailor-made research tool. The enormous variety in the applications of such methods illustrates their great potential.

A critical step in any MCA is the determination of political weights to be attached to the individual project effects. This can normally be done by interview methods, revealed preference methods, scenario approaches, and the like. Clearly, in general, a sensitivity analysis on the results in regard to the weights employed is always desirable. In all empirical applications, difficult analytical problems will be encountered regarding, for example, the precision of measurement, the identification of priorities, the demarcation of the impacts. Communication with all actors is then a sine qua non for the acceptance of the results of such techniques. Recursive or cyclical planning procedures are hence necessary for a structural and generally accepted evaluation method.

3. Evaluation tools for integrated transport plans

3.1 Park-and-Ride plans

In recent years, there has been a great interest among transport policy makers in Park-and-Ride concepts as a multimodal solution: the combined use of car and public transport in one trip. From a policy maker's point of view, Park-and-Ride might be the best of both worlds: avoiding the need to finance public transport in less densely populated areas, as well as relieving the pressure on congested roads in more urban environments. However, an integrated policy like this can only be a success if enough consumers share the feeling of 'best of both worlds' from their individual perspectives. The question in this sub-section is whether mainstream demand forecasting methods are sufficiently capable of handling multimodal transport.

The share of multimodal transport in mobility will vary from region to region and country to country. Rijkswaterstaat AVV (2002) gives figures on a nationwide basis for the Netherlands. These show that 2.7 per cent of all trips and 12.3 per cent of all kilometres travelled are multimodal, in the sense that two or more transport modes are used in one trip.[1] If only trips that combine car and train use are considered, no more than 0.1 per cent of all trips are multimodal: clearly a niche market. In the Netherlands, walking, cycling and public transport (bus, tram and metro) are more important in the access and egress of train trips than car use.

Until 20 years ago, in modelling passenger behaviour it was quite common to just compute total travel time. Nowadays, however, most models are refining the perceived costs of public transport trips by using transfer penalties. Iseki *et al.* (2006) (for a more detailed review of the evidence of transfer penalty see Chapter 14 in this volume) show that travellers are very sensitive to out-of-vehicle time, especially waiting time. People are averse to making transfers at interchanges. 'In practice the rule of thumb could be that walking and waiting time are valued twice as much as in-vehicle time for non-business trips' (Iseki *et al.*, 2006). Most values concern interchanges: bus-to-train, bus-to-bus, bus-to-metro, etc. Only two studies give values for car-to-rail (a transfer penalty equivalent to 15 minutes in-vehicle time) and car-to-bus (8.3 minutes). Looking at the general aversion to transfers and waiting times, it is clear that few travellers will consider integrated transport as long as driving the whole distance does not involve similar troubles.

There are several ways to give multimodal transport a better place in demand forecasting:

1. Stop the automatic assignment of one spatial zone to one particular public transport loading node.
2. Recognize that in some zones weighting factors or penalties for car use must be introduced in connection with parking problems. Introduce the possibility of an access/egress mode for car trips towards these zones in analogy to nested access-egress modal-split modules in public transport, like ProMiSe (Cohn *et al.*, 1996).
3. Consider multimodal trips as a separate mode from public transport and car, with its own disutilities. Li *et al.* (2007) and Molin and van Gelder (2008) took this route.

The empirical focus in Park-and-Ride ex-post policy evaluation is often restricted to the parking space occupation level and consumer satisfaction. A premature conclusion based on these indicators could be that policy goals are achieved because the Park-and-Ride park is full and the users are satisfied. But most policy makers have other goals in mind: Park-and-Ride parks should lead to a reduction of car use, creating environmental benefits. Or Park-and-Ride parks should relieve congested roads in urban areas, and relieve the pressure on city-centre parking space. Parkhurst (2000) and Mingardo (2008) show that Park-and-Ride facilities do not automatically serve these goals, and describe several side effects:

- Park-and-Ride sites attract other parkers from nearby locations. Not every occupied place is a new Park-and-Ride traveller.
- Park-and-Ride sites attract additional visitors to nearby destinations. This problem can be of importance in city-centre areas with scarce parking space and high parking fees.
- Park-and-Ride sites cause a switch from public transport and non-motorized trips to car trips in travel to and from the station.
- Park-and-Ride sites cause extra trip generation.
- Inner-city road capacity or parking space is refilled by latent demand coming from travellers who previously used public transport or non-motorized transport.

In general Park-and-Ride generates more car kilometres than it saves. It can contribute to a redistribution of environmental pressure on the city-centre area towards the outskirts of the city, as far as the latent demand from city-centre inhabitants is suppressed. Li *et al.* (2007) show that the parking charging level and the number of parking spaces supplied at the Park-and-Ride site and in the city-centre area, as well as the despatching frequency and fare of a metro line, significantly influence the commuters' choice behaviour and the network performance in terms of total realized travel demand and social welfare gain.

However, not serving the goals set by policy makers, or generating side effects, does not automatically mean that Park-and-Ride is undesirable from a welfare-economic point of view. Margail and Auzannet (1996) conducted a CBA on underground or covered Park-and-Ride facilities in the Paris region. They found that time savings were insignificant. The economic and social profitability of Park-and-Ride projects lies essentially in savings in parking provision in the centre, followed by the issue of the consumption of road space in heavily populated areas. How societal profits are redistributed depends on the parking tariffs (see Table 7.6).

3.2 Generalized cost-benefit analysis

Transport plans are often integrated, in the sense that they include land use policies (see Chapter 3 in this volume). Car use is associated with urban sprawl, while public transport is assumed to lead to more concentrated development. Litman (2009) shows that American cities with 'big rail' (subways, etc.) use much less space per capita than cities without 'big rail'. However, we may note that the spatial structure of a city changes very slowly over time. Therefore, the effect of any policy occurs in the very long run. In CBAs, the standard use of discounting factors for future benefits will reduce these long-run spatial benefits to a very small size. Therefore, the often-heard critique of CBA, that discounting does not take account of long-term environmental effects in a proper way, may also apply to spatial effects. The way future costs and benefits are discounted affects the distribution of environmental value, spatial value and welfare in general among generations. If we were to adopt sustainability as a fixed restriction, we should not use standard discounting.

Generalized CBA of land use and transport investment requires the analysis of both transport markets and the land market. This is particularly difficult for the land market, which is affected strongly by government regulation. Zoning laws effectively split up the land market into different segments: agricultural segments with low prices, and areas for industry, offices and housing with higher land prices. In effect, land for industry, offices and housing is rationed by government regulation. As economic activities and housing needs grow, agricultural zones are shifted to the industry/offices/housing segments. This creates a windfall profit which is divided among the original owners (farmers), project developers, the new owners of the land and local government.

Apart from creating windfall profits, the expansion of the built-up area reduces open space and often affects scenery, heritage, etc. The costs of these effects are very hard to quantify. Sometimes researchers assume that the value of open space is equal to the difference in land prices between agricultural land and built-up land. However, this is begging the question: do government policies adequately reflect the value of open space? Given that decision making is influenced not only by rational considerations but also by many other factors (as described at the beginning of this chapter), this assumption is hardly sustainable. A better way to value open space is to measure its value for people (willingness-to-pay), comparing

Table 7.6 Breakdown of costs and benefits for different scenarios of underground or covered Park-and-Ride in the Paris region, in 1996 French francs (millions)

Assumption about number and origin of park and ride user	*60 'ex-all car' 100 'ex-bus+RER' 100 'ex-car+RER'*	*100 'ex-all car' 140 'ex-bus+RER' 100 'ex-car+RER'*	*100 'ex-all car' 100 'ex-bus+RER' 80 'ex-car+RER'*	*100 'ex-all car' 100 'ex-bus+RER' 100 'ex-car+RER'*
COSTS				
– Investments	15.05	19.67	16.17	17.36
– Operations	0.65	0.84	0.69	0.74
BENEFITS				
– Time savings	0.01 (0.6%)	0.02 (0.7%)	0.01 (0.3%)	0.01 (0.3%)
– Reduction in car use	1.35	2.46	2.79	2.79
• consumption of highway space	0.66 (39.5%)	1.16 (41.6%)	1.25 (40.2%)	1.25 (40.2%)
• externalities	0.12 (7.2%)	0.21 (7.5%)	0.24 (7.7%)	0.24 (7.7%)
• direct costs of car use	0.39 (23.3%)	0.79 (28.3%)	1.00 (32.2%)	1.00 (32.2%)
• parking savings	0.18 (10.8%)	0.30 (10.8%)	0.30 (9.6%)	0.30 (9.6%)
– Variation in public transport use	0.31 (18.6%)	0.31 (11.1%)	0.31 (10%)	0.31 (10%)
Total yearly benefits	1.67 (100%)	2.79 (100%)	3.11 (100%)	3.11 (100%)
Parking costs avoided	5.85	9.35	9.23	9.35
• in the city centre	5.25	8.75	8.75	8.75
• on the outskirts	0.60	0.60	0.48	0.60
Net present value	3.34	13.50	22.64	20.95
Internal rate of return	11.08%	18.86%	34.85%	29.56%

Source: Margail and Auzannet, 1996

Notes:
RER is a regional train system in the Paris region. 'Ex-all car': Park-and-Ride users formerly travelling the whole trip by car; 'ex-bus+RER': users formerly travelling by bus in combination with RER; 'ex-car+RER': users formerly travelling by car in combination with RER.

prices of housing and offices which are located near open space with similar houses and offices which are not.

The forces which induce dispersion (population growth, economic growth) are partly countered by the benefits of agglomeration. Integrated transport plans often aim to increase spatial density, especially near train and subway stations. Therefore, agglomeration benefits are an important topic for transport CBAs, and especially for integrated transport plans. Unfortunately, the research into agglomeration benefits (which appear to be substantial) and the CBA literature is not yet fully integrated. An exception is Graham (2006). He computes an elasticity of productivity with respect to 'effective density' (defined as the employment that can be reached from locations) of 0.125 for the UK. That is, if, in a certain area, an (integrated) transport policy makes it possible to reach 10 per cent more people (at the same costs), this leads to a 1.25 per cent increase in the average productivity of the firms in that area. We may note that standard transport appraisals already evaluate travel time savings. Therefore, it is important to avoid double-counting.

In the joint analysis of spatial and transport options, it is important not to assume ex-ante that housing (or offices) require new infrastructure, or vice versa. Sometimes policy makers plan spatial developments 'to support public transport'. Here, means (public transport) turn into goals, on the implicit assumption that public transport is – without further research – a good thing. In addition, this approach mixes up political choices (to support public transport) with research (which should be neutral). Therefore, it is important to perform CBAs for all options, looking at the real benefits without prior restrictions.

As an example, we present a CBA of urbanization policy options in the Netherlands. Until 2005, the Netherlands had the goal of realizing at least 40 per cent of new housing within the city limits. The objectives of this policy were to preserve open space, and to support public transport use. Furthermore, a large part of housing outside of the large cities was built in commuter towns at distances of 15–20 kilometres, in order to preserve green areas near the cities. 'High quality public transport' was built to transport commuters. In practice, however, most commuters use private cars, causing serious congestion and heated debate on road expansion and road pricing. In 2005, the government commissioned a study of alternative policies (Ecorys, 2005), concentrating on the five largest urban areas (Amsterdam, Rotterdam, The Hague, Utrecht, Arnhem/Nijmegen). Three spatial alternatives were compared:

- Base case: the existing policy, 40 per cent of new housing within city limits.
- Higher density: 55 per cent of new housing within city limits.
- Controlled sprawl: 25 per cent of new housing within city limits.

Within these alternatives, public transport options were discerned.

Table 7.7 presents the main results. The value of open space was estimated in two ways: a (lower) estimate based on costs of compensation, and a (higher) estimate based on the assumption that government policies correctly reflect the benefits of open space. A higher density appears to have more costs than benefits, although a few intangible effects were not monetized. Allowing more sprawl,

Table 7.7 CBA of urbanization in the Netherlands, differences compared with existing policy (net present values in 2004, in million Euros)

	Higher density	*Controlled sprawl*
Investment and running costs	–264	489
Land revenues	32	59
Benefits for housing consumers	3	8
Windfall profits to land owners	–106	97
User benefits of 'high-quality public transport'	7	–13
Quality and profitability of existing public transport	+PM	–PM
Congestion (travel times)	11	–120
Open space (including 'green spots' in cities)	58 to 211	–45 to –194
Support for and variety of amenities	PM (+/–)	PM (–/+)
Quality of existing housing	20	–25
Synergy with urban improvement policies	PM (+)	PM (–)
Safety	PM	PM
Environmental effects of traffic	–3	–25
Net benefits	–88 to –242 +/–PM	275 to 424 +/–PM

Note: PM = Pro Memorie (not monetized).

however, shows more benefits than costs. The revenues from selling agricultural land for housing are reflected in negative investment costs (net benefits) in the controlled sprawl alternative. These benefits appear to be larger than the value of open space, especially if open space is valued using compensation costs. Also, the costs of building within city limits rise over time, as cheap locations are already filled up and expensive redevelopment areas are the next option. Another result is that 'high quality public transport' (dedicated bus lanes) is not viable, even to spatially concentrated commuter towns. Normal buses appear to be a better alternative in terms of costs and benefits.

From this example, we may conclude that strict zoning laws in densely populated areas may incur considerable costs, which are reduced by urban sprawl. However, as urbanization progresses, open space may become more scarce and more valuable. Furthermore, we also see that assumptions about an automatic need for expensive public transport may be refuted by a CBA which looks at a full set of alternatives.

4. Epilogue

In a modern society, spatial movements (mobility of people, transport of goods) are a basic feature. Intense spatial interaction and large volumes of transport flows put a heavy stress on the accommodation possibilities of regions and cities. We increasingly witness a 'struggle for space', where in a given area a multiplicity of spatial actors compete for a 'place under the sun'. This territorial competition – with many interactions between actors with different trip motives and spatial behaviour

– is accompanied by various positive and negative externalities, impacting on, inter alia, throughflows in transport (e.g. congestion), safety, ecological quality, land use, access to transport systems, etc. Transportation planning is not in the first place an engineering activity, but a multifaceted rational investigation and organization of scarce space. This calls for a more integrated perspective on transportation planning. As a consequence, there is a need – and scope – for new, broader-based approaches that are able to include a wide variety of different types of impacts, as illustrated in this chapter. Of course, extensions of the currently available state-of-the-art methodologies would be desirable. Interactive decision-aid methods could be one direction; while GIS-based assessment might provide new departures for operational research in the transport field.

In this context, the use of modern evaluation methods is extremely useful. They offer a rational basis for discussion and provide more systematics and transparency on the ins and outs of planning and decision making. In this way various trade-offs can be made more visible, while policy priorities can explicitly be taken into consideration as well.

Note

1 This is a very wide definition. As regular train transport inherently demands access and egress trips with other modes, the majority of all multimodal trips concern train trips (67 per cent of the multimodal trips, and 81 per cent of all travelled kilometres in multimodal trips). About 14 per cent of all train trips combine train with car use (as driver or passenger) at the home end of the trip, and about 6 per cent at the activity end (almost entirely car passengers) (see Chapter 11 in this volume).

References

Annema, J.A., Koopmans, C. and Van Wee, B. (2007) 'Evaluating Transport Infrastructure Investments: The Dutch Experience with a Standardized Approach', *Transport Reviews*, 27(2): 125–150.

Bakker, P. and Zwaneveld, P. (2009) *Het Belang van Openbaar Vervoer, de maatschappelijke effecten op een rij*, The Hague, Netherlands: CPB Netherlands Bureau for Economic Policy Analysis and KiM Netherlands Institute for Transport Policy Analysis.

Black, A. (1995) *Urban Mass Transportation Planning*, New York: McGraw-Hill.

Chorus, C.G. (2007) 'Traveler Response to Information', T2007/2, TRAIL Thesis Series, Delft, The Netherlands.

Cohn, N., Daly, A., Rohr, C., Dam, A.F., Oosterwijk, W. and Van der Star, T. (1996) 'ProMiSe: Policy Sensitive Rail Passenger Forecasting for the Netherlands Railways', paper presented at the PTRC European Transport Forum, Cambridge.

Deakin, M., Vreeker, R. and Nijkamp, P. (eds) (2008) *Toolbox for Sustainable Urban Planning*, London: Spon.

Debrezion, G., Rietveld, P. and Pels, E. (2007) 'The Impact of Railway Stations on Residential and Commercial Property Value: A Meta-analysis', *The Journal of Real Estate Finance and Economics*, 35(2): 161–180.

DfT (2009) *Transport Analysis Guidance*, Department for Transport. Available online at: http://www.dft.gov.uk/webtag/ (accessed 1 February 2010).

Dickey, J.W, Stuart, R.C. and Walker, R.D. (1983) *Metropolitan Transportation Planning*, Bristol: Taylor & Francis.

Ecorys (2005) *Maatschappelijke kosten en baten IBO Verstedelijking*, Rotterdam: Ecorys.

Fox, J. (2005) 'Modelling Park-and-Ride in the West Midlands Region', paper presented at the European Transport Conference, Strasbourg, France.

Graham, D.J. (2006) 'Wider Economic Benefits of Transport Improvements: Link Between Agglomeration and Productivity', Stage 2 Report, London: Centre for Transport Studies, Imperial College.

Hayashi, Y. and Morisugi, H. (2000) 'International Comparison of Background Concept and Methodology of Transportation Project Appraisal', *Transport Policy*, 7(1): 73–88.

Iseki, H., Taylor, B.D. and Miller, M. (2006) *The Effects of Out-of-Vehicle Time on Travel Behavior: Implications for Transit Transfers*, Los Angeles, Institute of Transportation Studies, University of California.

Koopmans, C.C. (2006) '"Zachte" beleidseffecten in maatschappelijke kosten-batenanalyses: hoe kan het beter?', *Tijdschrift voor Politieke Economie*, 27(3): 4–19.

—— (2004) *Een heldere presentatie van OEI resultaten; Aanvulling op de Leidraad OEI*, The Hague: Ministries of Transport and Economic Affairs.

Li, Z., Lam, W.H.K., Wong, S.C., Zhu, D.L. and Huang, H. (2007) 'Modeling Park-and-Ride Services in a Multimodal Transport Network with Elastic Demand', *Transportation Research Record: Journal of the Transportation Research Board*, No. 1994. Washington: Transportation Research Board.

Litman, T. (2009) 'Rail Transit in America: A Comprehensive Evaluation of Benefits', Victoria, Canada: Transport Policy Institute.

Margail, F. and Auzannet, P. (1996) 'Evaluation of the Economic and Social Effectiveness of Park-and-Ride Facilities', Proceedings of the European Transport Conference (ETC) 1996.

Meyer, M.D. and Millar, E.J. (1984) *Urban Transportation Planning: A Decision-Oriented Approach*, New York: McGraw-Hill.

Mingardo, G. (2008) *Effecten van Park and Ride in Rotterdam*, Rotterdam: EURICUR/EUR, Rotterdam.

Molin, E. and Van Gelder, M. (2008) *Freeway Access to Public Transport: A Hierarchical Multimodal Choice Model*, Delft, the Netherlands: Transport Policy Section, University of Technology.

Nijkamp, P. and Blaas, E. (1994) *Impact Assessment in Transportation Planning*, Dordrecht: Kluwer.

Nijkamp, P., Rietveld, P. and Voogd, H. (1992) *Multicriteria Analysis and Physical Planning*, Amsterdam: Elsevier.

Nijkamp, P., Ubbels, B. and Verhoef, E.T. (2003) 'Transport Investment Appraisal and the Environment', in D.A. Hensher and K.J. Button (eds) *Handbook of Transport and Environment, Handbooks in Transport*, Elsevier, Amsterdam, pp. 333–354.

Parkhurst, G. (2000) 'Influence of Bus-based Park and Ride Facilities on Users' Car Traffic', *Transport Policy*, 7: 159–172.

Pearce, D., Atkinson, D. and Mourato, S. (2006) *Cost-Benefit Analysis and the Environment: Recent Developments*, Paris: OECD.

Quinet, E. and Vickerman, R. (2004) *Principles of Transport Economics*, Cheltenham, UK: Edward Elgar.

Rijkswaterstaat AVV (2002) *De markt voor multimodaal personenvervoer, OVG Analyse*, Rotterdam: Rijkswaterstaat, AVV.

Simon, H.A. (1955) 'A Behavioral Model of Rational Choice', *Quarterly Journal of Economics*, 69(1): 99–118.

Van de Riet, O.A.W.T. (2003) 'Policy Analysis in a Multi-actor Policy Settings: Navigating Between Negotiated Nonsense and Superfluous Knowledge', Dissertation, Delft University, Netherlands.

Zondag, B. and de Jong, G. (2005) 'The Development of the TIGRIS XL Model: A Bottom-up Approach to Transport, Land-use and the Economy', paper presented at the Economic Impacts of Changing Accessibility Conference at Napier University, 27–28 October 2005.

Part 2

Application of integrated transport policy

Chapter 8

Integrating individual travel desires in transport planning

What is too far and what is too close?

Yusak O. Susilo

1. Individual travel desires in urban and transport planning

There has been a large amount of research over the last decade that has explored the relationship between travel behaviour and urban form, and how land use characteristics may influence individual travel behaviours (e.g. Banister, 1992; Cervero and Kockelman, 1997; Ewing, 1995; Frank and Pivo, 1994; Naess and Sandberg, 1996; Newman and Kenworthy, 1989; Schwanen *et al.*, 2004; Susilo and Axhausen, 2007). For example, a general consensus among researchers is that the 'right' urban design can stimulate the use of public transport, resulting in a reduction of car use. There is a vast amount of empirical evidence which points to car use being lower in traditional-style neighbourhoods, which are often characterized by higher density and a mixture of land use. Accessibility is often better in this type of neighbourhood with more pedestrian-oriented design features which encourage greater use of non-motorized modes (e.g. Snellen *et al.*, 2002; Handy *et al.*, 2005; Williams, 2005; Susilo and Maat, 2007). Some other studies have also shown that providing activity locations close to residential locations is likely to reduce travel time of residents and encourage non-motorized transport (Kenworthy and Laube, 1999; Naess, 2006). Based on these notions and empirical evidence the calls for integrating land use and transport planning (as discussed in Chapter 3 in this volume) are made.

However, some recent work shows that the definition of the 'right' urban design may not be so easy to determine. While one may expect that certain neighbourhood designs may encourage residents to travel more sustainably than the rest of the population, this may not actually be the case. A recent study by Williams *et al.* (2009) found that the residents of schemes designed with many sustainable design features in the UK did not seem to travel any more sustainably than the rest of the population. In fact, they found that the residents of 'sustainable' schemes travelled less sustainably than the rest of the UK population in general (for some trips) and owned more cars. Presumably these results may relate to the residents' attitude towards travel and their spatio-temporal constraints, which can be a reflection of their socio-economic class. This is in line with recent studies, (e.g. Stead, 2001; Mokhtarian and Salomon, 2001; Boarnet and Crane, 2001; Schwanen *et al.*, 2003; Susilo and Maat, 2007) that have shown that individual factors are more crucial in determining commuters' travel behaviour than place accessibility and built environment factors. Lyons *et al.* (2008) noted that while individuals may consider travel distance, time and cost in making their journey, these variables are measured on a perceptual basis rather than an absolute basis. In addition, car drivers are often motivated by their desire for independence and freedom, provided by personal mobility directly under their control. This can sometimes have a negative effect by turning car use into a habit as individuals become more dependent on the car themselves. This can in turn limit the independence sought and thus potentially reduce the effectiveness of any sustainable development schemes that are promoted by the government (Susilo *et al.*, 2009).

These perceptions, attitudes and constraints have been overlooked in many transportation studies on the impacts of the built environment. In most cases, the transport network is still regarded as a combination of different sub-networks, distinguished mainly by mode, and analysed based on a certain logic scheme without taking into account travellers' unique perceptions, attitudes and constraints in planning their daily travel and choosing their activity locations. This approach may provide a biased description in analysing how individuals are willing to use public transport and non-motorized modes in reaching their travel destinations and it will undermine the outcome of measures designed to better integrate these networks. It is important to understand the influences of individual characteristics, along with various sustainable development measurements and opportunities provided by each travel mode, on individuals' travel behaviour and the way they choose their activity locations. For example, situating local shops close to the residential location may reduce travel time of residents and encourage them to walk to do their shopping. However, in many cases individuals choose a location further away to do shopping, for reasons such as being willing to spend more time on travel or wanting to go to shops which seem more attractive. Moreover, while providing too many shopping locations within a given area may cause fierce business competition and may not be good for the profitability of the shops, providing local shops at a distance from residential locations may encourage individuals to start using their private car. So, what is too far and what is too close for an individual to participate in an activity? In order to answer these questions, we need to understand how far individuals are willing to travel, given their various spatio-temporal constraints. It is reasonable to assume that given spatio-temporal constraints people balance time spent on activities and total travel time (Dijst and Vidakovic, 2000). This balancing process is a bounded rational search not for an optimal value but rather for a sub-optimal but acceptable outcome, and this outcome will mainly depend on the way activities are spatially distributed and the nature of the transport system provided to reach them. Therefore, it is important to understand the relationship between travel time individuals spent to reach the activity location and the activity duration at the location, as this will influence travel behaviour through better integrating spatial planning with the provision of transport infrastructure and services.

In order to explore how the relationship between travel time and activity duration is influenced by travel mode in this study we use the concept of Travel-Time Ratio (TTR) introduced by Dijst and Vidakovic (2000). This is based on the idea that *where* an individual will participate in an out-of-home activity is, among other things, related to the travel time investment needed to reach the activity place and the activity duration at that destination. Although this concept has been tested for commuting and some other activities in general (Schwanen and Dijst, 2002; Susilo and Dijst, 2009, 2010), a comprehensive and detailed analysis of the influence of mode choice within the trip-chain on different types of non-discretionary activities' TTR is lacking. This will be the focus of the discussion in this chapter.

A better understanding of how travel mode choice can affect individuals' TTR for a large variety of activities will help urban and transport planners to optimize spatial distribution of activity locations in relation to infrastructure for various population categories: for example, when the 11 themes suggested in Chapter 3 of this volume are considered. It will also inform policy makers about the optimum spatial configuration that may encourage individuals to shift from private car to public transport and non-motorized modes. The next section discusses the concept of individual TTR, followed by a description of the Dutch NTS data set used in this study. A brief description of the TTR values and how it varies across different individuals and activity types and travel modes is then presented. A synthesis analysis using multilevel regression is provided. The chapter ends with conclusions and a discussion of the results.

2. The concept of the Travel-Time Ratio (TTR)

Individuals' daily lives consist of activities in space and time, such as personal care, family interaction, work, shopping, recreation and socializing, and these occur at a relatively few geographic locations and for limited durations of time. In order to take part in activities, individuals often have to travel between different places. Their movement ability in space and time depends in part on the resources available to them, such as time, and the level of accessibility of these places (Susilo and Kitamura, 2005).

While individuals will try to maximize their 'utility' within their limited perception of opportunities, they are at the same time constrained in their trade-offs between travel time and activity duration (Hägerstrand, 1970; Deserpa, 1971; Evans, 1972; Lenntorp, 1976; Dijst, 2009). Hägerstrand (1970) defined three constraints that shape individual activity-travel engagements: *capability* constraints (biological, mental and instrumental), *coupling* constraints (synchronization of individuals, equipment and materials) and *authority* constraints (regulation of access to places). These constraints restrict the set of spatio-temporal opportunities individuals have for travelling to activity places and to participate in activities. For example, the probability that an individual would be able to engage in a certain activity depends on whether he/she would have time to travel to the location and spend a reasonable amount of time there on the activity. Shorter travel times (resulting from shorter distances and/or higher speeds) could mean that more time is allocated to activities. Conversely, longer travel times mean that less time will be allocated to activities. If a person does not have sufficient time to travel and to participate in an activity in a satisfactory manner, then he/she may either re-schedule the activity or change the activity location (Susilo and Dijst, 2010).

To empirically study the relationship between travel time and activity duration Dijst and Vidakovic (2000) proposed the TTR concept, which is defined as the ratio obtained by dividing the travel time to a particular activity place by the sum of travel time and activity duration for the same activity location:

$$\tau = \frac{T_t}{T_t + T_a} \quad (1)$$

In this formula τ stands for TTR, T_t indicates the travel time spent to reach the activity location and T_a denotes the activity duration. Given that T_a will always be more than zero, τ will always be higher than zero and lower than one.[1,2]

Dijst and Vidakovic (2000) have applied the TTR concept to various non-work activities for a sample which consisted of double income families residing in a medium-sized city and a suburban community in the Netherlands. As an extension of Dijst and Vidakovic's study, Schwanen and Dijst (2002) analysed the TTR for the commuting activity in the Netherlands, and then Susilo and Dijst (2009, 2010) used a large dataset (comprised of various population categories and various urban forms) to analyse the TTR values for both mandatory and discretionary activities. They found that the value of the TTR varies according to the nature of the journey and also trip purpose. Important variables include the type of activities, individual commitments, available travel modes and the selection of activity locations. The results also show that the TTR value is not solely influenced by individuals' activity commitments, resources and constraints – there is also a trade-off within household commitments and constraints. However, whether the influences of different travel modes on the TTR values are similar among different types of activity is still unknown. This is the issue which will be explored in this chapter.

It is important to explore TTR values for different types of travel modes separately, because the distance individuals are willing to travel by foot will be very different to the distance individuals are willing to travel by private car. Moreover, it is also very important to understand the impacts of various trips and travel-mode chaining on the ratio of travel time and activity duration that an individual is willing to spend in various different activities. This would help us to better understand the impacts of mode- and trip-chaining in scheduling the activity participation and choosing the activity locations. The end result would enable deeper insights which may help provide a better transport system that can lead to more use of sustainable transport modes, like walking and cycling. Again, this kind of insights are important in informing the planning process, for example that of walking and cycling networks (see Chapter 9 in this volume).

3. Research design

This chapter draws on data from the 2005 Dutch National Travel Survey (NTS) which provides detailed information about individuals, households and their journeys (for more information about the dataset, see van Evert *et al.*, 2006). The dataset consists of 49,583 individuals from 21,743 households who made 182,797 trips on the given day. The NTS classified the individual activities into 45 different categories. Within this study, these 45 different activities were reclassified into 15 broader categories, which are:

1. work
2. business trip
3. transport as occupation: e.g. taxi and public transport drivers
4. regular school and nursery activities: e.g. primary and secondary schools, universities and nurseries
5. other education: e.g. school trips, training or short courses
6. daily shopping activities: e.g. visiting a bakery, grocery, butcher and supermarket
7. non-daily shopping activities: e.g. visiting clothes and shoe stores, department store and domestic decoration store
8. medical services: e.g. pharmacy, hospital, doctor or therapist
9. other services: e.g. bank, post office, city hall, or tax office
10. chauffeuring a person: e.g. to pick-up and drop-off person to/from school or station
11. visit café and restaurant: e.g. dining out
12. culture: e.g. visiting a museum, cinema or concert
13. sport/recreation: e.g. walking, playing football, or going to the beach
14. social visits: e.g. seeing friends and relatives and attending a community centre
15. other leisure: e.g. attending church service and various meetings

The NTS provides information on a range of personal and household attributes, which were used to create a set of variables at the individual and household level. In addition to these socio-demographics attributes, we have added travel attributes at the journey level and built environment attributes at the municipal level. It is important to note here that the classification of daily and non-daily shopping trips is based on respondents' reports.

3. Travel-Time Ratio index

Since the nature of the activity and its duration also depends on the nature of the trip-chain, the methods for calculating the TTR and the distribution of travel time values T_t within a multi-activity journey are based on the number of bases (e.g. home only or home and work) and the number of visits/activities within the journey (Figure 8.1). In this study the base location is either an individual's home or work location.

I. Journey has only one base location B (either home or work)

Journey type 1: Only 1 visit journey

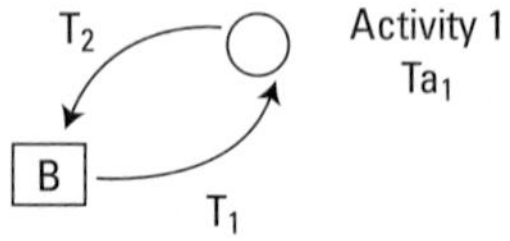

$T_t = T_1 + T_2$

$T_a = T_{a1}$

$$\tau = \frac{T_t}{T_t + T_a}$$

Journey type 2: More than 1 visit journey

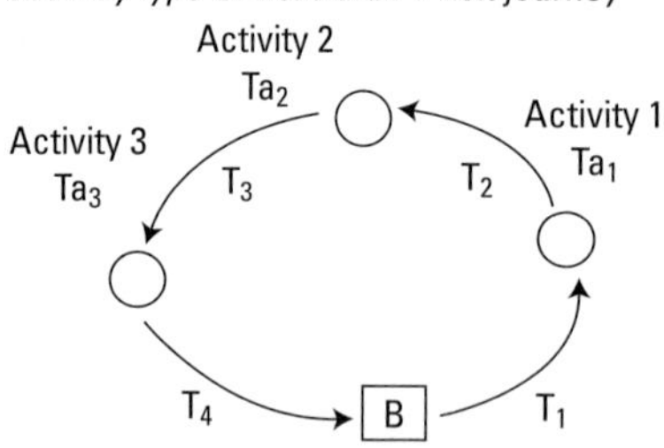

For a pattern with more than 1 visit, the travel time is allocated proportionally to the stay times in activity places (for discussion of the assumption, see Dijst, 1995). For example:

If $T_{a1} = 1$, $T_{a2} = 12$ and $T_{a3} = 1$

and $T_1 + T_2 + T_3 + T_4 = 24$

then $T_1' = 2$, $T_2' = 20$ and $T_3' = 2^*$

$$\tau = \frac{T'}{T' + T_a}$$

II. Journey has two base locations B (home and work)

Journey type 3: No additional activity location (simple commute pattern)

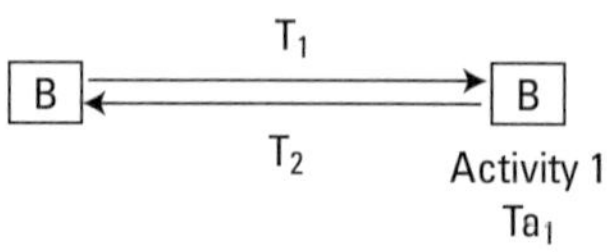

$T_t = T_1 + T_2$

$T_a = T_{a1}$

$$\tau = \frac{T_t}{T_t + T_a}$$

Journey type 4: With 1 additional activity location

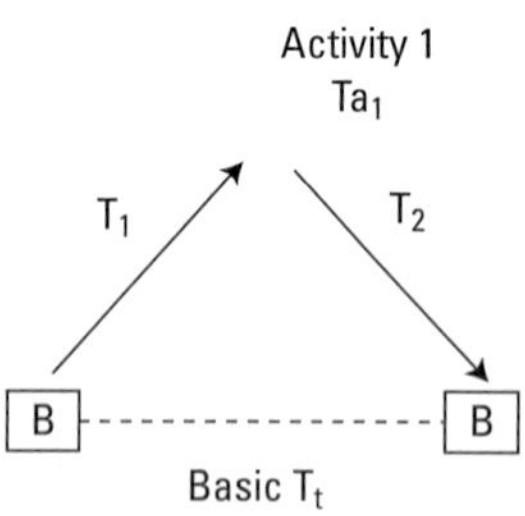

For a pattern with two bases, the travel time spent on the visited activity place is computed by subtracting 'the travel time needed for a direct trip between bases' from 'the actual total travel time'. The steps are:

1. Determine basic travel time (basic T_t) between base locations (using BasNet 2000 database)
2. Calculate travel time spent to visit additional activity, T'' = actual total travel time ($T_1 + T_2$) – basic T_t
3. The TTR index, $\tau = \frac{T''}{T'' + T_a}$

Journey type 5: With more than 1 additional activity location

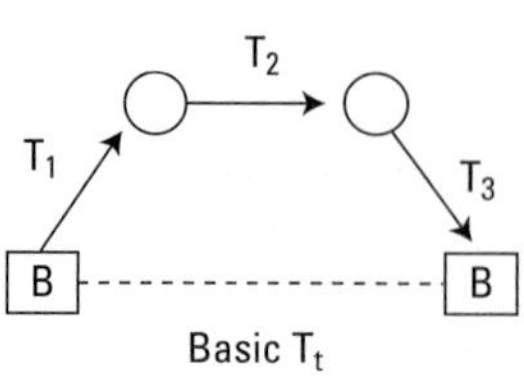

Similar to the journey type 4 approach, combined with journey type 2 approach:

1. Calculate T_t'' (time spent to visit additional activity as journey type 4)
2. Distribute T_t' for each activity, weighted by the amount of T_a, as journey type 2.

(continued)

Figure 8.1
TTR calculation methods by type of journey

4. Travel-Time Ratio values by main travel modes

The average TTR values by journey type (Figure 8.1) and transport mode are presented in Table 8.1. As expected, the value of TTR varies according to the type of journey and also by mode. It is clear from Table 8.1 that those journeys carried out in tighter time windows, such as work base activities (w-w in journey types 1 and 2) and morning commutes (h-w in journey types 4 and 5) have higher TTR values than their counterparts. Adding an extra trip during the morning commute and in a tight lunchtime window will leave travellers less time to participate in the activity; whereas adding extra non-work activities during the afternoon commute will leave the individual sufficient time to engage in these activities. The balance of this trade-off might also, for example, depend on the type of activity, individual commitments, the availability of a time-window to do the activity, available travel mode, and the selection of activity locations. Except for journey type 1, within the same type of journey, non-motorized travellers constantly have lower TTR values than private car users. This means that given the same amount of activity duration, private car users spend more time in travelling to reach activity locations (by selecting farther locations) than non-motorized travellers. Table 8.1 also shows that there is not a clear pattern for public transport travellers' TTRs compared to other modes. This might be due to the small size of the samples, which may cause rather different distributions of TTR values.

It is also reasonable to assume that different activity types would have different TTRs because each activity has a different nature and a different level of urgency and needs. For example, an individual may be willing to travel an hour to do his/her business travel, while he/she may not be keen to spend the same amount of time to just buy milk. These differences are shown in Figure 8.2, which indicates that the activity types of 'chauffeuring a person' and 'transport as occupation' have the highest TTR values, which means that in these two activities people spent quite a considerable amount of travel time in relation to the activity duration at the destination location; this is understandable given the nature of the trip purposes. On the other hand, activities which on average have a long participation duration (such as education/nursery and work) have lower TTRs than others. The TTR of sport/recreation for non-motorized travellers is very high compared to other activities. Presumably this is because, among non-motorized travellers, this activity is dominated by morning/afternoon walks to the nearby parks and the walking activity is treated as the trip purpose (with no significant amount of activity duration at the destination location).

Figure 8.1 (*cont.*)

Note:

Journeys without any additional activity (e.g. jogging) and which did not return to either home or work are excluded. The respondents who have such journeys in their daily activity pattern on the given day were excluded from the analysis.

* $T'_{1,2,3}$ is used instead of $T_{1,2,3}$ because in choosing activity locations within a trip-chain, an individual would not choose activity 1 solely under consideration of T_1. It is reasonable to assume that individuals would do a balancing process of activity time and travel time within their trip-chains. He/she would choose the locations based on the effectiveness of the whole chain (in this case, three different activities) given his/her constraints. Therefore it is reasonable to measure travel time proportional to the stay times in activity places.

Table 8.1 The average TTR values by type of journey and by main transport mode

Type of journey	*Nature of the journey*	*Private car*		*Public transport*		*Non-motorized*	
		Number of activities	*TTR mean*	*Number of activities*	*TTR mean*	*Number of activities*	*TTR mean*
Journey type 1	home base	21459	0.324	991	0.280	28437	0.409
(N = 52,413)	work base	498	0.391	10	0.429	370	0.571
Journey type 2	home base	11628	0.302	383	0.307	6068	0.266
(N = 8,043)	work base	365	0.385	3	0.707	40	0.282
Journey type 3	simple commute (h-w-h)	7726	0.133	817	0.166	3483	0.106
(N = 13,823)	only h-w/w-h journey	890	0.171	33	0.229	671	0.162
Journey type 4	h-w journey	528	0.544	19	0.359	172	0.490
(N = 2,237)	w-h journey	1060	0.366	60	0.390	383	0.325
Journey type 5	h-w journey	348	0.426	2	0.802	38	0.256
(N = 746)	w-h journey	1080	0.288	19	0.321	144	0.209
All samples	**77,262 journeys**	**45,582**		**2,337**		**39,806**	

Note:
h-w-h = home-work-home; h-w = home-work; w-h = work-home; 'journey' is defined based on the definitions found in Figure 8.1. Individuals who used 'other modes' are not shown.

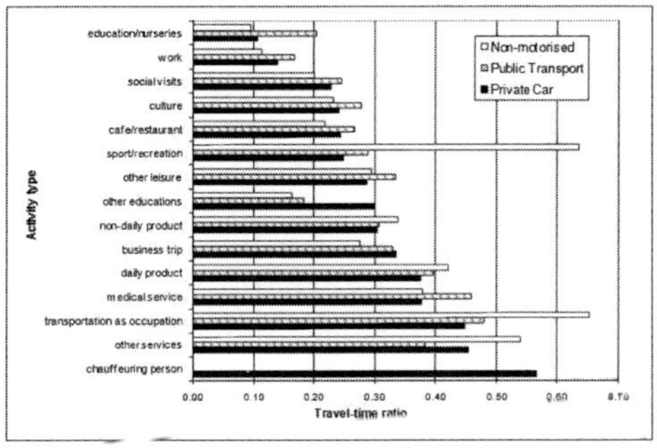

Figure 8.2 Travel time ratios by activity type and by main transport mode

Overall, TTRs of public transport users are higher than those of private car users. This suggests that public transport users travel longer to spend the same amount of time on an activity compared to private car travellers. This may be due to slower average travel speed of public transport compared to private cars in the Netherlands (Susilo and Kees, 2007). On the other hand, non-motorized travellers have lower TTRs compared to private car users. However, this is not the case for sports/recreation and leisure trips and other services. These higher values might be caused by concentration of these facilities at larger distances from the home.

For both daily and non-daily product shopping, non-motorized travellers have higher TTRs than private car travellers. This may mean that, given the same amount of travel time, non-motorized travellers tend to spend less time in activities. This result is in line with some previous studies (e.g. Susilo and Axhausen, 2007) which found that residents who live closer to shopping locations do their shopping on foot, in much shorter duration, but with more frequent visits during the week.

When the TTR values of each activity were plotted based on the proportion of their travel-mode chain, there were not any clear trends due to the changes in mode proportion. It may be because a very small proportion of respondents (only about 2 per cent) mixed their main travel mode within a trip-chain journey. Moreover, when the TTR values are grouped based on different types of journeys (single/dual base location, single or more extra trips within a journey), the results do not show that the TTRs are lower for activity places visited in trip-chains than in single trips. Nevertheless, it is understandable because TTR values are a result of a complex trade-off of various aspects of individual activity-travel participations and are neither solely influenced by the purpose of the activity nor the type of journey. In order to explore this complexity further, a multilevel regression analysis is used in the next section.

5. Multilevel regression analysis

In order to further explore the impacts of different travel modes on individuals' TTRs when engaging in various different activities, a multilevel regression model is used. This model will help us to understand the impacts of individual socio-demographic conditions, journey characteristics and built environment factors, along with travel mode constraints on the size of the TTRs for each different activity. In this study,

MLwiN v. 2.10 (Rasbash *et al.*, 2005) is used for the multilevel regression analysis. The general form of the model is:

$$y_{ijkl} = \beta_0 + \beta' X_{ijkl} + (f_{0l} = v_{0kl} + u_{0jkl} + e_{0ijkl})\ i = 1,2,\ldots,N,$$
$$j = 1,2,\ldots,M, \quad k = 1,2,\ldots,H, \quad l = 1,2,\ldots,R, \tag{2}$$
$$f_{0l} \sim N(0,\sigma^2_{f0}) \quad v_{0kl} \sim N(0,\sigma^2_{v0}) \quad u_{0jkl} \sim N(0,\sigma^2_{u0}) \quad e_{0ijkl} \sim N(0,\sigma^2_{e})$$

Where i refers to the journey episode (from base to base), j to the individual, k to the household, and l to the particular spatial location. N is the number of the journey episode, M is the number of individuals, H is the number of households, R is the number of spatial locations, and y_{ijkl} is the TTR of journey episode i of individual j of household k at location l, with an overall mean β_0, X_{ijkl} is a vector of explanatory variable e_{0ijkl} is journey episode random term u_{0jkl} is individual random term v_{0kl} is household random term, and f_{0l} is residential area random term. These random terms are assumed to have a mean of 0 and be mutually independent. This formulation shows that journey episodes are assumed to be nested within individuals in households which are grouped in a particular residential environment.

In this chapter we will focus on three discretionary activities: 'daily product shopping', 'non-daily product shopping' and 'sport and recreation'. The transport modes included are 'private car', 'public transport' and 'non-motorized modes'. Although it is very interesting to analyse the impact of a combination of transport modes on the size of TTR, given that almost 98 per cent of the samples only used a single transport mode within the journey, the impact of mode-chaining will not be carried out in this analysis.

Results

The estimation results are presented in Tables 8.2, 8.3 and 8.4 for daily shopping, non-daily shopping and sport activities, respectively. Since this study aims to understand the individual behaviours underlying the value of TTRs, some insignificant variables were still included in this estimation in order to show and to explore the behaviour variability among different groups of people. The salient results are as follows.

It is shown in Tables 8.2, 8.3 and 8.4 that individual socio-demographics, journey characteristics and built environment factors have different impacts on the TTR values of activities with different travel modes. For example, while private car travellers' TTRs are significantly influenced by the time of the journey, this is not the case with public transport and non-motorized travellers. Presumably it is because private car travellers have more choice and flexibility in chaining their trips and are also highly influenced by traffic conditions at peak and off-peak periods (Kitamura and Susilo, 2006). The estimation results show that in the morning period (6 am to 12 noon) private car travellers prefer to choose a closer location to home/work locations to do shopping than they do in the afternoon (12 noon to 6 pm) or evening (from 6 pm until midnight) periods. For sport and leisure activities, however, private car users travel to a further location in the afternoon than they do if they perform the activity in the morning or evening.

More activity engagements within a trip-chain would encourage private car and non-motorized travellers to find a closer place to do their shopping and encourage them to reduce the duration of their sport/leisure activities. However, that is not the case with public transport travellers. Less flexibility in routes and location choices among public transport users seems to be the reason.

Given the same amount of travel time, regardless of travel mode, males have significantly shorter durations for shopping than females, for both daily and non-daily product shopping. However, there is no significant gender difference (at $\alpha = 5$ per cent) for sports and recreational activities.

In doing shopping activities, private car and non-motorized travellers who are younger tend to find a closer shopping location to their home/work locations than their older counterparts, but this is not the case among public transport travellers. On the other hand, for sport and leisure activities, public transport and non-motorized travellers who are younger choose a closer activity location than older travellers. No significant age-related difference was found among private car travellers in choosing their sport activity locations.

More trip-chains within a given day encourage private car and non-motorized travellers to do their shopping, sport and leisure activities in shorter durations. However, similar impacts are not found among public transport users. Dependency of public transport users on the service route and its coverage reduces the ability of the users to increase the effectiveness of their activity-travel patterns.

The presence of dependent children encourages all travellers to reduce their daily shopping duration, but this seems to increase the sport and recreation activity duration for non-motorized travellers.

The TTR values increase with the decrease of the urbanization level, which shows that for the same amount of activity time, less urbanized residents would spend longer time travelling than their more urbanized counterparts. Jobs, accessibility to population centres, and being located in the Randstad area (the main metropolitan area of the Netherlands) were not consistently significant across different types of activities, especially for public transport travellers. This supports the earlier suggestion that public transport users tend to be less flexible in choosing their activity locations and adjust their activity duration according to the constraints that are imposed by the travel mode. Their TTRs and activity locations were not influenced by journey or built environment characteristics. On the other hand, non-motorized and private car travellers show more flexibility in choosing their activity locations and adjusting their activity duration. Land use characteristics, journey periods and socio-demographics play an important role in influencing private car and non-motorized travellers' activity locations.

6. Summary and discussion

Using the concept of TTR as proposed by Dijst and Vidakovic (2000) and the 2005 Dutch National Travel Survey dataset, this chapter explored the influence of travellers' socio-demographics, journey patterns, built environment factors, and (main) travel mode constraints on the ratio of their travel time and activity duration when

Table 8.2 Multilevel regression analyses of TTRs for daily shopping activities

	Random-intercept model							
	(i) All modes		*(ii) Private car*		*(iii) Public transport*		*(iv) Non-motorized*	
	Coefficient	*t-stats*	*Coefficient*	*t-stats*	*Coefficient*	*t-stats*	*Coefficient*	*t-stats*
Constant	0.3360	20.21	0.3709	16.43	0.3275	2.26	0.3146	12.11
Journey characteristics:								
The journey departed between 06–12 h			–0.0184	–2.03				
The journey departed between 12–18 h								
The journey is done on weekend	–0.0151	–3.10					–0.0144	–2.03
The journey only has one-base (home/office)	0.0921	7.63	0.0322	1.91			0.1429	7.75
Number of activities on this particular journey	–0.0401	–16.95	–0.0252	–8.45			–0.0473	–11.92
The journey started from office location	0.0799	6.84					0.1178	7.12
Individual characteristics:								
Male travellers	0.0379	12.44	0.0242	6.80	0.0310	1.66	0.0494	10.39
Travellers younger than 25 years old	–0.0231	–2.89	–0.0336	–3.18				
Travellers between 25 and 40 years old	–0.0217	–3.23	–0.0256	–2.82			–0.0185	–1.89
Travellers between 40 and 64 years old	–0.0101	–1.88						
High education level travellers							0.0183	2.07
Medium education level travellers								
High income group travellers							0.0272	2.49
Medium income group travellers								

	Random-intercept model							
	(i) All modes		*(ii) Private car*		*(iii) Public transport*		*(iv) Non-motorized*	
	Coefficient	*t-stats*	*Coefficient*	*t-stats*	*Coefficient*	*t-stats*	*Coefficient*	*t-stats*
Total amount of time spent on obligation commitments	0.0000	–2.80					–0.0001	–3.79
Number of journeys on the given day	0.0201	10.69	0.0174	7.77			0.0208	7.09
Household characteristics:								
Number of household members	0.0068	2.81	0.0074	2.29			0.0078	2.17
Presence of dependent children (less than 15 years old)	–0.0193	–3.16	–0.0218	–2.67	–0.0787	–1.91	–0.0157	–1.74
Travellers come from single earner household	–0.0130	–2.35			–0.0924	–2.16	–0.0159	–2.08
Two or more earner household					–0.1207	–2.37		
Total number of journeys within household	–0.0056	–3.73	–0.0046	–2.36			–0.0051	–2.28
Total number of activities within household	0.0035	3.03	0.0033	2.24				
Land use characteristics:								
Number of retail per km^2								
Number of jobs that can be reached within 30 minutes travel								
Number of population that can be reached within 30 minutes travel								
Home municipality has ≥ 2,500 addresses per km^2	–0.0189	–2.12	–0.0317	–2.39			–0.0333	–2.65

(*continued*)

Table 8.2 (cont.)

	Random-intercept model							
	(i) All modes		*(ii) Private car*		*(iii) Public transport*		*(iv) Non-motorized*	
	Coefficient	*t-stats*	*Coefficient*	*t-stats*	*Coefficient*	*t-stats*	*Coefficient*	*t-stats*
Home municipality has 1,500–< 2,500 addresses per km^2	–0.0267	–4.22	–0.0341	–3.89			–0.0304	–3.29
Home municipality has 1,000–< 1,500 addresses per km^2	–0.0274	–4.38	–0.0326	–3.83			–0.0296	–3.24
Home municipality has 500–< 1,000 addresses per km^2	–0.0231	–3.75	–0.0251	–3.04			–0.0225	–2.47
The residents of Randstad area (the main metropolitan area in the Netherlands)								
The respondent resides next to Randstad area								
Random part:								
Variance at level land use (level 4)			0.0092	37.35				
Variance at level household (level 3)	0.0167	22.77	0.0203	20.41	0.0222	5.63	0.0176	15.27
Variance at level individual (level 2)								
Variance at level journey (level 1)	0.0158	52.67	0.0092	37.25	0.0028	4.06	0.0164	30.04
–2 * log-likelihood:	–10112		–6690		–232		–4085	

Note: Only variables significant at $\alpha = 5\%$ are shown in the table.

Table 8.3 Multilevel regression analyses of TTRs for non-daily shopping activities

	Random-intercept model							
	(i) All modes		*(ii) Private car*		*(iii) Public transport*		*(iv) Non-motorized*	
	Coefficient	*t-stats*	*Coefficient*	*t-stats*	*Coefficient*	*t-stats*	*Coefficient*	*t-stats*
Constant	0.3496	13.58	0.3409	10.33	0.3914	2.68	0.3637	8.68
Journey characteristics:								
The journey departed between 06–12 h	–0.0360	–3.37	–0.0422	–3.36				
The journey departed between 12–18 h								
The journey is done on weekend	–0.0200	–3.30	–0.0180	–2.43				
The journey only has one-base (home/office)	0.0447	2.14						
Number of activities on this particular journey	–0.0162	–4.56	–0.0125	–3.07			–0.0171	–2.56
The journey started from office location								
Individual characteristics:								
Male travellers	0.0209	5.99	0.0120	3.16	0.0076	1.81	0.0344	5.55
Travellers younger than 25 years old	–0.0428	–4.44	–0.0318	–2.67			–0.0624	–4.00
Travellers between 25 and 40 years old	–0.0391	–4.65	–0.0312	–3.02			–0.0441	–3.25
Travellers between 40 and 64 years old	–0.0238	–3.54	–0.0204	–2.38			–0.0204	–1.99
High education level travellers								
Medium education level travellers								
High income group travellers								
Medium income group travellers								

(continued)

Table 8.3 (cont.)

	Random-intercept model							
	(i) All modes		*(ii) Private car*		*(iii) Public transport*		*(iv) Non-motorized*	
	Coefficient	*t-stats*	*Coefficient*	*t-stats*	*Coefficient*	*t-stats*	*Coefficient*	*t-stats*
Total amount of time spent on obligation commitments							–0.0001	–2.80
Number of journeys on the given day	0.0160	6.53	0.0113	4.10			0.0200	4.74
Household characteristics:								
Number of household members								
Presence of dependent children (less than 15 years old)							0.0262	2.07
Travellers come from single earner household								
Two or more earner household								
Total number of journeys within household			–0.0057	–2.30				
Total number of activities within household			0.0052	2.62				
Land use characteristics:								
Number of retail per km^2	9.15E–07	2.31					1.43E–06	2.37
Number of jobs that can be reached within 30 minutes travel								
Number of population that can be reached within 30 minutes travel								

	Random-intercept model							
	(i) All modes		(ii) Private car		(iii) Public transport		(iv) Non-motorized	
	Coefficient	*t-stats*	*Coefficient*	*t-stats*	*Coefficient*	*t-stats*	*Coefficient*	*t-stats*
Home municipality has ≥ 2,500 addresses per km²	–0.0302	–2.71					–0.0655	–3.65
Home municipality has 1,500–< 2,500 addresses per km²	–0.0194	–2.32			0.1612	3.07	–0.0525	–3.66
Home municipality has 1,000–< 1,500 addresses per km²							–0.0453	–3.16
Home municipality has 500–< 1,000 addresses per km²					0.1694	2.80	–0.0303	–2.06
The residents of Randstad area (the main metropolitan area in the Netherlands)								
The respondent resides next to Randstad area			–0.0172	–2.13				
Random part:								
Variance at level land use (level 4)								
Variance at level household (level 3)	0.0160	19.15	0.0153	25.41	0.0180	8.21	0.0200	20.97
Variance at level individual (level 2)								
Variance at level journey (level 1)	0.0085	33.61	0.0054	25.38	0.0002	4.42	0.0090	18.26
–2 * log-likelihood:	–6890		–4725		–365		–2348	

Note: Only variables significant at $\alpha = 5\%$ are shown in the table.

Table 8.4 Multilevel regression analyses of TTRs for sport and leisure activities

	Random-intercept model							
	(i) All modes		*(ii) Private car*		*(iii) Public transport*		*(iv) Non-motorized*	
	Coefficient	*t-stats*	*Coefficient*	*t-stats*	*Coefficient*	*t-stats*	*Coefficient*	*t-stats*
Constant	0.2146	5.50	0.1721	4.80			0.3480	6.19
Journey characteristics:								
The journey departed between 06–12 h			0.0154	2.03				
The journey departed between 12–18 h	0.0461	7.15	0.0608	8.76			0.0235	3.22
The journey is done on weekend	0.0723	7.37	0.0172	2.08			0.0917	7.34
The journey only has one-base (home/office)	0.3084	9.14					0.3780	7.54
Number of activities on this particular journey	–0.0490	–10.02	0.0121	3.25	0.0927	2.63	–0.0776	–10.80
The journey started from office location	0.2105	7.22					0.1950	5.98
Individual characteristics:								
Male travellers							–0.0127	–1.92
Travellers younger than 25 years old	–0.1613	–9.94			–0.2976	–3.63	–0.2047	–10.24
Travellers between 25 and 40 years old	–0.0736	–5.08			–0.1509	–2.38	–0.0644	–3.57
Travellers between 40 and 64 years old					–0.1508	–2.38		
High education level travellers	–0.0431	–3.76			–0.1507	–2.37	–0.0442	–3.09
Medium education level travellers	–0.0281	–3.05			–0.1446	–2.61	–0.0281	–2.46
High income group travellers								
Medium income group travellers					–0.1059	–2.76		
Total amount of time spent on obligation commitments	–0.0002	–9.76	–0.0001	–2.50			–0.0002	–8.11

	Random-intercept model							
	(i) All modes		*(ii) Private car*		*(iii) Public transport*		*(iv) Non-motorized*	
	Coefficient	*t-stats*	*Coefficient*	*t-stats*	*Coefficient*	*t-stats*	*Coefficient*	*t-stats*
Number of journeys on the given day	0.0532	17.97	0.0191	6.99			0.0401	11.48
Household characteristics:								
Number of household members	–0.0253	–5.17					–0.0377	–6.06
Presence of dependent children (less than 15 years old)								
Travellers come from single earner household								
Two or more earner household	–0.0260	–1.91						
Total number of journeys within household	0.0082	2.88	–0.0056	–2.53			0.0097	2.59
Total number of activity within household			0.0054	3.26	–0.0247	–2.27		
Land use characteristics:								
Number of retail per km^2	1.80E–06	2.44					2.10E–06	2.15
Number of jobs that can be reached within 30 minutes travel							1.14E–07	2.57
Number of population can be reached within 30 minutes travel							–2.22E–07	–2.60
Home municipality has ≥ 2,500 addresses per km^2	–0.0481	–2.54					–0.0802	–3.27
Home municipality has 1,500–< 2,500 addresses per km^2								
Home municipality has 1,000–< 1,500 addresses per km^2								

(continued)

Table 8.4 (cont.)

	Random-intercept model							
	(i) All modes		*(ii) Private car*		*(iii) Public transport*		*(iv) Non-motorized*	
	Coefficient	*t-stats*	*Coefficient*	*t-stats*	*Coefficient*	*t-stats*	*Coefficient*	*t-stats*
Home municipality has 500–< 1,000 addresses per km^2								
The residents of Randstad area (the main metropolitan area in the Netherlands)	–0.0278	–2.40						
The respondent resides next to Randstad area	–0.0489	–4.69					–0.0461	–3.46
Random part								
Variance at level land use (level 4)								
Variance at level household (level 3)	0.0713	24.74			0.0140	5.38	0.0813	8.37
Variance at level individual (level 2)			0.0191	3.02				
Variance at level journey (level 1)	0.0415	55.51	0.0089	28.25	0.0001	2.12	0.0340	42.83
–2 * log-likelihood:	4701		–4177		–175		2795	

Note: Only variables significant at α = 5% are shown in the table.

1 If T_t remains constant, τ will decrease if T_a increases and τ will increase if T_ε decreases. On the other hand, if T_t decreases (with constant T_a), the impact on τ may appear ambiguous at first since the decrease in the numerator points to a decrease in τ, while the decrease in the denominator points to an increase in τ. Given that T_a will always be more than zero, it is clear that for a given decrease in T_t, the denominator will decrease by a smaller proportion than will the numerator, so the net outcome is that the denominator is relatively larger than before, and τ decreases. Conversely, τ will increase if T_t increases.

2 It needs to be noted here that in this study we are focusing on out-of-home activities. Therefore, in-home activities are not included in the analysis.

engaging in daily and non-daily shopping and sport/recreation activities. The results show that the value of TTR differs according to trip purpose and travellers' main modes. While private car travellers' activity locations and activity durations are significantly influenced by the time of the journey, this is not the case with public transport and non-motorized travellers. Presumably this is because private car users have more flexibility in chaining their trips and are also highly influenced by the uncertainty of traffic conditions. Overall, the results show that public transport users tend to be less influenced by their socio-economic and built environment factors in choosing their activity locations and adjusting their activity duration. On the other hand, land use characteristics, journey periods and socio-demographics play important roles in influencing private car and non-motorized travellers' activity locations.

The understanding of these unique trade-off indices provides us with an insight of how far individuals with specific travel mode constraints would travel to participate in activities. This, in turn, will help inform planning, with consideration of spatio-temporal distances and of activity locations from individuals' home/work locations in order to encourage more sustainable transport modes. For instance, on average, the shopping duration of a female in the Netherlands is 34.9 minutes (Schwanen, 2004) and (after controlling for all other factors) the TTR of a high income female aged 25 to 40 years with dependent children for grocery activities with a non-motorized mode is 0.308. Therefore, a reasonable location for a simple home-based grocery shopping trip for her by a non-motorized mode is about 15.5 minutes from her house (31 minutes for a roundtrip). Shops that are located 30 minutes from her house may be too far for her to access by foot, although providing many stores that are located 10 minutes from each house may not bring the optimum land use configuration. Of course this also depends on the nature and the time of the journey, such as whether it is part of a complex trip-chaining or not; whether she travels with an infant or not; whether the activity locations are accessible at that particular time; and also on her socio-demographic factors. Nevertheless, it could provide a bridge between sustainable urban design and transport planning at an individual level which may provide a better forecast of travel mode impacts in predicting individual travel destinations.

Some important factors have not been taken into account in this analysis, such as the impact of mode-sharing, multimodal journeys and how this ratio will evolve in short- and long-term periods. Given that the way individuals carry out their activity-travel behaviour is evolving over time (Susilo and Kitamura, 2008), the stability and acceptability of TTR values will evolve over time as well.

Acknowledgements

This chapter is an extension of the author's previous research papers with Martin Dijst of Utrecht University, the Netherlands, which were presented at the NECTAR Cluster 1 workshop at Oxford University, at the 88th TRB Annual Meeting in Washington, DC, and at the 41st UTSG Annual Meeting in London. The author would like to thank Martin Dijst for his comments on an earlier version of this manuscript and Tom de Jong of Utrecht University, the Netherlands, for his help in providing the travel distances of various different locations from the BasNet 2000 database. The author also would like to thank Hebba Haddad of the Centre for Transport and Society, University of the West of England, for her help in proof-reading the manuscript, and also two anonymous reviewers of this chapter for their constructive comments.

Notes

1 If T_t remains constant, τ will decrease if T_a increases and τ will increase if T_a decreases. On the other hand, if T_t decreases (with constant T_a), the impact on τ may appear ambiguous at first since the decrease in the numerator points to a decrease in τ, while the decrease in the denominator points to an increase in τ. Given that T_a will always be more than zero, it is clear that for a given decrease in T_t, the denominator will decrease by a smaller proportion than will the numerator, so the net outcome is that the denominator is relatively larger than before, and τ decreases. Conversely, τ will increase if T_t increases.

2 Need to be noted in here that in this study we are focusing on out-of-home activities. Therefore, in-home activities are not included in the analysis.

References

Banister, D. (1992) 'Energy use, transport and settlement patterns', in M.J. Breheny (ed.) *Sustainable Development and Urban Form*, London: Pion, pp. 160–181.

Boarnet, M. and Crane, R. (2001) *Travel by Design: The Influence of Urban Form on Travel*, Oxford: Oxford University Press.

Cervero, R. and Kockleman, K. (1997) 'Travel demand and the 3Ds: Density, diversity, and design', *Transportation Research Part D*, 2, 199–219.

Deserpa, A. (1971) 'A theory of the economic of time', *The Economic Journal*, 81, 826–846.

Dijst, M.J. (1995) Het elliptisch leven: actieruimte als integrale maat voor bereik en mobiliteit [An Elliptical Life: Action Space as Integral Measure for Accessibility and Mobility], Delft: Technische Universiteit Delft, Faculteit Bouwkunde (Ph.D. thesis).

Dijst, M. (2009) 'Time geographical analysis', in R. Kitchin and N. Thrift (eds) *International Encyclopedia of Human Geography*, Oxford: Elsevier, pp. 266–278.

Dijst, M. and Vidakovic, V. (2000) 'Travel time ratio: The key factor of spatial research', *Transportation*, 27, 179–199.

Evans, A.W. (1972) 'On the theory of the valuation and allocation of time', *Scottish Journal of Political Economy*, 19, 1: 1–17.

Ewing, R. (1995) 'Beyond density, mode choice, and single-purpose trips', *Transportation Quarterly*, 49, 15–24.

Frank, L. and Pivo, G. (1994) 'Impacts of mixed use and density on utilization of three modes of travel: Single occupant vehicle, transit and walking', *Transportation Research Record*, 1466, 44–52.

Hägerstrand, T. (1970) 'What about people in regional science?', *Papers of the Regional Science Association*, 24, 7–21.

Handy, S., Cao, X. and Mokhtarian, P. (2005) 'Correlation or causality between the built environment and travel behavior? Evidence from Northern California', *Transportation Research Part D*, 10, 427–444.

Kenworthy, J.R. and Laube, F.B. (1999) 'Patterns of automobile dependence in cities: An international overview of key physical and economic dimensions with some implications for urban policy', *Transportation Research A*, 33, 691–723.

Kitamura, R. and Susilo, Y.O. (2006) 'Does a grande latte really stir up gridlock? Stops in commute journeys and incremental travel', *Transportation Research Record*, 1985, 198–206.

Lenntorp, B. (1976) 'Paths in space-time environment', *Lund Studies in Geography*, Series B, 44, CWK Gleerup, Lund, Sweden.

Lyons, G., Goodwin, P., Hanly, M., Dudley, G., Chatterjee, K., Anable, J., Wiltshire, P. and Susilo, Y.O.

(2008) *Public Attitudes to Transport: Knowledge Review of Existing Evidence*, Final Report for Department for Transport, United Kingdom.

Mokhtarian, P. and Salomon, I. (2001) 'How derived is the demand for travel? Some conceptual and measurement considerations', *Transportation Research A*, 35, 695–719.

Naess, P. (2006) *Urban Structure Matters: Residential Location, Car Dependence and Travel Behaviour*, London: RTPI Library Series, Routledge.

Naess, P. and Sanberg, S.L. (1996) 'Workplace location, modal split and energy use for commuting trips', *Urban Studies*, 33, 557–580.

Newman, P.W.G. and Kenworthy, J.R. (1989) 'Gasoline consumption and cities. A comparison of US cities in a global survey', *Journal of the American Planning Association*, 55, 24–36.

Rasbash, J., Steele, F., Browne, W. and Prosser, B. (2005) *A User's Guide to MLwin, Centre for Multilevel Modelling*, University of Bristol, United Kingdom.

Schwanen, T. (2004) 'The determinant of shopping duration on workdays in the Netherlands', *Journal of Transport Geography*, 12, 35–48.

Schwanen, T., Dieleman, F.M. and Dijst, M. (2003) 'Car use in Netherlands daily urban systems: Does polycentrism result in lower commute time?', *Urban Studies*, 24, 410–430.

Schwanen, T. and Dijst, M. (2002) 'Travel-time ratios for visits to the workplace: The relationship between commuting time and work duration', *Transportation Research A*, 36, 573–592.

Schwanen, T., Dijst, M.J. and Dieleman, F.M. (2004) 'Policies for urban form and their impact on travel: The Netherlands experience', *Urban Studies*, 41, 579–603.

Snellen, D., Borgers, A. and Timmermans, H. (2002) 'Urban form, road network type, and mode choice for frequently conducted activities: Multilevel analysis using quasi-experimental design data', *Environment and Planning*, 34, 1207–1220.

Stead, D. (2001) 'Relationships between land use, socio-economic factors and travel patterns in Britain', *Environment and Planning B*, 28, 499–528.

Susilo, Y.O. and Axhausen, K.W. (2007) 'How firm are you? A study of the stability of individual activity-travel-location pattern using Herfindahl Index', *The 11th World Conference on Transport Research (WCTR)*, Berkeley, California.

Susilo, Y.O. and Dijst, M. (2009) 'How far is too far? Travel time ratios for activity participations in the Netherlands', *Transportation Research Record: Journal of Transportation Research Board*, 2134, 89–98.

—— (2010) 'Behavioural decisions of travel-time ratio for work, maintenance and leisure activities in the Netherlands', *Journal of Transportation Planning and Technology: UTSG Special Issue*, 33, 19–34

Susilo, Y.O. and Kitamura, R. (2005) 'On an analysis of the day-to-day variability in the individual's action space: An exploration of the six-week *Mobidrive* travel diary data', *Transportation Research Record*, 1902, 124–133.

—— (2008) 'Structural changes in commuters' daily travel: The case of auto and transit commuters in the Osaka metropolitan area of Japan, 1980 through 2000', *Transportation Research A*, 42, 95–115.

Susilo, Y.O. and Kees, M. (2007) 'The influence of built environment to the trends in commuting journeys in the Netherlands', *Transportation*, 34(5), 589–609.

Susilo, Y.O. and Maat, K. (2007) 'The influence of built environment to the trends in commuting journeys in the Netherlands', *Transportation*, 34, 589–609.

Susilo, Y.O., Williams, K., Dair, C. and Lindsay, M. (2009) 'Does green in mean green out? An exploration of individual travel patterns and the influences of their environmental preferences in the UK "sustainable neighbourhood"', paper submitted for publication at the *89th Transportation Research Board Annual Meeting*, Washington, DC.

van Evert, H., Brog, W. and Erl, E. (2006) 'Survey design: The past, the present and the future', in P. Stopher and C. Stecher (eds) *Travel Survey Methods – Quality and Future Directions*, London: Elsevier.

Williams, K. (2005) 'Spatial planning, urban form and sustainable transport: An introduction', in K. Williams (ed.), *Spatial Planning, Urban Form and Sustainable Transport*, Hampshire: Ashgate Publishing Ltd.
Williams, K., Dair, C. and Lindsay, M. (2009) 'Neighbourhood design and sustainable lifestyles', in C. Jones and M. Jenks (eds) *Dimensions of the Sustainable City*, New York: Springer.

Chapter 9

Planning walking networks and cycling networks

John Parkin

1. Introduction

Good quality walking networks and, separately, good quality cycling networks, are able to provide end-to-end journey possibilities for many journeys. However, both types of network will inevitably be physically integrated with the public realm, which includes highways and other public rights of way. The physical integration of these networks is the main subject matter of this chapter. An important additional aspect is the integration of walking and cycling with public transport, particularly at public transport nodes. In order to achieve integration, good quality planning is required and methods for doing this, based on good practice guidance and research evidence, are presented.

The chapter begins by considering walking and cycling planning guidance offered to practitioners. It reviews the ways that walkers and cyclists are characterized, summarizes network characteristics deemed to be important, and discusses the difficulties of integrating networks for different purposes. It also evaluates guidance on integration with public transport.

Drawing on research evidence, a summary is presented of the important features of transport networks for walkers and cyclists, and consideration is given to both the traditional issues of time and distance, and also to the important issues of accounting for effort and risk. Issues which are relevant for walkers and cyclists at the point of interchange to public transport are also considered.

The chapter concludes by summarizing the methodological approaches for walking and cycling network planning to achieve physical integration with the highway network and public realm, and with public transport.

2. Guidance on planning for walking and cycling

There was a lot of international activity on the development of guidance for the planning and engineering of networks and facilities for walking and cycling in the closing stages of the twentieth century. The seminal document for cycling, and to which many others refer, is the Dutch guidance *Sign up for the bike* (CROW, 1993). The guidance was a by-product of the Bicycle Master Plan which was a part of the second Structure Plan for Traffic and Transport. Planning and engineering needs to be based on a rationale which, in the first instance, is determined by the characteristics of the users.

2.1 Characteristics of walkers and cyclists

Revised Dutch guidance for cycling (CROW, 2006) notes a number of important distinguishing features of the bicycle as a vehicle, including that it: is powered by muscle; is inherently unstable; has no protection;[1] has virtually no suspension; and that it is open to the air. It makes the point that the skills and limitations of the rider need to be taken into account. Equally important, it suggests that cycling can be a sociable activity. English cycle design guidance (DfT, 2008) caricatures cyclists by type (fast commuter, utility, inexperienced and/or leisure, child, and users of specialized equipment). While usefully reinforcing the variability in types of cyclist, it arguably only differentiates and does not resolve design issues concerning the characteristics.

Other guidance (e.g. Scottish Executive, 1999; Lancashire County Council, 2005; Highways Agency, 2005) chooses to emphasize the distinctions between types of cyclist based on the speed capabilities of the rider with a view to properly designing geometry and sight lines for cycle traffic. New Zealand guidance (LTNZ, 2005) emphasizes the training needs of cyclists and suggests that, where it is not possible for plans to accommodate those with only basic skills, further training is required to the point where cyclists may be regarded as experienced. The guidance suggests that experienced cyclists do not require specific facilities, and that they will be able to 'defend' their space in a mixed traffic road and 'judge' faster traffic. The need to undertake these tasks requires a high level of confidence and ability; these are attributes which many potential cycle users may not wish to develop or adopt.

Insofar as walkers are concerned, guidance tends not to offer comments on their characteristics other than their speed and space requirements. Restrictions on mobility because of, for example, encumbrance with baggage or accompaniment by children in pushchairs, or resulting from a sight or other physical disability, are important to consider and are given prominence in guidance (DfT, 2002). Clearly, walkers share a number of characteristics with cyclists, including the fact that walking can be a sociable activity. It is also self-evidently powered by muscle, has no protection from the bodywork of a vehicle and is open to the surrounding environment. The difference in mean speed between cycling and walking (22 kph for cycling compared with 4 to 5 kph for walking) is, however, so great that these two modes should never be treated as an homogenous whole.[2]

The guidance summarizes appropriately the characteristics of walkers and cycle users relevant for the design of transport networks, to which the discussion now turns.

2.2 Network characteristics for walkers and cyclists

The Dutch guidance (CROW, 2006) for design for cycle traffic suggests five main requirements for planning networks for cycle traffic. Again, these five network characteristics have been adopted with zeal by the writers of many other design guidance documents, particularly in Europe and Australasia.

The first requirement is that the network is 'coherent' and merely confirms that, at a network level, connections are in place to link a cyclist's origin and destination. At face value, this would seem to be so self-evident as not to need stating, particularly as a network of rights of way, especially in urban areas, is likely to be well established and relatively dense. The fact that it is stated, and that it has been repeated by many others as a requirement, indicates that planners consider existing highway networks insufficient for the needs of cycle traffic. Good coherence, or integration, with other networks, particularly public transport and pedestrian precincts, is also noted as being important and, it is suggested, is exemplified by adequate capacity and location of bicycle parking.

The second requirement is 'directness', which may be measured by the difference between a crow's flight distance and the network distance. Time also becomes relevant, particularly where excess time is caused by junction delay. The third requirement is 'safety', which is defined as the avoidance of conflicts with other traffic and ensuring uniformity in design parameters so that riders' expectations are met.

The final two network characteristics recognize more fully some of the distinguishing characteristics of bicycles as distinct from other types of vehicles. The requirement for 'comfort' demands that nuisance (proximity, noise and fumes) from motor traffic is minimized. Comfort is also defined as including the 'comprehensibility' of the network based on good signage and making best use of wayside features to help create a 'mental map'. These aspects would appear to be more properly linked with 'coherence' and 'directness' of the network, and omit the very important issues of gradient, surface quality and the eradication of steps in the running surface (e.g. at kerb faces), as noted in the UK guidance (DfT, 2008). Finally, and very importantly, the route needs to be attractive and this requires thoughtful and creative public realm design which encourages people to enjoy the network, and which in turn can enhance personal security.

Insofar as walking is concerned, five network attributes, which also alliterate and parallel the five requirements for cycle networks, have been identified in the UK (DfT, 2000). The first attribute is 'connectedness' and requires that walkers can travel from their origin to their destination without encountering major obstacles or severance. The second attribute is 'conviviality', which, it is suggested, relies on the interaction of the walker, not only with other people, but also with the built and natural environment, and other network users. The third attribute is suggested as being 'conspicuousness', the net result of which should be that public spaces feel safe and inviting. Routes should have high quality surfaces, be well landscaped and provide appropriate shelter and rest facilities, and these are described as contributing to 'comfortableness', the fourth attribute. The final attribute is that the routes need to be 'convenient' by virtue of their layout, and lead to minimal delay.

Schemas which define the characteristics of networks provide useful descriptions of the objectives of planning for walking and cycling. The process by which such objectives are implemented, however, will rely on a methodology, or a design approach. Such approaches have usually been defined in terms of a 'hierarchy of provision' (for example DfT, 2004a). Such hierarchies encourage transport planners and engineers first of all to consider reducing motor traffic volumes, then to consider reducing motor traffic speeds. If neither of these are achievable, or do not produce the benefits desired, particular treatment with appropriate special facilities at junctions and particular hazard sites may be considered and may be coupled with traffic management measures at the site. Further on down the hierarchy, consideration may then be given to re-allocating carriageway space and finally, the construction of routes away from the highway.

Such a hierarchy is meant to provide a process by which appropriate networks and provision for cycle traffic and walkers may be achieved. The difficulty with such a hierarchy, however, is that it is not self-evident that its application will create routes suitable for walking and cycling traffic because it does not focus on providing the attributes of walking and cycling networks as previously defined. The hierarchy appears to lead the designer into making amendments to route networks without undertaking the higher-level introductory processes of understanding demand, its origins and destinations, and the shortest path-routing that demand would like to take. It also does not address the additional design issues more specific to walkers and cyclists based on their particular characteristics as previously described. It is perhaps German national policy (BVBW, undated) which best emphasizes the need for network planning to create interconnected, safe, speedy and extensive networks for cycle traffic. This is what may lead a 'hierarchy of provision' in a Northern European context to more appropriately mean a hierarchy of primary, secondary and leisure routes for cycle traffic as part of a whole network, and which takes account of the speed and connectivity needs of such traffic.

2.3 Difficulties of integration of networks for different purposes

Design documents used within the engineering community tend to be prescriptive because they need to specify requirements for adequate performance without failure. This is unsurprising, given the specific and measurable characteristics of materials, and the precise nature of scientific analysis which underpins engineering design. A high degree of prescription is also required for transport networks where the dominating feature is the speed of the traffic using the network. In some transport networks, such as railways, the speeds are such that direct control of the system is required at all times. For highway networks, there are elements of control imposed, for example by using signal control at junctions, but mostly the system operates adequately so long as the network, the vehicle and the behaviour of the driver are well aligned. The variability of the behaviour of the driver is such that the system itself needs to be designed to closely defined rules so that drivers may adequately predict, and thus respond to, conditions as they unfold.

Over-prescription, however, is not well suited to the design of transport networks where some of their essential features are their intrinsic comfort and attractiveness, which is the case for walking and cycling networks. The challenge of integration is to merge the technical design requirements, made necessary because of the linkage with other parts of the transport network, with those creative design aspects which do not require prescription. It is not appropriate or possible to adopt 'templates' for such designs. This aspect calls for 'educating' engineers and urban designers about each other's approach, or in a wider sense educating for integration (see discussion on this in Chapter 19 in this volume).

Of fundamental importance in design for both cycling and walking networks is the ability for the users of these networks to be able to establish eye contact with others using the system and with users of parallel transport systems. Such eye contact leads to human interaction and hence to human behaviour dominating the interaction. Such interactions, rather than rule-based control mechanisms, will then predominate.

It is very important that the quality of provision for walkers and cyclists is not compromised by the different approach that is needed. The use of different design approaches could easily lead to a mismatch in standards of provision when some aspects need to be rigorously related to standards, while others are not subject to the same demanding constraints. To overcome this problem, the use of case studies as means of sharing design ideas is now common (e.g. Danish Road Directorate, 2000; DfT, 2004b and 2007) and brings the benefit of experience in the use of different solutions to the attention of a wide range of designers. The alternative is to ensure that users of guidance documents understand their purpose and recognize that they need at all times to exercise their professional judgement, and this is emphasized in the London Cycling Design Standards (TfL, 2005).

Overall, it can be seen that the design approach for integration within the highway and public realm of walking networks and, separately, of networks for cycle traffic requires the designer to apply appropriately both engineering-based design rules as well as a wider range of good design practice.

2.4 Guidance on integration with public transport

Journeys based solely on walking or cycling are local in nature. Longer distance journeys are also possible when coupled with a bus or a rail journey, but the apparent complexity of the interchange needs to be minimized and eased. Cycling and walking as access modes are important for public transport journeys, and additional attention to providing for them appropriately, and integrating them appropriately with public transport are necessary as part of any package of measures designed to shift travel away from the motor vehicle (see Chapter 11 in this volume for a broader analysis of integrating rail with access to rail station modes).

The design and ease of flow through the point of interchange is critical to the success of such integrated journeys. Particular issues present themselves for those with a bicycle. Sufficient quantities of well-managed and easily accessible parking which is safe from the weather and which offers confidence of a low risk of theft and vandalism needs to be made available (CA/DfT, 2004). The parking needs to have adequate lighting, be overseen by being in a busy area, and abandoned and vandalized bicycles need to be removed regularly. In some cases it may be appropriate to offer reserved spaces, but any additional management should not create additional delay. Routes within the interchange need to be well signed with level differences accommodated by ramps, or wheeling channels on stairs or lifts.

It may also be appropriate to offer and advertise bike hire and breakdown rescue and assistance after theft or vandalism. With antecedence in Amsterdam in the 1960s, bicycle hire schemes have met with varying degrees of success in recent times in Copenhagen (beginning in 1995), Munich from 2000 (run by Deutsche Bahn) and most extensively in Paris since 2007. Many other cities are now exploring the potential benefits of such schemes. Such bike hire provision could take the edge off the debate about the need for bicycle carriage on trains. Rail companies recognize the primacy of the need to move people, rather than bicycles, on busy urban commuter routes at peak times. On the other hand, cycle advocacy groups recognize the benefits of cycle access at both ends of a public transport journey and for leisure journeys.

3. Important features for integrated walking and cycling networks

While guidance can distil recognized good practice, good network planning also needs to be informed by evidence, and this section discusses available research into the important features of networks for walking and cycling.

Transport mode and route choice modelling has traditionally focused on assessing time and distance. Other attributes of the journey which may be considered, particularly for public transport, include reliability and comfort. Consideration needs to be given to the features of walking and cycling journeys which, in addition to time and distance, are factors in determining mode and route choice. These include perceptions concerning effort and risk.

Cycling and walking are distinct from other forms of vehicle transport because they require human effort to provide the locomotive power. The effect is arguably heightened for cycling compared with walking because of the coupling of the rider and the machine. Effort may be equated with energy output, and the rate of energy output, power, for cycling has been investigated by Wilson (2004). The components include air resistance; inertia of the bicycle and rider and inertia of the wheels to rotation; potential energy gain (and loss); and rolling resistance and mechanical efficiency of the bicycle. Variation in hilliness and the number of stops and starts will affect the total energy consumption of cyclists and their journey times, and some evidence is presented on this under 'Cyclability' later in this section.

For walking, even with lower speeds, air resistance can be significant in windy conditions, while potential energy gain and loss remains important and other energy requirements are related to maintaining an erect position and moving the legs. Goodwin (1976) perhaps pioneered consideration of effort and walking and rightly differentiated between the quantifiable measures of energy expended in terms of watts and, secondly, heart rate, the latter also being influenced by psychological factors including stress and anxiety. In a consideration of London Underground stations, Collins *et al.* (1977), however, collapsed this extended thinking back into a single time-based measure. They identified a linear relationship between additional energy expended while walking as compared with sitting, and the additional value of time[3] of walking relative to sitting in a public transport vehicle. They then extrapolated this to the case of additional energy expended while ascending steps and added a commensurate weighting to the value of time. In an alternative approach, which still also maintains the metric solely in terms of time, the team monitoring the impacts of the Supertram in Sheffield used Naismith's hill walking rule of adding an hour for every 2,000 feet of ascent to assist in understanding the relative catchment areas of tram and bus stops (WS Atkins, 1996). Typically, however, the additional subtleties of effort are not fully considered in econometric models of walking, and a simple factor of 1.4 to 2.0 times the in-vehicle time is used as a weight on its value of time (Paulley *et al.*, 2005).

Existing and potential walkers and cycle users are generally aware of their vulnerability in a busy urban environment. The perceptions that exist about this vulnerability however are myriad and complex and an understanding of the cultural constructs surrounding risk leads people to react in different ways to different situations. Adams (1985) suggests that an individual will decide on action based on his or her inherent propensity to take risks, but is also influenced by rewards and experience of (accident) losses. Adams points out that science can never provide an objective measure of risk, as the management of a risk will modify the risk, thus he suggests an analogy with the Heisenberg Uncertainty Principle.[4]

The ways that risk and effort, and other relevant network attributes are perceived are now discussed for both walking and cycling.

3.1 Walkability

Arguably considerations of 'walkability' began with space syntax theory, as developed by Hillier and Hanson (1984), which suggests that spaces within a built environment that are better physically connected, including through lines of sight, will tend to display higher density of movement. In addition, the literature since then, which is well summarized by Millington *et al.* (2009), suggests that walkability can be measured through a wide range of attributes. These attributes include residential density, land use mix and street connectivity, and the presence of en route 'destination' facilities such as shops. The importance of safety from traffic and good maintenance of footpaths and street lighting is also recognized, as are aesthetics and access to green spaces and parks. Ewing and Handy (2009) suggest, however, that walkability audit tools, even if validated, do not capture people's overall complex and subtle perceptions of a street environment.

3.2 Cyclability

A stream of American work classifies 'cyclability' into a six-point 'level of service' scale (TRB, 2000) and includes the attributes of volume and speed of general traffic and highway pavement quality (Landis *et al.*, 1997); junction crossing distances

(Landis *et al.*, 2003); and width of facilities (Harkey *et al.*, 1998 in urban areas and Jones and Carlson, 2003 in rural areas). Using a ten-point scale in the UK, Guthrie *et al.* (2001) identified road width; traffic flow; speed of traffic; presence of heavy goods vehicles and buses; gradient; bumpiness (texture and potholes); lateral conflict (presence of minor junctions; accesses and parking) and aesthetics as being relevant and important.

None of these cyclability studies considered the effects of links and junctions (that is to say a complete network) together in the same model. The Risk Rating Model developed by Parkin *et al.* (2007) has, however, sought to provide a respondent-based rating for a whole journey, and thus comes closer to a method that could be extended to include other attributes and hence plan whole networks. The risk scale ranged from 1 to 10, with 1 representing a low risk and 10 a high risk, and responses were regressed to the S-shaped logit functional form suitably constrained to the bounds 1 and 10. Link factors which reduced the risk rating included: traffic calmed roads; bicycle routes adjacent to the motor traffic carriageway; routes through park; bicycle-only routes in urban centres; and carriageways with cycle or bus lane. Link features which increased the risk rating included heavily trafficked roads and roads with on-street parking. Junctions generally contributed to higher risk ratings, with special facilities for cycle traffic not having a significant effect. The model, which explained most variation among the respondents, did not account for the duration of time on links of different types, but only included a dummy variable to represent their presence in a journey. The model has been extended to create an area-wide measure for the risk of cycling and has been successfully used to validate cycle review and audit guidelines (IHT, 1998; Parkin and Coward, 2009).

Insofar as effort is concerned, the results from an experimental survey of cyclists (Parkin, 2008; Parkin and Rotheram, forthcoming), using journey time data collected by global positioning system devices, has been used to assess the power outputs and ultimate total energy consumptions for assessment of routes in Bolton, UK.[5]

Two estimates of journey time and energy consumption have been calculated: one assuming that the cyclist has to stop at every 'give-way' and traffic signal controlled junction; and the second assuming that the cyclist is able to proceed without stopping. In the non-stop scenario, the difference in time between the inbound and outbound journey is larger for larger altitude differences and rises to as much as 30 per cent. The difference in energy consumption between the inbound and outbound journey is, however, proportionally much larger and requires up to 250 per cent more energy. This significant difference, while manifest to a cyclist, would not, using simple time-based accessibility modelling tools, be manifest to a transport planner.

In the stopping scenario, travelling times (excluding stopped time) are broadly 10–15 per cent longer and result from time at slower speeds during the acceleration and deceleration phases. This difference is again relatively large and points to the importance of minimizing the number of stops for cycle traffic on short urban journeys. The increase in energy consumption for the scenario with stops was found to be generally between 6 per cent and 10 per cent more than the scenario without stops. The additional power required to overcome inertia during the acceleration phase is to some extent offset by the lower power required to overcome air resistance at lower speeds.

These examples of research on risk and effort and other network attributes demonstrate the variables which need to be adopted in network planning and design, and estimate some of their parameters.

3.3 Integration with public transport

The dislike of interchange by passengers is well recognized and has been frequently considered as a fixed additional penalty including walking and waiting time, with mean figures for Great Britain of 37 minutes of in-vehicle time for train to train interchange (Wardman, 2001). Many studies have evaluated the important attributes connected with waiting as being comfort, cleanliness, safety and protection from the weather (Paulley *et al.*, 2005), and the value of guarded bicycle parking (Givoni and Rietveld, 2007).

Rietveld (2000) rightly suggests that adequate provision for walking and cycling as access modes to transport terminals is as important as the speed and frequency of public transport service (see also Chapter 11 in this volume). Work of this kind is well exemplified by Ghebreegziabiher *et al.* (2008), who studied access mode choice to Dutch railway stations and choice between railway stations using econometric models, including a measure for rail service quality, access distance, availability of car and bicycle parking facilities and, for public transport as an access mode, journey time. They found that the utility of accessing the railway station at shorter distances is greater for walking and cycling than for car and public transport. However, the utility for the non-motorized modes declined with distance at a steeper rate than for car and public transport.

Coming from a different perspective, Sherwin (2008) suggests that, among other transport modelling applications, adoption of bicycle rail integration might be best explained by one of a number of socio-psychological theories. Adoption of the theory of planned behaviour (Ajzen, 1985) by Bamberg and Schmidt (1994) has shown that such a methodology allows for a deeper insight into travel decision-making processes.

4. Network planning for walking and cycling

4.1 The essentials of network planning

The fundamental nature of walkers and cyclists is that they are on journeys from origins to destinations. To be successful they need to complete the journey swiftly, securely and comfortably. The special characteristics of walkers and cyclists have been noted, but these should not cloud the understanding that a network of routes needs to be provided which is coherent at a fine level of detail and should offer many journey possibilities which minimize time, distance and effort. For cycle traffic, the level of coherence due to the issues of effort needs to be at a finer grain than for motor traffic, which, for practical reasons to do with space requirements and flow control, needs to be diverted around traffic management systems and by-passes.

Such networks should provide, as far as is possible, an advantage for walkers and cyclists relative to other road users in order to maximize use for practical, health and environmental reasons. Priority relative to public transport is more arguable, but, based on the much lower environmental impacts of walking and cycling, there is a case for priority over public transport as well.

It should be noted that the maximum speed of cycling and the maximum speed of walking are often higher than has been considered in design for these two modes in the past. Obstacles which cause delay and additional effort should be minimized and the networks should be integrated with public transport interchanges.

The majority of the network available to cycle traffic is the already existing highway network. This network has, however, particularly in urban areas, been the subject of extensive traffic management since the 1960s. Such traffic management has been concerned with expediting the movement of motor traffic and reducing its impact. With the assistance of route assignment models, it has involved, for example, the construction of inner ring roads, the use of one-way streets, turning restrictions, point closures and vehicle restricted areas. These constraints on movement for motor traffic are not always reasonable for cycle traffic, yet their impact on cycle traffic can be significant. Good cycle planning guidance should therefore emphasize the need to redefine a network for cycle traffic which reconsiders the traffic regulations and restrictions brought about to control motor traffic. Such redefinition may include exemptions to some of the restrictions and, in addition, include the construction of links and routes specifically for cycle traffic, for example through parks and other public realm spaces, and bridging across geographical obstacles such as railways and rivers. Such routes may be permissive, that is, the passage of traffic is permitted by the landowner, or may be created as a right of way. In the UK, the process of rights of way improvement planning under the Countryside and Rights of Way Act 2002 (DEFRA, 2002) provides to network planners a wide remit and additional duties in terms of network planning for non-motor traffic routes and should help in this regard.

4.2. Methods of network planning

Effort is expended by cyclists during a cycle journey. Using evidence of the actual acceleration and final speed characteristics of cyclists, and applying these to a route network, it has been shown that effort varies significantly and in a way independent of journey time. On this basis, and in order to ensure that the disutility to cyclists is considered and reduced as much as possible, effort is a factor which needs to be considered in planning and engineering for cycle traffic.

The personal effort that a traveller puts into a journey is something very intrinsic to them and their body and, on this basis, it is quite conceivable that the additional disutility that they may place on an increase in effort may be more highly valued than a commensurate proportional increase in journey time. It will also be important to attempt to balance the perceived disutility of energy required for a journey against the ambience of the journey: the value of additional energy expenditure may be acceptable if it occurs while travelling in an attractive environment. Conversely, it may be argued that, for some, the concept of effort being a disutility may be misplaced, and this may be evidenced by those willing to take a detour on a commute journey in order to satisfy a desire for a certain level of exercise. A fusion between work on perceptions of risk and effort would create, when combined with time, an overall metric of the disutility of a human-powered journey.

Both walking and cycling journeys are often adjuncts to journeys on public transport and a feature of such journeys is the interchange. For both modes, and particularly for the cyclist, this may mean that particular effort is required to carry and place a bicycle in a secure position (either off or on the public transport vehicle) and may require the cyclist to pack and unpack baggage and other accoutrements of cycling. All of these actions require effort which may be in excess of that for travellers interchanging from car to public transport or from public transport to public transport. In a similar manner, as for the journey itself, a detailed consideration of the energy requirements of interchange will provide a more accurate explanation of the perceived disutility of interchanging than a simple measure based on time alone.

In summary, the two main points that this chapter makes are that cycling and, separately, walking, need to be better integrated within the public realm and the current highway network and, secondly, more attention needs to be paid to cycling and, separately, walking, as access modes to public transport. The chapter suggests a methodology for this based on network planning, which should include consideration of risk and effort as well as time. Integration is therefore required at several levels, and these are with respect to design requirements based on users as human beings, engineering requirements arising from the need for systematic design, and links with public transport networks.

Notes

1 The guidance actually says it has no equivalent of the protective 'crumple zone' of the body of a motor vehicle.

2 The difference in speed between cycle traffic and walking is greater in relative terms than the difference, at least in urban areas, between the speed of cycle traffic and motor traffic, which is usually limited to 48 kph (30 mph) or sometimes 32 kph (20 mph).

3 The 'cost' of a journey in modelling is conventionally expressed in 'generalized cost minutes', and includes the in-vehicle time and time for other legs of the journey factored according to their perceived disutility compared with in-vehicle time. The fare and other costs for a journey are converted to generalized cost minutes by dividing them by a value of time.

4 With its origin in quantum physics, Heisenberg states that two related properties, such as momentum and position, may not be known with precision.

5 Bolton is the 293rd most hilly district out of 376 districts in England and Wales, with a population of 267,000.

References

Adams, J.G.U. (1985) *Risk and freedom: The record of road safety regulation*, London: Transport Publishing Projects.

Ajzen, I. (1985) 'From intentions to actions: A theory of planned behaviour', in Kuhl, J. and Beckmann, J. (eds) *Action control from cognition to behaviour*, Berlin: Springer Verlag.

Bamberg, S. and Schmidt, P. (1994) 'Auto oder Fahrrad? Empirischer Test einer Handlungstheorie zur Erklärung der Verkehrsmittelwahl' [Car or bicycle? An empirical test of theory of mode choice], *Kölner Zeitschrift für soziologie und social psychologie* 46, 1: 80–102.

BVBW (undated) *Fahrad! Nationaler Radverkehrsplan 2002–2012: Maßnahmen zur Förderung des radverkehrs in Deutschland* [Bicycle! National bicycle traffic plan 2002–2012: Measures for the promotion of cycling in Germany], Berlin: Bundesministerium für Verkehr, Bau- und Wohnungswesen.

CA/DfT (2004) *Bike and rail: A good practice guide*, London: Countryside Agency Publications/ Department for Transport.

Collins, P.H., Weston, J.G. and Smith, R.S. (1977) *An approach to measuring the disutility of walking up stairs and escalators at L.T. stations*, Technical Note TN 123. London: London Underground (unpublished).

CROW (1993) *Sign up for the bike – design manual for a cycle friendly infrastructure*, Publication 74, Ede, the Netherlands: Centre for Research and Contract Standardization in Civil Engineering.

CROW (2006) *Design manual for bicycle traffic* (Dutch version April 2006, English version June 2007), Ede, the Netherlands: Centre for Research and Contract Standardization in Civil Engineering.

Danish Road Directorate (2000) *Collection of cycle concepts*, Copenhagen: Danish Road Directorate.

DEFRA (2002) *Rights of way improvement plans: Statutory guidance to local highway authorities in England*, London: Department of Environment, Food and Rural Affairs.

DfT (2000) *Encouraging walking and cycling: Advice to local authorities*, London: Department for Transport (formerly Department of Environment Transport and the Regions).

—— (2002) *Inclusive mobility: A guide to best practice on access to pedestrian and transport infrastructure*, London: Department for Transport.

—— (2004a) *Policy, planning and design for walking and cycling: Consultation draft, April 2004*, London: Department for Transport.

—— (2004b) *Encouraging walking and cycling: Success stories*, London: Department for Transport.

—— (2007) *Manual for streets*, London: Thomas Telford Publishing on behalf of the Department for Transport, Welsh Assembly Government, Department of Communities and Local Government.

—— (2008) *Cycle infrastructure design*, Local Transport Note 2/08, London: Department for Transport.

Ewing, R. and Handy, S. (2009) 'Measuring the immeasurable: Urban design qualities related to walkability', *Journal of Urban Design* 14, 1: 65–84.

Ghebreegziabiher, D., Pels, E. and Rietveld, P. (2008) 'Modelling the joint access mode and railway station choice', *Transportation Research Part E: Logistics and Transportation Review* 45, 1: 270–283.

Givoni, M. and Rietveld, P. (2007) 'The access journey to the railway station and its role in passengers' satisfaction with rail travel', *Transport Policy* 14, 5: 357–365.

Goodwin, P.B. (1976) 'Human effort and the value of time', *Journal of Transport Economics and Policy* 10, 1: 3–15.

Guthrie, N., Davies, D.G. and Gardner, G. (2001) *Cyclists' assessments of road and traffic conditions: The development of a cyclability index*, TRL Report 490, Crowthorne: Transport Research Laboratory.

Harkey, D.L., Reinfurt, D.W. and Knuiman, M. (1998) 'Development of the bicycle compatibility index', *Transportation Research Record* 1636, Paper No. 98–1073: 13–20.

Highways Agency (2005) 'The geometric design of pedestrian, cyclist and equestrian routes', *Design manual for roads and bridges*, vol. 6, Section 3, Part 5, TA90/05, London: Highways Agency.

Hillier, B. and Hanson, J. (1984) *The social logic of space*, Cambridge: Cambridge University Press.

IHT (1998) *Guidelines for cycle audit and cycle review*, London: Institution of Highways and Transportation.

Jones, E.G. and Carlson, T.D. (2003) 'Development of a bicycle compatibility index for rural roads in Nebraska', *Transportation Research Record* 1828, Paper No. 03–3911: 124–132.

Lancashire County Council (2005) *Lancashire – the cyclists' county. A code for planning, designing and maintaining roads and tracks for cyclists*, Preston: Lancashire County Council Environment Directorate.

Landis, B.W., Vattikuti, V.R. and Brannick, M.T. (1997) 'Real-time human perceptions toward a bicycle level of service', *Transportation Research Record* 1578: 119–126.

Landis, B.W., Vattikuti, V.R., Ottenberg, R.M., Petritsch, T.A., Guttenplan, M. and Crider, L.B. (2003) 'Intersection level of service for the bicycle through movement', *Transportation Research Record* 1828, Paper No. 03-3292: 101–106.

LTNZ (2005) *Cycle network and route planning guide*, Wellington: Land Transport New Zealand.

Millington, C., Ward-Thompson, C., Rowe, D., Aspinall, P., Fitzsimmons, C., Nelson, N. and Mutries, N. (2009) 'Development of the Scottish walkability assessment tool (SWAT)', *Health and Place*, 15, 2: 474–481.

Parkin, J. (2008) *The importance of human effort in planning networks.* NECTAR workshop, University of Oxford, 12–13 September 2008

Parkin, J. and Coward, A. (2009) 'Comparison of methods of assessing routes for cycle traffic', *Proceedings of the Institution of Civil Engineers, Municipal Engineer*, 162, 1: 7–14.

Parkin, J. and Rotheram, J. (in press) 'Design speeds and acceleration characteristics of bicycle traffic for use in planning, design and appraisal'. *Transport Policy.*

Parkin, J., Wardman, M. and Page, M. (2007) 'Models of perceived cycling risk and route acceptability', *Accident Analysis and Prevention*, 39, 2: 364–371.

Paulley, N., Balcombe, R., Mackett, R., Titheridge, H., Preston, J., Wardman, M., Shires, J. and White, P. (2005) 'The demand for public transport: the effects of fares, quality of service, income and car ownership', *Transport Policy*, 13, 4: 295–306.

Rietveld, P. (2000) 'Non-motorized modes in transport systems: A multimodal chain perspective for the Netherlands', *Transportation Research Part D: Transport and Environment*, 5, 1:31–36.

Scottish Executive (1999) *Cycling by design: A consultation document*, Edinburgh: Scottish Executive.

Sherwin, H. (2008) 'Travel planning at railway stations: An examination of the potential for bike-rail integration', *Universities' Transport Study Group Conference,* University of Southampton, January 3rd to 5th. (unpublished).

TfL (2005) *London cycling design standards*, London: Transport for London.

TRB (2000) *Highway capacity manual 2000*, Washington, DC: Transportation Research Board.

Wardman, M. (2001) 'A review of British evidence on time and service quality variations', *Transportation Research Part E: Logistics and Transportation Review*, 37, 2–3: 107–128.

Wilson, D.G. (2004) *Bicycling science*, 3rd edn, Cambridge, Massachusetts: The MIT Press.

WS Atkins (1996) *Supertram monitoring study. Report of 'after' surveys.* Prepared for South Yorkshire Passenger Transport Executive and the Department of Transport (unpublished).

Chapter 10

The role of ICT in achieving integrated transport networks

Neil Hoose

1. Introduction

This chapter discusses the role of modern Information and Communications Technology (ICT) in delivering a better, more integrated approach to transport management, particularly real-time network management. Greater integration within the transport system, both physical and organizational, is needed to meet the challenge of continued demand for mobility of people and goods while, at the same time, there is the opportunity to achieve greater integration of services through exploiting recent developments in information technology.

The urban road network provides a common physical infrastructure that is used by several different modes: walking, cycling, private car, buses and trams and commercial traffic. All these modes have differing requirements, albeit within a common framework of being safe, environmentally acceptable and cost-effective, and the challenge is to balance these needs. The balance itself is not static and will be different at each location (e.g. residential streets, industrial areas) and may vary with time (e.g. rush hour, school terms). Therefore, we need to be able to dynamically change the balance according to circumstances using information, management and control techniques.

Network management in real time is a complex field. The number of ways of interfacing with vehicle drivers and travellers is increasing and the internet and mobile phone now provide a wealth of information. This creates an opportunity to harness the power and reach of ICT for transport operations in the form of what has become known as Intelligent Transport Systems (ITS).

The use of information and control techniques to influence and manage the movement of people and goods on the road network has become the norm and is very much part of everyday life. However, within an individual network (e.g. an urban road network) there is a lack of coherence between different functions such as bus operations, traffic signals, pollution monitoring, car park information, etc. In the UK, the Traffic Management Act sets out a network management duty to 'expedite the movement of all traffic', including non-motorized modes such as walking and cycling. Although the guidance makes it clear that this duty is subsidiary to other factors, such as road safety, it is explicit that the Highway Authority must take account of congestion both on the network under their jurisdiction and on adjacent networks. A holistic approach is needed that attempts to balance the competing needs of different modes and different policy goals such as safety, air quality and climate change.

In this chapter information is presented on the technologies now available and their use in the transport sector. Before looking at some of the techniques being used there is discussion about the need for a strategic, top-down approach to ensure that solutions are not being led by the technology but by the needs of the users and the management systems. The succeeding sections give an overview of the types of techniques used for managing urban road networks and outline some of the issues involved in balancing the needs of different modes, including walking and cycling. The chapter also considers recent developments in communications that bring the vehicles themselves into the data management system. The potential to integrate transport management with management of external impacts, notably in respect of air pollution, is described before some thoughts are offered on how effective integrated management can be achieved through use of ICT.

2. Technology context

Technology development is rapid and unrelenting. Communications systems are now available that will enable access to broadband internet services while on the move and allow mobile computers to talk to each other across *ad hoc* wireless networks. New technology offers the capability for rapid collection, real-time processing and dissemination of data and information that enables different strands of network management to be used collectively and coherently.

There has been a significant amount of research and development into the use of combinations of electronics, notably electronic display technology, and sensing, computing and communication technologies for a wide variety of transport functions (Chen and Miles, 1999, McDonald *et al.*, 2006, Taylor, 2004). Many of these intelligent transport systems have now been deployed on arterial highways and urban streets.

Sensor technologies have also been developed to enhance the amount of data that can be collected. Systems to count vehicles and to classify them according to length or number of axles, or some combination of both, are available and these can operate continuously, 24 hours a day, seven days a week. Speed at an individual location can be measured. Automatic reading of vehicle registration plates (licence plates) using computer vision enables matching of plate reads and thus the calculation of travel time and average speed over a distance. Data from vehicles fitted with satellite tracking provides information on speed and travel time, and traffic data can be derived from mobile phone location data. The latter marks an increasing trend for data to be provided from a mobile source (e.g. a vehicle or a person with a portable device).

A management system uses data collection and analysis to generate information in order to influence how the process being managed will be controlled and to report on its performance. ICT enables processes to be monitored using advanced sensors and the data collected to be communicated to one or more locations for storage and analysis. Computer technology can manipulate and store large and complex data sets. The process can be controlled automatically while providing supervisory information to the human component of the system, be they traveller, vehicle driver or system operator.

Broadband communications is changing the way computer and data storage resources are distributed and owned. Because large amounts of data can be moved very quickly the location of data can be remote from the computer processors used to analyse it, and both can be remote from the user of the results or the providers of 'live' input data. Hence, real-time and historic data can be enhanced, linked to other relevant data and processed through a powerful computing environment. The business model is that computing resources will be owned by a third party and processing time will be bought 'on demand'. Data may also be stored in professionally owned 'data warehouse' facilities. The consequence of this is that capital investment in IT is no longer required by either the end user or the service provider. Computing resources can be scaled according to the needs of the application and cost will remain proportional to the demand on those resources. It means that any organization with access to data can generate applications and services. Hence, access and ownership of data becomes increasingly important as access to IT becomes a service commodity.

An integrated data collection, management and processing environment can be created to parallel the integrated transport environment. The different organizations involved in delivering transport can share relevant data while purchasing computer processing power to meet their own needs. Commodity-based computing has the potential to empower community groups that can organize their own data collection, perhaps using sensors carried by an individual and linked using mobile phone data transmission (Kanjo and Landshoff, 2008). The Mobile Environmental Sensing Systems Across a GRID Environment Project (MESSAGE) has demonstrated these concepts using air pollution data from roadside sensors linked to microscopic traffic simulation, vehicle emission and pollution dispersion models (North *et al.*, 2008). Such changes to the availability of data and computing are quite revolutionary because they shift the basis of ICT from a capital intensive resource owned and maintained by the organization and individual users to a commodity that is purchased and used according to need.

The question of data ownership and access becomes critical in creating the integrated system environment. There are a variety of actors involved in urban network management some of whom are involved in operating and maintaining the fixed asset (e.g. the highway) and some who operate the vehicles using that asset (e.g. a bus fleet operator). However, particularly in a deregulated public transport environment, the organizational link between the fleet operator and the road network manager may be weak. The fleet operator may regard the real-time and performance information as commercially confidential and be unwilling to allow the road operator to access it in case it became available to competitor or even client organizations. This can be resolved technically by additional processing to remove, or make anonymous, sensitive data.

Electronic displays are now a common feature of public transport systems even to the extent of being placed at open-air bus stops. The displays show the destination and timing of the next departures. They can also be used to give additional information on delays or disruption. The more advanced systems have devices to track vehicles and use the information to keep estimates of arrival and journey times up to date. Tracking can be done using satellite positioning (GPS) on the bus combined with mobile phone communications (GSM or GPRS) or by the bus exchanging data with the central system at fixed communication points (the so-called 'tag and beacon') (Mintsis *et al.*, 2004).

Electronic displays are also used for more general traffic management. They enable speed limits to be varied or advisory speeds to be displayed. They can be used to warn of congestion, either current or at some time in the future, and perhaps advise an alternative route. They can be linked with systems that monitor the available capacity in car parks to advise drivers where best to park and hence minimize the amount of mileage associated with hunting for an available space. The same technology can be used to direct travellers from highways to park-and-ride sites, showing available parking, travel time to the city centre and next train or bus departure. One example of such a system has been in operation in Munich since the mid-1990s (Cervero, 1998).

The internet and mobile phone now provide a wealth of information sources that can be used to plan a journey and to get up-to-date information while en route. Such systems provide considerable information for fleet network managers. The advent of affordable and reliable technologies for measuring traffic, identifying and tracking vehicles and real-time, wireless communication of the associated data opens up a range of techniques for traffic management. Some, such as congestion charging and distance-based road use charging, are well publicized and politically controversial. Others, for example pay-as-you-drive insurance, are in their infancy. A variety of ideas in respect of improved safety at junctions and on highways using vehicle-to-vehicle and vehicle-to-roadside communications are being developed and prototypes have been demonstrated at a variety of locations around the world.

This embedding of technology into everyday life comes about through a combination of usefulness, ease of use, convenience and reliability. Electronic payment cards, commonly referred to as 'smart cards', remove the need for specific tickets and customers do not have to queue at ticket booths or machines (Blythe, 2004). They allow customers to use different modes and benefit from multi-use discount schemes, and use pre-payment or post-payment accounts as appropriate. They can reduce operational costs for the public transport system operator and improve revenue collection. It is now possible to use mobile phones as an electronic ticket. Customers do not need to obtain an additional device and the mobile phone operator's payment system can be used to collect fares. The introduction into the market of handsets that can work out their own location using satellite and wireless ranging techniques introduces the possibility of 'location specific' services. It could be possible to register for a computer service to recognize the current mode and route of the owner and provide relevant and up-to-the-minute travel information.

It is important to realize that the physical manifestation that the user sees – the handset, smart card and the electronic reader – are only the tip of the iceberg. Behind the scenes there will be a complex network of interacting communications and data services. These will authenticate the users' transaction, charge the correct amount, support clearance of payments between different operators and provide full user account and system management services. In some cases it is the difficulty of creating the organizational cooperation between system owners that is the barrier to deployment. For example, the OYSTER card system in London is not compliant with the ITSO smart card standard being deployed elsewhere in the UK. This prevents linking the smart card systems for long distance coach and rail services to London into the Transport for London Underground, bus and local rail services.

All these systems involve some form of data collection from a variety of sensors linked to back-office central systems for data processing and storage. However, these systems are generally delivered as complete, stand-alone applications with little integration. For example, traffic signal control systems are designed and implemented as separate entities from bus passenger information systems. Those responsible for specifying and operating such systems are continually faced by rapid changes in technology and external policy that impacts on functionality. Demand management, traffic control and travel information are frequently seen as separate entities. As a result, most urban areas have a number of 'technology islands' that have similar functional elements but cannot share data or functionality. This leads to duplication and weakens the business case because each application has to fund all its components. For example, data from the bus tracking system may not be available in real time to support the local traffic information services.

There are methods in place, such as the Urban Traffic Management and Control (UTMC) protocols in the UK and the ITS Systems Architecture and NTCIP protocols in the USA, that enable data to be shared between systems and encourage 'plug and play' system components from multiple vendors. However, these only address the technical issue of exchanging bits and bytes and do not address the higher issues of how to obtain additional value from the integration. The move to commodity-based ICT changes the cost profile of the business case and increases the accessibility of information about the state of the transport network and the options available for a given trip. This increase in data and technology accessibility results in a step change in system complexity. System interactions become many-to-many instead of few-to-many and change to data 'pull' (e.g. use of search engines to find data across the internet) from data 'push' (e.g. provision of information on a electronic display).

3. Strategy development

Technology should not be the driving force in ITS. Just as transport demand is a 'derived' demand to enable people to go about their daily lives, so technology can be seen as a second order derived demand to enable transport systems to deliver more effectively. What is meant by 'more effectively' is determined by the overall policy context, which may be broader than transport, for example the climate change agenda. Therefore it is important to consider the higher level strategy before developing the technological approach. The potential power of these new approaches can only be realized if they are harnessed into a coherent framework that enables the demand management, control and information to be used together as part of a single system. This can only be achieved from a strong understanding of the policy goals for the network and ways of resolving the inevitable trade-offs. A top-down, strategic approach is needed to ensure that technology is used in an appropriate and effective way.

Technology develops rapidly and over the long term it is the functionality that is being delivered that is the key factor. The role of integrated transport management is to deliver those aspects of transport policy that require management and information, including real-time management of the available capacity. One technique for capturing the strategic context is to develop a set of high-level goals. These can be used as a reference point against which the broad aspects of both transport and technology system designs can be compared to ensure that the implementation will deliver the strategic goals. This assists in keeping the importance of integration at the fore and in mitigating the risk that systems become fragmented once they are in the detailed design and implementation stages.

Such an approach has been developed in the London Traffic Systems Vision (LTSV) project. The objective of the LTSV project was to develop a long-term, strategic vision for the development of systems that support the on-street operations of the Directorate of Traffic Operations (Transport for London, 2006). It became apparent at an early stage that the context of the use of the systems was important and the project identified the set of high-level goals set out in Table 10.1.

This set of goals aimed to capture the broad thrust of what the organization had to achieve over the long term. At first sight they seem to be a statement of the obvious. However, the technical complexity of the 'day job' is such that it is very easy to lose sight of what you are trying to achieve.

Table 10.1 LTSV high-level goals

Goal	*Key words*
A centre of excellence	Best practice, knowledge base, partnerships, research and development
In-depth understanding of the network	Data collection, network description, shared data access, knowledge tools
Policy responsiveness and flexibility	Changing policy, geographic diversity, impact of policy
Means to manage street space	Coordinated actions, advanced control, dynamic allocation of street space
Informed choice and compliance	Enable choices of mode, route; understanding of rules and restrictions

Source: Transport for London, 2006

4. Complexity of traffic control and management

Virtually every urban settlement of any size in the developed world has some form of traffic control and management system, and every part of the transport network falls within the scope of traveller information systems. Figure 10.1 shows the major components of network management, namely *transport network state descriptions* based on *monitoring* leading to *interventions* which in turn feedback to change the state of the network.

The monitoring, transport network state description and interventions elements shown in Figure 10.1 can be related to the high-level goals of LTSV. The first two relate to the 'understanding the network' goal and the latter to the goal 'means to manage street space'. Technology can also be an enabler for integration between all modes as well as for the network management components for individual modes.

Figure 10.1 illustrates that there are multiple linked components required for an integrated approach to network management and that there are feedback loops between the effects of interventions as measured by the monitoring component, which updates the current and potential future 'States of the Network' components. Furthermore, interventions in respect of one mode or through any one area of intervention (e.g. signal control, public information) will result in changes elsewhere in the model. ICT provides the tools to measure and trace these interactions and communicate the related data and information to multiple agencies that have an interest in network management (e.g. bus fleet manager, urban traffic control centre).

The scale of the challenge for traffic management has increased because there is rarely a steady state across the network. Somewhere within an urban network there will be circumstances that are outside the range of 'normal' conditions. In the case of the road network, for example, this affects signal-green time efficiency. Hence real-time operations are required to supervise the operations of the systems, identify faults that require attention, identify incidents, and if necessary take action to mitigate their impact. There is a question of how these responses relate to the different users. Should the response be targeted at getting the bus network back

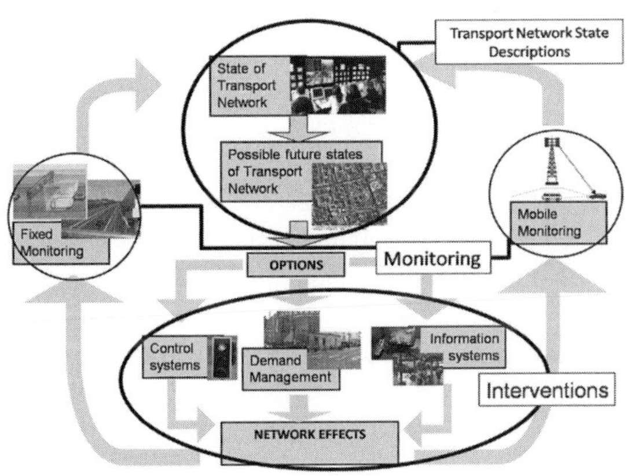

Figure 10.1 Conceptual model of road network management

into normal operation as soon as possible even if that means that other traffic has to queue for longer and create more pollution?

5. Managing the network using ICT

The red, amber and green traffic light has been around for a long time and its basic form has changed little since the 1920s. However, the methods for determining the timings of the lights have become more sophisticated. Algorithms have been developed to maximize the throughput of vehicles across a junction or to minimize the delay to vehicles passing through it. Timings at adjacent junctions can be synchronized to allow the same technique of maximizing capacity or minimizing delay across a whole section of the network. Use of automatic vehicle detectors allows these timings to be adjusted in response to the actual traffic flow and fully dynamic systems such as SCOOT (Split Cycle Offset Optimization Technique) and SCATS (Sydney Central Area Traffic System) are found in many cities around the world. Some cities have developed their own techniques to meet their own specific needs, for example the UTOPIA system in Turin that incorporates trams into the traffic control algorithms (Hounsell and McDonald, 2001).

The scale of some of these systems is quite large. For example, London has more than 2,000 junctions under SCOOT control and another 4,000 under planned fixed-time control (i.e. signal timings vary according to predetermined plans for the time of day, week and year). Each junction will have its own local control electronics and is linked to a central computing facility that provides the timing updates and monitors the health of the system. The associated numbers of signal heads, vehicle detectors, poles, etc. numbers tens of thousands. Most cities do not have systems of this scale but the majority of dynamic systems will be controlling a hundred or more intersections, 24 hours a day, 365 days a year.

The main goal of traffic signal control is to optimize the throughput of vehicles through an intersection, or minimize their delay, so that the junction is both safe and efficient. SCOOT and SCATS extend the same approach to networks of junctions. This treats all streams of traffic equally but this may not be in line with the local transport policy. In recent years there has been a growth in techniques to give greater priority to public transport (Hounsell *et al.*, 2008). Although such techniques can be used to reduce bus journey times, the main objective is usually to improve the reliability by keeping the bus to its published schedule or by maintaining headway between buses to prevent 'bunching'. There are several approaches in common use ranging from biasing the optimization towards traffic streams that include buses to the provision of dedicated 'bus only' lanes to the junction stop line combined with real-time detection of an approaching bus. In the last example the lights will either be held at green for the approaching bus or the lights will change so that there is a green signal for the bus before it reaches the stop line.

6. Integrating vehicles and users into the network management system

The use of satellite tracking (GPS) and real-time wireless communications across the mobile phone network (GPRS) to manage bus fleets represents a first step in integrating the vehicles into the wider systems. Instead of being entities that the management system acts upon, technology now enables vehicles, and even individuals, to be an integral part of the overall system.

There is much research interest into greater integration between individual vehicles (Vehicle-to-Vehicle or V2V) and between vehicles and roadside systems (Vehicle-to-Infrastructure or V2I). V2V enables the information and control systems on different vehicles to cooperate with each other, for example passing data about the rate of deceleration from a leading vehicle to those following. Such 'cooperative vehicle infrastructure (highway) systems' (referred to as IntelliDrive, in the USA, formerly Vehicle-Infrastructure Integration or VII (Sharafsaleh *et al.*, 2008)) have the potential to improve safety and reduce congestion and environmental impacts, and offer the potential for a single 'system' to address multiple policy goals (Ward and Hoose, 2006).

It is advances in wireless communications and computing that enable V2I and V2V systems to be created. The key development is for the mobile units to be networked together to create *ad hoc* mesh networks and for individual vehicles to be linked to the infrastructure systems by the best available connection. In some cases a mobile unit may talk to the infrastructure using the mobile *ad hoc* network to reach the best location to connect to the roadside systems. Hence, mobile units can enter and leave the network (that is why it is called an *ad hoc* network) and they are 'always connected' by the best available wireless medium. This can provide both more direct access to the driver via the Human-Machine Interface (HMI) and also direct links to vehicle systems without involvement of the driver.

There is a huge range of applications including improved safety (e.g. through 'over the horizon' sensing beyond the range of sensors mounted on the vehicle), travel information both driving related and non-driving related, and improved techniques for network management and control. The latter could include providing road sign and location- and context-based highway code information through the HMI. Potentially this could reduce the amount of physical infrastructure, such as road signs and speed humps.

There is significant potential for cooperative systems to support transport integration. The data sets created are very detailed in both time and space. This enables the pattern of demand to be determined to a much greater degree of detail. Local fluctuations can be identified. This should lead to better matching of infrastructure and services to demand patterns and the ability to design improved resilience into the transport network.

The EU, USA and Japan are all exploring the potential of cooperative systems. Cooperative Vehicle-Infrastructure Systems (CVIS) is an EU-funded project under FP7 which is developing a technical reference model for multiple services over multiple wireless communications bearers, with demonstrations in 2009 of several information oriented applications. The target domains include urban, inter-urban, network monitoring and freight fleet operations.

Cooperative systems comprise a whole spectrum of applications from *ad hoc* information networks through to a highly controlled and managed network. At present this spectrum is not well defined but the information driven end is developing rapidly. Dynamic route guidance based on a central system providing frequent route updates without the need for an extensive map database in the vehicle is likely to become commercially available during 2010. Location-aware mobile phones, such as Apple's iPhone, provide personal navigation services that can be used for all modes within a trip. This marks the start of new services based on user communities; in the case of navigation perhaps based around a supplier. Cooperative systems provide the capability to extend internet-based information sharing into the mobile environment. This may lead to multiple communities serving their own needs and these may not necessarily be in the best interests of the broader community. An open question is: at what point will there be a need for a centralized approach in certain locations? Either way, cooperative technology points to a potential revolution in the way the travellers are informed and in the way the network is managed.

A UK Department for Transport study (Ward and Hoose, 2006) included an initial analysis of the business case for a range of cooperative applications. This concluded that there were very significant benefits available, of the order of £2 billion net present value over 25 years, from automated driving as it provides safety benefits and additional capacity without additional construction costs. For less ambitious schemes, such as Intelligent Speed Adaption (ISA) or collision warning, the benefits are lower but still worthwhile, although the situation is complicated because the same application can be implemented using autonomous equipment on each vehicle.

Cooperative systems in the CVIS project are based on the underlying concept of 'always connected', meaning that the mobile unit is always linked to a wider computing environment, sometimes referred to as the 'cloud'. As well as providing connectivity for vehicles and drivers these technologies can also provide general internet services for bus and train passengers. Wide-area 3G can be used to link a bus or train to the internet and this connection can be shared within the vehicle or carriages using local Wi-Fi. The provision of internet access on board public transport vehicles can be particularly attractive for customers. Although clearly more attractive for longer, inter-urban journeys (e.g. in the UK the West Coast train service from London to Newcastle and Edinburgh, and the London–Oxford coach service both have on board Wi-Fi), it is being deployed on longer commuting routes such as those provided by AC Transit in San Francisco (Twichell and Minoofar, 2006).

The most far reaching developments create vehicles where the driving task is fully automated. While there is some considerable way to go before automated driving will be accepted in mixed-use public roads, the technology can be used now for vehicles working on their own dedicated, segregated tracks. Such systems can provide smaller vehicles, say 2–20 passengers, operating at frequent headways. They can provide high frequency transport across areas where the land use is relatively less dense and travel demand more dispersed. Hence, in such areas, they provide a more economic service than buses, trams or trains. A number of alternative systems have been developed and are being trialled, including the NBP in Germany, Posco Vectus in Sweden and ULTra in the UK (Buchanan, 2008). Such systems could have a significant impact in reducing private car use and associated climate change emissions. Modelling work for Gothenburg predicts a halving of car use but as there is no significant practical experience with such a system it is impossible to say how near to such predictions the actual performance will be. Public perceptions of personal safety, reliability and freedom from graffiti and vandalism will be key factors.

7. Managing the environmental impact

It is well understood that transport systems have impacts outside of their primary function of moving people and goods. Noise, air pollution (e.g. CO, NO, NO_2, SO_2 and fine particulate matter) and visual intrusion are all produced by vehicular based surface transport modes (Schwela, 1998). An integrated transport system should include measures to manage and mitigate the adverse external impacts.

Advanced sensor technologies enable air pollution and noise to be measured at street level in real time. A range of diverse, low-cost sensors are able to provide data for the planning, management and control of the environmental impacts of transport activity at urban, regional and national levels. Sensors can be fitted on vehicles and people to act as mobile, real-time sensor platforms. Using existing sensors combined with location, local processing and wireless networking means data can be captured continuously, in real time, but with a much finer granularity in space and time than is the current norm. In particular, the use of mobile sensors changes the nature of the spatial component and close spacing of sensors allows them to exchange data with each other.

Buses make an attractive option as mobile 'probes' to collect pollution data. They follow prescribed routes at known frequencies and pass through areas with high densities of pedestrians. The unit on board the bus, sampling air from outside the vehicle and well away from any exhaust system, can provide data at high temporal and spatial frequency along the route. Data sampled every 5 seconds would give a spatial frequency of 50 metres or less for most urban bus routes. These readings would be updated at approximately the bus headways, say every 2–30 minutes depending on the route.

The academic research and demonstration project MESSAGE (Cohen *et al.*, 2009) mentioned above has integrated sensing, positioning, wireless communications and GRID computing. GRID computing is the technology that allows computing tasks to be spread across a network of computers, thereby creating a virtual parallel computer with very large computational capacity. Data analysis and modelling takes place in a GRID computing environment making use of internet technology to partition data processing across available computing resources. This allows computation to be carried out as a series of parallel tasks.

The use of such an integrated data collection and processing environment is an evolution in terms of technology because most of the components have been around for some time. However, bringing them together and using them to support the real-time management of transport, air quality and potentially personal healthcare is a revolutionary step.

8. Effective network management enabled by ICT

Two key conclusions can be drawn. First, network management needs to have clear, far reaching objectives in order to provide a context for the systems that will be implemented. Second, developments in technology will enable these objectives to be addressed in innovative ways. However, technology continues to evolve rapidly and there is a need for organizational and financial structures that can change at the same pace. Otherwise there is a risk that maintaining obsolescent technology will become a drain on resources and the ability to deliver the broader objectives will be diluted.

Evolution in network management is inevitable as the need for and use of the road network will evolve anyway. Mechanisms are needed to enable the operational systems to evolve in tandem and not to lag behind. For example, the ability to obtain information while on the move via mobile phones has been around for several years (e.g. Transport Direct), but the underlying data and information systems are still weak, particularly in respect of dynamic, real-time information.

There is also a need for revolution involving two step changes. The first involves changing the focus to moving people and goods, with more emphasis on behaviour, purpose and value of trips. Much of today's operational performance is based on that of the vehicle and not on achieving the best overall balance and efficiency of movement. In particular, there needs to be a better understanding of the factors that influence travel decisions at the microscopic level. The second step change is integrating vehicles and drivers, public transport passengers and non-motorized road users more closely into network management. This involves making use of the pervasive sensing and control devices and ubiquitous communications technology that is now available. Utilizing such technology creates challenges in policy, user acceptability, privacy, security and business viability, as well as the whole question of trust and reliability of technology. However, by harnessing that potential it should be possible to get better value from fixed assets and improve resilience of the networks. The Eddington Review (Eddington, 2006) points out the dangers of running networks to their limits where any disruption quickly becomes catastrophic and recovery becomes more lengthy and costly.

Flexibility and scalability are today's issues: how can network operators deploy the systems currently available so as to increase coverage and be able to respond to changes in policy and the consequent public reaction and behaviour? A focus on people and goods mobility has consequences for the skills needed within the organizations involved in network management, and for the data systems. The nature of the data needed will have to change in order to grow the understanding of how the network behaves and how it may respond to different interventions. New technology creates opportunities for innovative approaches and for new players to be involved. For example, the volume of data needed to run micro-scale models over complete networks will require access to significantly more computational power than a local authority could own and operate. This creates an opportunity for a third party to provide computational and data processing services, perhaps including public dissemination, and therefore the public sector may not have the monopoly on influencing travel behaviour. The Highways Agency National Traffic Control Centre for England has some features of this business model as it is provided and run by a private company via a ten-year concession.

In conclusion there is likely to be some degree of revolution, that is, step change brought about by changing the management approach and by taking advantage of technical possibilities, but that will result in an evolution in the way urban travel is managed. The sum total of this will be to consider all urban travel as part of the same continuum and to create management techniques that are integrated across modes as well as across geographic areas.

References

Blythe P. T. (2004) 'Improving public transport ticketing through smart cards', *Proceedings of the Institution of Civil Engineers, Municipal Engineer 157*, March, Issue ME1: 47–54.

Buchanan M. (2008) 'Sustainable public transport to compete with the car', *7th European Congress on Intelligent Transport Systems and Services*, Geneva, 3–6 June.

Cervero R. (1998) *The Transit Metropolis: A Global Inquiry*, Chicago: University of Chicago Press.

Chen K. and Miles J. C. (1999) *ITS Handbook 2000: Recommendations from the World Road Association (PIARC)*, Boston: Artech House.

Cohen J., Fuchs B., North R., Hoose N. and Polak J. (2009) 'Computational grid-based data management and analysis for a mobile sensor network deployment', *16th World Congress on Intelligent Transport Systems and Services*, Stockholm, Sweden, September.

Eddington R. (2006) *Eddington Transport Study*. London: HMSO. Available online at: http://www.dft.gov.uk/about/strategy/transportstrategy/eddingtonstudy/ (accessed February 2010).

Hounsell N. B. and McDonald M. (2001) 'Urban network traffic control', *Proceedings of the Institution of Mechanical Engineers, Part I: Journal of Systems and Control Engineering*, Professional Engineering Publishing, Vol. 215, No. 4: 325–334.

Hounsell N. B., Shrestha B. P., Bretherton R. D., Brown T. and Souza C. D. (2008) 'Exploring priority strategies at traffic signals for London's iBUS', *7th European Congress on Intelligent Transport Systems and Services*, Geneva, 3–6 June.

Kanjo E. and Landshoff P. (2008) 'Fresh: cell-ID based mobile forum for community environmental awareness', *Tenth International Conference on Ubiquitous Computing, COEX*, Seoul, South Korea, 21–24 September.

McDonald M., Hall R. and Keller H. (2006) *Intelligent Transport Systems in Europe: Opportunities for Future Research*, Singapore: World Scientific Publishing Company.

Mintsis G., Basbas S., Papaioannou P., Taxiltaris C. and Tziavos I. N. (2004) 'Applications of GPS technology in the land transportation system', *European Journal of Operational Research*, Vol. 152, No. 2: 399–409.

North R., Richards M., Cohen J., Hoose N., Hassard J. and Polak J. (2008) 'A mobile environmental sensing system to manage transportation and urban air quality', *Proceedings of the 2008 IEEE International Symposium on Circuits and Systems (ISCAS 2008)*, Seattle, WA, May: 1994–1997.

Schwela D. (1998) *Urban Traffic Pollution*, Oxford: Taylor & Francis.

Sharafsaleh A., VanderWerf J., Misener J. and Shladover S. (2008) 'Implementing vehicle infrastructure integration (VII): real world challenges', *Intellimotion*, Vol. 14, No. 1. Available online at: http://www.path.berkeley.edu/PATH/Intellimotion/IM_14-1.pdf (accessed February 2010).

Taylor M. A. P. (2004) 'Intelligent transport systems: emerging technologies and methods in transportation and traffic', *Transportation Research Part C: Emerging Technologies*, Vol. 12, Nos 3–4 (June–August): 167–169.

Transport for London (2006) *London Traffic Systems Vision: A Prospectus*. Available online at: http://www.tfl.gov.uk/assets/downloads/LTSV-prospectus.pdf (accessed 2 February 2010).

Twichell J. and Minoofar C. P. E. (2006) 'The "NETBUS" WiFi project: delivering internet access to

AC TRANSIT bus riders', *13th World Congress on Intelligent Transport Systems and Services*, London, 8–12 October.

Ward C. and Hoose N. (2006) 'A policy orientated approach to the development of co-operative vehicle highway systems', *13th World Congress on Intelligent Transport Systems and Services*, London, 8–12 October.

Chapter 11

Developing the rail network through better access to railway stations

The need for integration

Moshe Givoni and Piet Rietveld

1. Introduction

At the heart of the EU transport policy lies the goal to revitalize the railways in order to shift the balance between transport modes, especially from private to public modes, and from car and plane to rail. The 2001 Transport White Paper explicitly stated that "rail transport is literally the strategic sector, on which the success of the efforts to shift the balance [between the modes] will depend" (CEC, 2001, p. 13).

A rail journey is rarely an end in itself; it is almost always part of a journey "chain" that includes a journey to, and later from, the railway station by different modes of transport. The *integration* of these components is essential to achieve a continuous travel, door-to-door journey when using rail, and in order to make rail a viable and attractive alternative to the car. Furthermore, these before and after (the rail journey) parts of the journey could be an important part in the decision whether to use rail at all. National modal shares suggest that in most cases when rail is not the chosen for the journey the car will be. It is therefore necessary to also look outside "rail" element of a rail journey to continue the revitalization of rail transport in Europe.[1]

Moreover, the need to change between transport modes and services is associated with great inconvenience for passengers. This is evident in the high Value of Time (VOT) passengers assign to interchanging between services (Wardman, 2001). For this reason, the EU, in its call for integration between modes, emphasizes the argument that integration depends on the extent to which the interchange between transport modes and services is seamless (CEC, 2001). The above suggests that there is scope for increasing rail use by improving the accessibility of rail stations and by improving the interchange between the mode used to access the station and the rail service. For this to happen, integration at several levels is required.

The need for integration in this respect very much relates to the physical level (the quality and location of the infrastructure required for seamless transfer between the access and the rail modes) and the operational level (timetable coordination, through ticketing, etc.). To achieve integration at these levels, very often integration at higher levels is a prerequisite (i.e. integration at the institutional level). For example, integration is required between the service operators and/or the authorities contracting and regulating them as well as those influencing land use, especially around railway stations.

This chapter provides a summary of research which examines the potential to increase rail use by improving the access to railway stations.[2] Its main aim is to demonstrate the need and potential benefits of integration, in this case when it comes to promoting rail use.

To carry out the research the Netherlands was taken as a case study. The Dutch rail network consists of over 350 stations covering most of the country, as can be seen in Figure 11.1. In terms of density, it has 68 metres of lines for every square kilometre, which is higher than the EU15 and EU25 average of 50 and 47 m/km^2 respectively, but lower than countries like, for example, the Czech Republic (122) and Belgium (115) (EC, 2005). The wide coverage of the Dutch rail network is represented by the fact that just 8.4 percent of the population lives further than 10 km away from the nearest railway station (Keijer and Rietveld, 2000). Despite these qualities, rail is not the main mode for passenger transport in the Netherlands. Only 8.2 percent of the passenger kilometres traveled in the Netherlands by passenger car, rail, bus and coach, and tram and metro are carried by rail,[3] which is still better than the EU25 and EU15 average of 6.5 percent and 6.3 percent respectively (EC, 2005).

First, the modal share on access to and egress from rail stations in the Netherlands was analyzed (described in Section 2). Then, using the Dutch Railways (NS) customer satisfaction survey, the extent to which the quality of the interchange facilities, as passengers perceived them, plays a role in their overall satisfaction with the rail journey was investigated (Section 3). A broader approach was then adopted and the factors determining the propensity to use rail were estimated, focusing on the role of the access to station characteristics (Section 4). Finally, some conclusions are drawn (Section 5).

2. Access mode to stations in the Netherlands

Table 11.1 presents rail passengers' choice of mode when accessing the rail station at the home end and when egressing it at the activity end, based on the NS customer satisfaction survey. It is assumed that passengers get to and from each of the stations (the home end station or the activity end station) using the same mode. Therefore, the analysis refers to the access journey to the home end station and the egress journey from the activity end station. This assumption does not imply that the journeys to and from the station are similar; indeed each of these journeys might involve additional and different activities when traveling between home and station. The assumption is that the same mode of transport will be used for these journeys.

Cycling, public transport (the reference in the NS questionnaire is to "Bus/Tram/Metro") and walking are the main modes used in the Netherlands to get to or from the railway station, together they account for about 85 percent of the trips at the home end or 99 percent when including the car (with the traveler as driver or a passenger). At the activity end station, public transport and walking dominate the modal share with 82 percent (the share of these modes at the home end station is only 47 percent). Together with bicycle and car (mainly when travelers are passengers) these modes account for 97 percent of the journeys from the railway station at the activity end. This makes the use of other modes of transport available, like taxi and motorcycle, negligible. Furthermore, initiatives like the "Traintaxi"

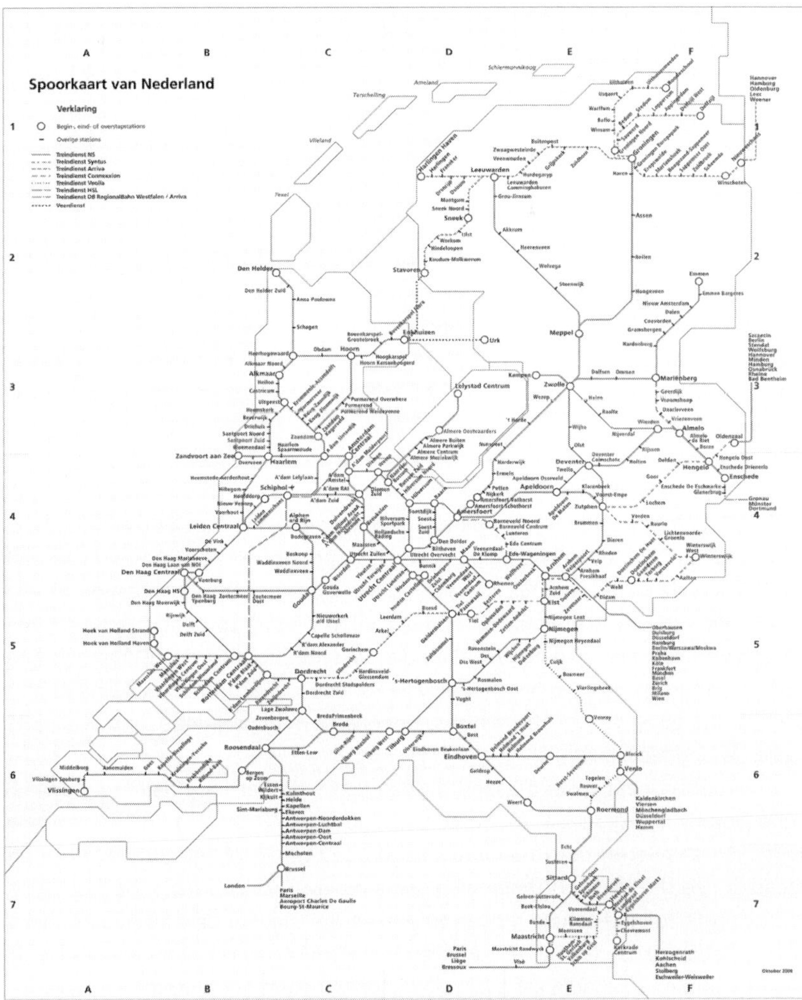

Figure 11.1 The Dutch rail network
Source: Dutch Railways. Used with permission.

Table 11.1 Mode choice on the access journey to the home end station and the egress journey from the activity end station (%)

	Access at the home end station			*Egress at the activity end station*
		Distance to station		
		< 3 km	> 3 km	
Bicycle	38.3	46.3	22.8	9.5
Bus/Tram/Metro	26.7	16.4	50.0	34.6
(Only) walking	20.1	27.0	4.6	47.2
Car (driver)	7.2	4.1	13.6	0.9
Car (passenger)	6.6	5.1	8.1	4.6
Taxi	0.2			0.9
Motorcycle	0.1			0.1
Train taxi	0.1			0.0
Other	0.7			2.2
Total	100	98.9	99.1	100
Valid answers	1,203			1,196

Source: Givoni and Rietveld, 2007

Note: Based on a survey carried out between 26 and 30 September 2005 (Monday to Friday).

scheme (where passengers can share taxi services with other rail passengers at a fixed price within a specified area around the railway station), which are designed to promote rail use, achieve only a negligible share at both the home end and the activity end stations. It is clear therefore that in order to improve stations' accessibility focus in most cases must be on the main four modes: walking, cycling, public transport and car.

It is apparent that in the Netherlands most passengers use non-motorized modes and public transport before or after a railway journey. The car is only the fourth most popular mode used to get to/from a railway station. The distinction between driving a car to the station and being driven is important, as the former requires the use of parking facilities and this is mainly relevant at the home end station. The low share of the bicycle at the activity end is also important to notice, as it suggests that relatively few passengers take their bicycle with them on board the train. More commonly they opt to own another bicycle for use just at the activity end or rent a bicycle at the activity end. The last two options have implications for the supply of bicycle parking facilities.

The questionnaire asked passengers whether they had a car available for the journey. Interestingly, for home base journeys 43 percent of the passengers questioned answered that they could have used a car for the journey but opted instead to use the train. Of those who had a car available to make the journey, the majority preferred not to use it even to access the station, only 16 percent did so. Also the option to use the car to be driven to the station was not the most attractive option for those who had a car available, as only 9 percent opted for this option.

3. Passengers' satisfaction with access facilities at the railway station

A rail journey is part of a "chain" of journeys that include a journey to, and later from, the railway station, but are these journeys important in the eyes of the passengers when making a rail journey? The NS questionnaire asked passengers for their opinion on the quality of "connections between the railway and public transport," "the capacity of car parks" and the "quality of guarded bicycle parking" and "unguarded bicycle parking." The quality scale used was from 1 – "cannot be worse" to 10 – "excellent," 5 stood for "insufficient" and a score of 6 for "sufficient." Together with the question about the "general opinion of traveling by train," it was possible to assess how passengers' satisfaction with the access facilities influences their overall satisfaction with the rail journey. An OLS regression analysis was used to estimate this link.

Most passengers view the quality of the access infrastructure somewhere between insufficient and sufficient (between 5 and 6) and it was not found that the relevant infrastructure (e.g. car parks capacity) is viewed better or worse by those using it (those rail passengers driving to the station). Also, no apparent differences could be identified in the perception of the access facility between those who had a car available for the journey and those who did not.

In the regression analysis, the "general opinion of traveling by train" was assumed to be a linear function of the general cost of traveling (measured as passengers' satisfaction with the price/quality ratio), the perception of the railway station and the perception of the access mode facilities (a linear function of the different questions on the access facilities). The analysis refers to passengers which were identified as making a journey from home (their origin station is closer to their home address than the destination station). The results of the linear regression analysis are presented in Table 11.2; all variables are on the 1 to 10 quality scale discussed above.

The regression analysis shows that passengers' satisfaction with the rail journey is mainly influenced by their perception of the journey's value for money, followed by their perception of the quality of the station. The access mode facilities, on the other hand, appear to have a more modest effect when considered separately, but as a group, the overall effect of the access variables on the journey perception is similar to that of the station variable.[4] Even if not the most important, improving passengers' satisfaction with the access facilities is likely to have a positive effect on the overall satisfaction from the rail journey. The connections with public transport appear to be the most important access facility.

The same regression model was estimated for passengers using a specific access mode (e.g. passengers using public transport) and the respective access mode facilities (e.g. connections with public transport). This did not yield considerably different results or a much better fit. Surprisingly, the estimation for passengers using a bicycle to access the station did not yield significant results for the bicycle parking and the effect of these variables was also small.[5] It appears that cyclists do not care much about parking facilities at the railway station, which might be related to relatively poor conditions for the unguarded bicycle parking facilities and

Table 11.2 Regression results for customer satisfaction with the railway journey (only passengers making a journey from the home end station are included)

	Coefficient	*t*
Intercept	3.504	12.697
Price/Quality	0.246	9.321
Station in general	0.144	3.942
Connections with public transport	0.082	3.362
Car parking capacity	0.044	1.995
Bicycle parking (guarded)	0.064	2.568
Bicycle parking (unguarded)	–0.051	–1.896
R square: 0.312; N = 462		

Source: Givoni and Rietveld, 2007

the perception that the guarded bicycle parking is too expensive. Nevertheless, it appears that most travelers who use the bicycle as an access mode do not incorporate this in their overall valuation of the trip, since the regression results demonstrate that those passengers who experience better bicycle parking conditions do not have a higher valuation of the rail journey.

Another approach to estimate how passengers' satisfaction with the access facilities influences their overall satisfaction with the rail journey was used on a larger sample from the NS questionnaire, using the whole of 2005 (17,033 questionnaires). In this approach, the 37 questions asked by NS were grouped into ten dimensions of the rail journey using factor analysis. One of the dimensions defined is "accessibility" of the station. This dimension was only ranked 8th in the passengers' satisfaction score. Next, a regression analysis, similar to the one presented in Table 11.2, derived the importance of each of the ten dimensions in terms of the impact on the overall satisfaction with the rail journey (the coefficient for each dimension marked its importance). The accessibility dimension was ranked 7th in importance (coefficient 0.052, t-statistic 6.94). The most important dimension was "travel comfort" (coefficient 0.213, t-statistic 22.12) followed by "travel time reliability" (coefficient 0.201, t-statistic 35.89).

The satisfaction score for each dimension can be plotted against its importance to yield the impact of each dimension on the overall satisfaction score (Figure 11.2). The impact of a dimension is positive when the associated satisfaction score is higher than the overall satisfaction score and negative when it is lower. Furthermore, the impact (be it positive or negative) of a dimension is stronger when its weight is higher. Thus, rail operators aiming at improving the rail service must consider both the importance of each dimension for the passengers and their current satisfaction with its level.

In Figure 11.2, the thick vertical line represents the overall satisfaction score for the rail journey in 2005. The value of the impact indicator for the accessibility corresponds to the surface of the grey rectangle to the left of the vertical line in Figure 11.2 (for comparison purposes, the rectangle to the right of the vertical line represents the (positive) impact of travel comfort on the overall satisfaction).

The current level of satisfaction with the access infrastructure at rail stations in the Netherlands has a negative impact on the average overall satisfaction with the rail journey. In other words, this is one of the dimensions where improvement is most needed. The impact level for the accessibility dimension is ranked 3rd after the travel time reliability and price-quality ratio dimensions. Thus, in terms of impact, improving the accessibility might not be the first priority. Yet, rail operators must also include cost considerations, which can increase the attractiveness of improving station accessibility. In some cases and circumstances it can be assumed that improving access services to the station is cheaper than improving travel time reliability. When accounting for the impact of each dimension together with the likely costs involved in improving it, the scope for increasing rail use through improvements to the accessibility might be higher than suggested in Figure 11.2. The above analysis was repeated for infrequent travelers only, those who use the train less than five times per year. It was found that for infrequent travelers the accessibility is ranked second in terms of impact behind travel time reliability. This might imply that for current non-users of rail services the accessibility of the railway station is more important than for those already using it regularly.

Analysis of the NS customer satisfaction survey provided clear evidence that passengers' satisfaction with the rail journey is partly the result of their satisfaction with the access facilities provided to them. This means that improving the quality of the access to the rail station is likely to increase rail use. It is an important signal to rail operators that they must consider the whole rail journey, door-to-door, as within their responsibility if increasing rail use is one of their goals. The above supplements previous research on the characteristics of the access and egress journeys to and from railway stations, which mainly account for distance, time and other supply oriented variables (e.g. Kuby *et al.*, 2004) by including quality variables, and showing that these influence the general opinion of traveling by rail. For rail operators to be able to consider the whole rail journey, door-to-door, integration at several different levels, as suggested at the outset, is clearly required.

A clear limitation of the above analysis, in the context of investigating the potential to increase rail use by improving access to railway stations, is that only rail passengers were surveyed and not those who currently do not use it. Nevertheless, it can be assumed that to some extent the findings from surveying rail passengers

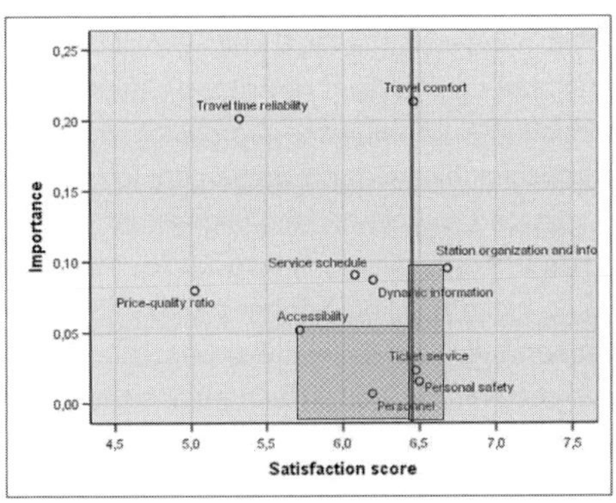

Figure 11.2 Satisfaction, importance and impact of each dimension of rail travel (all passengers)
Source: Brons et al., *2009*

would apply to current non-users, although some research finds that many current non-users do not even consider rail use (e.g. Wardman and Tyler, 2000). At least the findings clearly demonstrate that there is a potential to increase rail use by those currently using it infrequently if access is improved.

4. The role of access to rail stations in the propensity to travel by rail

From the above analysis it can be inferred that improving the access to railway stations and the interchange at the station between different modes and the train would increase rail use. This was directly examined in the analysis described below where the propensity to use rail (measured as the number of daily trips per person) was considered as a factor of a) the level and quality of the rail service provided, b) the level and quality of the access to the rail service, and c) the characteristics of the area and population served.

The analysis was carried out at the (four-digit) postcode level in the Netherlands. Out of the 3,961 postcode areas at this level in Netherlands, 1,440 postcodes were included in the analysis.[6] Data on the number of trips made by rail were only available at the municipality level (489 in the Dutch case) and therefore it was assumed that all postcodes within a municipality (eight on average) have a similar number and share of trips made by rail. The mean, median, maximum and minimum values for the number of daily trips, rail share (out of the total trips made by rail and car at each Dutch municipality), and other variables considered for the analysis are given in Table 11.3. The highest number and share of trips by rail are in the municipality of Diemen, south of Amsterdam, while in 17 municipalities in the Netherlands rail is not used at all.

To investigate how the number of trips made by rail is determined by the variables described in Table 11.3, and especially the accessibility variables, an OLS regression analysis was carried out. The linear model to be estimated is:

$$y_i = \beta_0 + S_i\beta_s + A_i\beta_A + PC_i\beta_{PC} + \varepsilon_i \qquad (1)$$

Where y_i is the number of rail trips in the postcode, i, and S_i, A_i and PC_i represent the rail service characteristics, accessibility characteristics and postcode characteristics of postcode i, respectively. The results of the best fitted model are presented in Table 11.4. The variables included in the regression explain 36 percent of the variation in the number of rail trips per person per day at the Dutch postcode level.

All the variables included in the regression analysis have the expected sign and were found to be significant at the 95 percent level, the exception being the dummy variables for guarded bicycle parking (significant just under the 90 percent level) and Park and Ride (P&R) facilities. The result for the guarded bicycle parking variable is not surprising given the findings described in Section 3. The explanation for the insignificant result for the P&R variable probably relates to the way these facilities are defined in the Netherlands, which means that in practice they are not, or at least not perceived to be, much different from car parks at rail stations which are not designated as P&R. Furthermore, car parks designated as P&R are provided

Table 11.3 Descriptive statistics of variables considered to estimate the number of trips made by rail

	Mean	*Median*	*Min.*	*Max.*
Trips by rail (per day per person)	**0.052**	**0.04**	**0.00**	**0.25**
Rail modal share (of car and rail trips)	**3.51%**	**2.57%**	**0.00%**	**15.24%**
Rail service variables				
RSQI per postcode*	0.67	0.64	0.05	1.57
Rail punctuality (% of trains delayed on departure)	0.15	0.15	0.01	0.50
Rail punctuality (average delay per train on departure in minutes)	1.52	1.60	–6.00	3.63
Access to rail station variables				
Average distance in km†	8.681	7.615	0.543	29.742
Public transport travel time in minutes†	25.41	26.11	5.51	50.23
Public transport service frequency per hour†	1.98	1.33	1.00	9.33
Guarded bicycle parking (in at least one station)	90.7%			
Park and Ride facilities (in at least one station)	59.4%			
Postcode characteristics				
Postcode land (hectare†)	982	517	15	13,606
Postcode population (2003)	6,468	6,140	170	22,850
Population density (people/ km†)	2,833	1,060	30	41,660
Population over 65	13.8%	13.0%	1.1%	53.7%
Share of immigrants	7.5%	3.8%	1.0%	83.0%
Population with higher education	10.0%	8.7%	0.0%	80.2%
Average income per inhabitant (Euro/year)	11,067	10,781	6,667	25,531
Number of cars per household	0.97	1.01	0.16	2.60
Valid N (list-wise): 1394				

Source: Brons *et al.*, 2009

Notes:

* RSQI stands for the Rail Service Quality Index. The RSQI for a departure station, calculated by Debrezion (2006), is a function of (i) the number of trips attracted to all other stations on the network, (ii) the generalized travel time from the departure station to all other stations on the network (accounting for service frequency, actual travel time and penalties for having to transfer), and (iii) the generalized travel time to distance ratio, which is used as a control for the effect of other modes of transport on the attractiveness of rail transport. The highest RSQI for a station on the Dutch network is 2.001 (Utrecht Centraal), followed by 1.832 (Duivendrecht in south-east Amsterdam), and 1.818 (Leiden Centraal). The average RSQI for all the stations on the Dutch network is 0.44 and lowest is 0.112. The RSQI for Amsterdam Centraal is 1.381 (ranked 8th).

† Between the postcode centroid and the stations.

in almost all the stations on the Dutch rail network (so in comparison, the reference group is small), overall implying the limitation in using this variable in the analysis. The punctuality of the rail service provided, the share of population with higher education in each postcode, and the share of immigrants were not included since they were not found to be significant or were correlated with other variables.

To illustrate the regression results, a 1 percent improvement (in relation to the current mean level of the variable) is considered for the significant variables. The change in daily trips per person per day is multiplied by the average population per postcode and the number of days per year to yield the number of additional trips per year for an average Dutch postcode (Table 11.4). The results show that the number of rail trips is most sensitive to the characteristics of the population, but these are difficult to change, and change is not necessarily desirable. What is possible and important to change is the level of car ownership, a 1 percent reduction in the average car ownership would lead to 607 additional rail trips per year for an average postcode.[7] However, changes in car ownership are mostly outside the control of rail operators and in the short term also outside the control of policy makers.

Rail operators are thus left with two main options to increase the number of trips: improving the rail service and/or the access to it. Improving public transport services to the station (rather than increasing the level of service or reducing distance to the station, by opening new stations) might result in a more modest increase in the number of rail trips, but at the same time might be less costly and require shorter lead time, making it more cost-effective in some circumstances. However,

Table 11.4 Regression results for the number of daily trips per person made by rail and the effect of 1% improvement[1] on the number of rail trips

	Linear model		ΔX^2	*Δ rail trips*[3]
Variable	*Coefficient*	*t-value*		
(Constant)	0.0647	7.41		
RSQI per postcode	0.0146	4.97	0.0067	231
Average distance	–0.0017	–6.79	–0.0868	350
Public transport travel time	–0.0004	–2.43	–0.2541	251
Public transport service frequency	0.0027	3.24	0.0198	125
Guarded bicycle parking	0.0048	1.63		
Park and Ride facilities	0.0013	0.78		
Population density (×1000 people/km^2)	0.0555	1.98	0.2833	37
Population over 65	–0.0691	–4.49	–0.0014	225
Average income per inhabitant	0.0024	4.13	0.1107	625
Number of cars per household	–0.0265	–6.96	–0.0097	607
Dependent variable	Rail trips per day			
N	1438			
Adjusted R square	0.359			

Source: Brons *et al.*, 2009

Notes:

1 Depending on the direction of the effect of the variable, this can be an increase or a decrease in the variable.

2 ΔX = 1% of mean.

3 Δ rail trips = additional trips in PC per year.

in most cases, the quality of public transport access to stations is controlled by other transport operators (not the rail operator) and/or by the local or regional municipality contracting them. Similarly, the land required to develop the access infrastructure around railway stations (e.g. parking facilities) very often is not owned by the railways. The need for integration is thus clear.

A similar regression analysis to the one presented in Table 11.4 was carried out but with the number of daily car trips per person as the dependent variable. Similar results were obtained (but with the opposite sign for the coefficients), showing that improving the rail service and the access to it can reduce car use.

The above analysis can be considered as presenting half of the picture, since only accessibility to the home end station is considered. The data used contained information on the passengers' home location and the mode used on the journey between home and station but it did not contain any information on activity location, only the mode used on journeys between activity location and the rail station. Improving the station accessibility is also important at the activity end and this requires a separate analysis since the balance between how different variables affect rail use would probably be different, with different implications for the development of the rail network depending on whether the station is a home or activity end station. This might imply differences in the requirements for integration at the physical level (as different access/egress facilities will be required) but not overall with respect to the general need for integration.

5. Conclusions

The above analysis provides a broader consideration of rail travel, one which considers the access to the station journey and the egress from it, as an integrated part of the rail journey. It emphasizes the need to consider a journey by rail as a door-to-door journey, and for rail operators to view the whole journey chain as part of the service they provide. The direct implication of this is that increasing rail use cannot be worked out within the rail industry alone and therefore requires integration (at several levels) with a variety of actors responsible for the provision of transport services. The analysis thus validates the need to keep "integrated transport" as the main policy goal, which can directly contribute to a more sustainable transport system. It also reinforces the conclusion reached by Rietveld (2000, p. 74) that "the market potential of railway services depends to a considerable extent on the quality of the total chain from residence to place of activity and vice versa." One implication of the above is that rail operators should consider providing the journey to/from the station themselves in order to provide a door-to-door service.

The above analysis revealed that there is still scope for improving the access to railway stations, and this will in most cases contribute to a higher satisfaction with railway journeys, which can be expected to translate to higher use rates. Many of those who use rail use it infrequently and irregularly, implying that they probably use the car more often. Those who do not use rail or seldom use it, represent a market for rail operators which can be attracted to use rail (more often) through improvements to the accessibility of stations. In this sense, the challenge lies not primarily in changing the way passengers get to or from the railway station (certainly in the Netherlands where "green modes" dominate), but in attracting new passengers to use the railway by making it more accessible. A change of the travel mode and not the access mode is the main target. Those who use the rail less frequently, but still have some experience of using it, will probably be more sensitive to improvements in the access to/egress from the station compared with those who do not use it at all; many of those who do not use the train at all do not even consider using it (at least in the UK; see Wardman and Tyler, 2000). In this respect, it is important to emphasize that the analysis accounted only for those who already use (even if infrequently) rail services. Still, it is likely that many current non-users and infrequent users would use rail if it was more accessible, implying that the potential benefits of integration, when it comes to access to railway stations, are probably greater than what is apparent from the above analysis.

Once a rail service is provided at a certain area, improving that service (e.g. higher frequency of service) and improving the access to it are substitutes when it comes to increasing rail use and decreasing car use. Improving station accessibility might be more cost-effective. Reducing the distance to the station was found to have an important and significant effect on rail use, more than other elements of accessibility, but it relates more to changes in the rail network and not the access to it. Furthermore, reducing the distance to the station by opening more railway stations will most likely require substantial investments and might not have a positive impact on the level of service across the network, as more stations will impose travel time penalties for passengers not using (but traveling through) the new stations. The implications are that distance to the station should be reduced not in the physical sense but in the real and perceived travel time sense. Thus, improving public transport services to railway stations seems to be one way to effectively increase rail use.

The possible substitution between improving the rail service and the access to it has an important spatial dimension. The rail network is more developed around and close to the major urban centers while it is less developed in the periphery. If promoting rail use is the policy the implications could be to expand the rail network to better serve the periphery. But this can be costly and not cost–effective, considering the relatively low demand for rail in the periphery. The analysis here suggests that an alternative approach could be to provide better access to already existing stations on the network. This alternative could prove a solution only if the access service and the rail service are fully integrated.

The reforms of the European railways which, among other things, aimed at increasing competition and private sector involvement in rail operation, led in several cases to large bus operating companies starting to also operate rail services and this facilitates integration at different levels. Some examples are given by Alexandersson (2009). In Sweden, BK Tåg, a new entrant to the rail market, introduced new and innovative practices to the rail sector by making use of its experience from the bus sector. Also in Sweden, and more importantly, the decentralized responsibility for regional passenger rail lines and the fact that they are under the same authorities as the public bus services appears to have brought about better coordination of regional train services with bus services. In the UK, companies related to the bus industry (such as Stagecoach, National Express and First Bus) were very successful in bidding for rail franchises when these were first introduced. To date there is no evidence on how this affected the level of integration between the modes and thus no evidence to indicate if introducing competition (e.g. through the bidding process) can also increase integration (or not). In the Netherlands, the demand for coordinated bus and railway services has led to the creation of new consortia of public and private firms such as NS and Arriva, and NS and Keolis (Alexandersoon, 2009). While privatization might facilitate integration in some way, it seems that integration is more important at the institutional level, with respect to who controls and governs land use (around railway stations) and the operation of different transport services, and it is integration at this level which can result in better accessibility to railway stations and therefore increased use of rail services.

Acknowledgment

This chapter summarizes the findings of the IBRAM research which was funded through a Marie Curie Intra-European Fellowship. The project was also affiliated with the TRANSUMO project "The Reliability of Transport Chains." We thank the Dutch Railways (NS) for providing the data and for their support throughout the research.

Notes

1 The use of the term "revitalization" is often taken from the 1996 *White Paper: A Strategy to Revitalize the Community's Railways* (CEC, 1996). This and the 2001 White Papers refer to the EU15 countries where rail use has been starting to increase after many years of decline. In contrast, most of the new EU countries (part of the EU27) have seen rail use decline over that period (see discussion in Givoni and Banister, 2010).

2 The "Integration between Rail and Access-to-railway-stations Modes" (IBRAM) research was financed by a Marie Curie Intra-European Fellowship. This chapter is mainly based on Givoni and Rietveld (2007) and Brons *et al.* (2009), where a detailed description of the methodology and the results can be found.

3 On some intercity routes in the Netherlands rail is the main mode of transport used.

4 An F test for the access mode facility variables shows that as a group these variables are significant at the 1 percent level and also that the bicycle parking variables are significant (but at the 5 percent level).

5 For the guarded bicycle parking variable the coefficient is 0.011 (t-statistic 0.315) and for the unguarded bicycle parking the coefficient is 0.007 (t-statistic 0.213).

6 For each postcode, the rail operator (NS) collects data for the three most used stations by people residing within that postcode. Only postcodes where these three stations serve more than 98 percent of the postcode's rail users are included in the analysis. The rail service and accessibility variables at the postcode level are then calculated as an average across these three most used stations.

7 Considering the statistics in Table 11.3, an average postcode would generate a total of about 123,000 rail trips per year. This will imply that reducing the number of cars per household by 1 percent will increase rail use by approximately 0.005 percent.

References

Alexandersson, G. (2009) Rail privatization and competitive tendering in Europe. *Built Environment*, 35, 1, 43–58.

Brons, M., Givoni, M. and Rietveld, P. (2009) Access to railway stations and its potential in increasing rail use. *Transportation Research Part A*, 43, 136–149.

CEC – Commission of the European Communities (1996) *White Paper – A Strategy to Revitalize the Community's Railways*. Commission of the European Communities, COM(96)421final, July.

—— (2001) *White Paper: European Transport Policy for 2010: Time to Decide*. Commission of the European Communities, COM(2001)370, Brussels, September. Available online at: http://europa.eu.int/comm/off/white/ index_en.htm (accessed February 15, 2002).

Debrezion, G. (2006) Railway impacts on real estate prices. PhD Thesis, Vrije Universiteit, Amsterdam.

EC – European Commission (2005) *European Union Energy and Transport in Figures 2005*. European Commission, Directorate-General for Energy and Transport.

Keijer, M. J. N. and Rietveld, P. (2000) How do people get to the railway station? The Dutch experience. *Transportation Planning and Technology*, 23, 215–235.

Kuby, M., Barranda, A. and Upchurch, C. (2004) Factors influencing light-rail station boardings in the United States. *Transportation Research Part A*, 38, 223–247.

Givoni, M. and Banister, D. (2010) Reinventing the wheel – planning the rail network to meet mobility needs of the 21st century. In Frenkel, A., Nijkamp, P and McCann, P. (eds) *Societies in Motion: Regional Development, Industrial Innovation and Spatial Mobility – Essays in Honor of Daniel Shefer*. Cheltenham: Edward Elgar (forthcoming).

Givoni, M. and Rietveld, P. (2007) The access journey to the railway station and its role in passengers' satisfaction with rail travel. *Transport Policy*, 14, 357–365.

Rietveld, P. (2000) The accessibility of railway stations: the role of the bicycle in the Netherlands. *Transportation Research Part D*, 5, 71–75.

Wardman, M. (2001) A review of British evidence on time and service quality valuations. *Transportation Research Part E*, 37, 107–128.

Wardman, M. and Tyler, J. (2000) Rail network accessibility and the demand for inter-urban rail travel. *Transport Reviews*, 20, 1, 3–24.

Part 3

Assessing the potential benefits of integrated transport policies

Chapter 12

Measuring the costs and benefits of integrated transport policies and schemes

John Preston

1. Introduction

Integration is a multifaceted concept that includes a number of stages. This has led some commentators to describe integrated transport as a scalar (Potter and Skinner, 2000), while others refer to the rungs of an integration ladder (Hull, 2005). For transport these rungs might include, in approximate ascending order of organizational difficulty:

i. The integration of fares, service patterns, terminals/stops and information within public transport.
ii. The integration of infrastructure provision, management and pricing for public and private transport.
iii. The integration of passenger and freight transport.
iv. The integration of (transport) authorities.
v. The integration between transport measures and land use planning polices.
vi. Integration between general transport policy and transport policies concerning the education, healthcare and social services sectors.
vii. The integration between transport policies and policies for the environment and for economic development.

For the purposes of this chapter, integration is defined as: *the organizational process through which the planning and delivery of elements of the transport system are brought together, across modes, sectors, operators and institutions, with the aim of increasing net social benefits* (NEA *et al.*, 2003, p. 17).

The zenith of integrated transport in the UK was the 1998 Transport White

Paper – *A New Deal for Transport: Better for Everyone* (DETR, 1998) – which established a framework for introducing integrated transport, although the more detailed delivery plan had to wait two years until the publication of the Ten Year Plan (DETR, 2000). However, a series of events occurred that were to effectively scupper the Ten Year Plan and the integrated transport policy that underpinned it. The fuel protests of September 2000 illustrated the power of the road lobby and the unpopularity of road user charging. The Hatfield rail accident of October 2000 led to a sequence of events that undermined rail finances, leading to Railtrack[1] being placed in administration in October 2001 and being replaced by Network Rail in October 2002. By early 2003, the Ten Year Plan was effectively dead, to be replaced by another Transport White Paper in 2004 (DfT, 2004). As a result most commentators would agree that New Labour's Integrated Transport agenda has largely failed (Docherty and Shaw, 2003, 2008). Nonetheless, integration as a concept continues to have some resonance and has featured in recent policy documents by bodies such as the Centre for Cities and the Institution of Civil Engineers (Preston *et al.*, 2008; ICE, 2008). The rest of this chapter will examine how integration has been appraised (Section 2) and the empirical evidence that has emerged from this appraisal (Section 3), with specific reference to the Eddington Transport Study in the UK and to work by the European Commission. It will then consider evidence on the impact of integration on city performance (Section 4), before drawing some conclusions about the efficacy of the integration concept (Section 5). However, before doing this it is necessary to clarify the integration concept.

In practice, the definition of integrated transport varies depending on disciplinary/theoretical perspectives, for example between engineering, micro-economics, management and political science viewpoints. Counterpoints are provided by the engineering and micro-economics perspectives. The engineering viewpoint, which is also shared by architects, planners and urban designers, is that certain aspects of integration can be based on best practice as determined by professional experts (see for example, Creswell, 1979 or Rogers, 1999). This is, of course, something of an exaggeration – best practice guides are often based on detailed case studies and analytical techniques such as the linear programming methods used to determine optimal public transport stop spacing (Newell and Vuchic, 1968).

The micro-economics perspective sees integration as a response to market failure. A neoclassical perspective is that coordination will be provided by competition. If consumers value integration, the free market will provide it (Hibbs, 2000). An opposite perspective highlights that transport exhibits a number of market failures of which the most important are wasteful competition, network failures and the presence of externalities (others include information imperfections and public good characteristics). With respect to wasteful competition it is argued that because transport operators have some monopoly in both space and time, deregulated markets will exhibit features of monopolistic (taxis) or oligopolistic (buses and trains) competition resulting in too much service being supplied at too high fares (Evans, 1987; Preston *et al.*, 1999). For public transport, there are benefits from having a network of services. The more users there are, the lower the time costs, as more frequent services will mean lower waiting times; more routes will mean lower walking time; and the greater scope for mixing stopping and express services will lead to lower in-vehicle time. This is referred to as the Mohring effect (Mohring, 1972) and provides 'first-best' arguments for an integrated public transport system that has lower fares, higher frequencies, smaller vehicles and higher levels of subsidy than has been the norm in Britain (see, for example, Nash, 1988). More recently, externalities have come to the fore. Initially emphasis was on the extent to which integrated transport could reduce road congestion but subsequently more emphasis has been placed on global and local environmental impacts and in particular reductions in carbon emissions, given the warnings of Stern (2006).

It will become evident that this chapter is largely informed by a micro-economic perspective, although we have taken account of other perspectives. In order for it to be made operational we need to define what we mean by net social benefit and how this can be measured. This will be considered in the next section.

2. Appraising integration[2]

An appropriate framework for appraising (ex-ante) and evaluating (ex-post) integrated transport is provided by the New Approach to Appraisal (NATA) and the accompanying web-based transport analysis guidance.[3] The thrust of this approach is that transport interventions are measured against five key criteria: Economy, Accessibility, Safety, Integration and Environment (sometimes referred to as the EASIE criteria). As Figure 12.1 shows, these are then split into a number of sub-objectives and the results displayed in an appraisal summary table (AST). NATA is in essence a combination of cost-benefit analysis and environmental impact assessment. Alternatively, it may be viewed as an unweighted multicriteria analysis (Glaister, 1999).

The cost-benefit analysis components of NATA are monetized and expressed in present value terms by discounting costs and benefits in years 1 to 30 by 3.5 per cent per annum and in years 31 to 60 by 3 per cent. The main monetized benefits of integration accrue as a result of generalized costs (time and money) savings to transport system users. These benefits are included in the Transport Economic Efficiency (TEE) calculations, and may be split between business users and commuters and leisure travellers (consumers). Transport providers will typically (but not always) have benefits of increased revenue but face additional costs as the result of the provision of new services. Where integration promotes mode shift from relatively unsafe modes (such as motorcycle and private car) to safer modes (such as bus and train), there will be safety benefits. More recently the benefits of reduced emissions of greenhouse gases and of reduced noise levels have been monetized, as has increased physical activity and improved journey ambience as a result of schemes promoting cycling and walking (see also Chapter 9 in this volume).

A key measure in the NATA is the Benefit Cost Ratio (BCR) which is simply the Present Value of Benefits (PVB) divided by the Present Value of Costs (PVC) – see Figure 12.1. Where there are no budget constraints, schemes with a BCR greater than 1 should be taken forward, as they have a net social benefit. However, in practice there are budget constraints and transport must compete with other sectors for government funding. As a result, the Department for Transport (DfT) initially suggested that a BCR less than 1 was poor, a BCR between 1 and 1.5 might

OBJECTIVE	SUB-OBJECTIVE	ASSESSMENT
ENVIRONMENT	Noise	Net population win/lose NPV £m
	Local air quality	Concentration weighted for exposure
	Greenhouse gases	PVB £m
	Landscape	Score
	Townscape	Score
	Heritage of historic resources	Score
	Biodiversity	Score
	Water environment	Score
	Physical fitness	Score
	Journey ambience	Score
SAFETY	Accidents	PVB £m
	Security	Score
ECONOMY	Public accounts	PVC £m
	Transport economic efficiency: business users and transport providers	PVB £m
	Transport economic efficiency: consumers	PVB £m
	Reliability	Score
	Wider economic impacts	Score
ACCESSIBILITY	Option values	Score
	Severance	Score
	Access to the transport system	Score
INTEGRATION	Transport interchange	Score
	Land use policy	Score
	Other government policies	Score

Figure 12.1
The NATA objectives and sub-objectives
Source: www.webtag.org.uk – TAG Unit 2.1, the Appraisal Process.

be considered low Value for Money (VFM), a BCR between 1.5 and 2 medium VFM, and a BCR greater than 2 high VFM. More recently an additional threshold of 4 has been introduced to denote very high VFM (DfT, 2009). This change is suggestive of tightening budget constraints.

With respect to non-monetized impacts, the influence of integration on local air quality is quantified by calculating the number of households affected by changes in exposure to particulate matter and nitrogen oxides, but it is not, as yet, monetized. All other sub-objectives are scored using semantic (verbal) scales, typically using a seven-point scale ranging from strongly beneficial to strongly adverse.

In this schema, integration is an objective in itself, which introduces an element of circularity, but the focus is on the provision of passenger and freight interchanges and consistency with land use policy and other government policies. For passenger interchanges, the emphasis is on the waiting environment, level of facilities, level of information, physical linkage for the next stage of the journey, and connection time and the risk of missing a connection. Land use planning includes consideration of the fit with Local Plans, Unitary Development Plans, Regional Planning Guidance and County Structure Plans (now subsumed within Regional Spatial Strategies – see also Chapter 3 in this volume). Other government policies highlighted include regeneration, housing, environmental protection, education, labour market flexibility, health, heritage, social inclusion, economic growth, and national and regional competitiveness.

The NATA has attracted criticism, largely on the basis of the somewhat uncritical combination of quantitative cost-benefit and more qualitative multicriteria analysis (Glaister, 1999), while other researchers have highlighted how only a subset of the criteria appear to be taken into account by decision makers (Nellthorp and Mackie, 2000). Furthermore, a number of sub-objectives have been subsequently monetized by the Eddington Review (Eddington, 2006), including local air quality, landscape and wider economic benefits; while the Stern Review (Stern, 2006) and accompanying work has suggested that carbon should be valued using shadow prices rather than social costs.[4] As a result the DfT has undertaken a NATA Refresh exercise that aims to make the guidance fully multimodal, to improve its use for non-infrastructure proposals and to align it with the DfT's new objectives (DfT, 2009). A particular weakness is the failure to assess the extent to which transport interventions can promote social inclusion, despite work in cities such as Belfast, Bradford, Bristol and Nottingham that has illustrated the importance of good public transport (Preston and Rajé, 2007; Hine, 2008).

An important issue is the extent to which integrated transport might promote wider economic benefits, primarily through the agglomeration benefits achieved by extending labour and land markets but also through introducing more competition into imperfectly competitive transport using sectors. The inclusion of wider economic benefits has been estimated to increase the BCR of Crossrail from 1.8 to 2.6 and the Jubilee Line extension from 1.75 to 2.75 (Banister, 2007). For these London rail projects, this suggests a multiplier of around 1.5 (i.e. the inclusion of wider economic benefits increases the benefit cost ratio by 50 per cent). For urban networks as a whole Eddington (2006) suggests a multiplier closer to 1.25, while for international gateways and inter-urban routes this may be 1.1 or lower.

3. Evidence base on the benefits of integration

Theoretical analysis suggests transport integration will not occur autonomously as free market provision is likely to be affected by service instabilities and schedule matching (van Reeven, 2003) and, with multiple operators across a number of jurisdictions, integration of fares, services and information is unlikely to occur (Roumboutsos and Kapros, 2008). Integrated transport usually requires public intervention. In the UK such interventions are assessed by NATA-type BCR calculations as outlined in Figure 12.1.

3.1 The Eddington Transport Study

A useful evidence base is that put together for the Eddington Transport Study (see also Dodgson, 2009), although a major constraint is that this study tended to examine individual measures rather than packages of measures. Given that improving public transport constitutes the first steps of the integration ladder, Table 12.1 summarizes the appraisal of over 30 public transport schemes, with mean BCRs in the range 2.1 to 2.6, although there are a number of schemes that have BCRs below 1.5. These BCRs are calculated by summing the PVBs in Figure 12.1 and dividing through by the PVC. For most of these schemes by far the biggest element of benefit is the time saving to users, although for schemes with dedicated rights of way, time savings to non-users are also important. Non-monetized impacts tend to be small but in general positive.

It is interesting to note that with respect to rail (both heavy and light), investments in devolved administrations such as London have gone ahead with relatively low BCRs, while in England outside London a number of schemes have been rejected despite high BCRs. For example, the Docklands Light Rail Woolwich extension went ahead despite a BCR of only 1.1. By contrast, the Leeds Supertram and South Hampshire Rapid Transit have been cancelled, despite BCRs of 2.3 and 3.6 respectively. Table 12.1 also includes three local road schemes that include public transport (such as bus priority and bus-based park and ride) and other elements (such as pro-walking and cycling measures) and are described as integrated transport schemes. These schemes seem to have relatively high BCRs (mean 4.97) and are worthy of further investigation.

Dodgson (2009, Table 2) shows that local road schemes have higher mean BCRs than public transport schemes – computed as 4.23 for a sample of 48 schemes. However, in the case of roads it might be argued that the non-monetized impacts are less favourable. For some schemes, a modified VFM calculation has been undertaken by the DfT which takes into account these non-monetized impacts and places a scheme in one of four qualitative categories.[5] There were 43 local roads schemes for which both BCR and modified VFM calculations were made. For 39, the BCR and VFM categorizations were the same, for two the modified VFM gave a higher category and for two BCR gave a higher category. By contrast, both BCR and

Table 12.1 Eddington evidence on public transport integration BCRs

Type	*Number*	*Locations*	*Mean*	*Range*
Integrated transport schemes	3	Reading, Scarborough, w2emms*	4.97	2.7–7.7
Public transport interchanges	6	Altrincham, Bradford, Coleshill, Mansfield, Warrington, Wolverhampton	2.57	1.0–4.8
Light rapid transit	6	Coventry, Leeds, London, Nottingham, South Hampshire	2.10	1.1–3.6
Urban rail	6	Glasgow, London, York	2.16	1.1–3.0
Bus schemes	13	Bletchley, Bristol, Cambridge, Doncaster, Leeds, Sheffield, Taunton, Warwick	2.51	1.7–4.0

Source: DfT, 2006

* West to East Midlands Multimodal Study.

the modified VFM calculations were made for 16 local public transport schemes. For 11 of these schemes, the categorizations were the same but for five cases the modified VFM analysis produced a higher category. This suggests that at the margins the inclusion of non-monetized impacts strengthens the case for local public transport compared to roads. However, given Dodgson estimated the mean BCR of the 25 local public transport schemes he examined as 1.71 compared to 4.23 for local road schemes, there is a large gap to be reconciled. There are other factors that could explain this gap, including the treatment of small time savings (which are particularly important for road schemes – if these were treated as zero then BCRs for local road schemes would reduce); the treatment of tax revenues (public transport schemes will attract some motorists leading to a reduction in taxation revenue[6]); and missing impacts (most notably related to social inclusion).

3.2 European Commission study

Table 12.2 includes some BCRs for public transport integration schemes examined in work for the Directorate General Transport and Energy (DGTREN) of the European Commission (NEA *et al.*, 2003). These schemes have been quantified in slightly different and not strictly comparable ways but broadly include the monetized elements of Figure 12.1, with particular emphasis on construction and operating costs, revenues, accident reductions and time savings. It is apparent that there are few examples where integration benefits are quantified in BCR terms outside the British Isles. This reflects the dominance of the micro-economics approach in the British Isles and of engineering approaches to integration in continental Europe. However, it can also be seen that all the schemes listed have BCRs in excess of 2, including network integration in Bucharest, based on a contra-flow bus lane in the city centre, and a Metro extension in Rotterdam. Indeed, in the case of the latter this permitted large reductions in bus operating costs and ultimately reductions in bus subsidy, which means that this scheme could lead to savings to government which in turn could result in a negative BCR because it saved government expenditure and the costs were thus negative (i.e. there was a financial benefit). Other examples of commercial integration include the Jutland Regional and Interregional express bus network in Denmark. This scheme resulted in a 35 per cent increase in patronage and the whole system was reported as being self-financing, with a cost recovery ratio of 107 per cent.[7] However, services between the main towns (e.g. Aalborg, Vejle and Viborg) had a cost recovery ratio of 180 per cent, whereas smaller feeder routes only had a cost recovery ratio of 55 per cent. It is unlikely that such cross-subsidy could occur in a deregulated system such as that which exists in Great Britain outside London.

Table 12.2 also illustrates the benefits of integrated ticketing and information and of area-wide integration. It should be noted that the Dublin scheme, based on the 2001 Platform for Change document by the Dublin Transportation Office includes a major programme of rail investments including extensions to the suburban rail (DART) network, on-street trams (Luas) and a segregated, higher capacity light rail network (Metro). In Greater Manchester, area-wide integration was based largely around Quality Bus Corridors and some extensions to the light rail (Metrolink)

Table 12.2 Public transport integration schemes

Location	*Description*	*BCR*
Bucharest	Network integration	2.90
Rotterdam	Bus integration with Metro extension	4.10
Greater Manchester	Tariff and information integration	2.53
	Local Transport Plan 2001/2–2005/6	4.86
	Major Schemes 2001/2–2005/6	3.71
West Yorkshire*	Integrated ticketing	5.40
London	Information integration	7.67
Dublin	Area-wide integration	2.75

Source: NEA *et al.*, 2003

* Based on Cottham, 1985.

Table 12.3 Public transport integration demand impacts

Location	*Dates*	*Overall % change*	*Annual % change*
Greater Manchester	1999–2001	4	2.0
Hamburg	1967–2002	19	0.5
Stockholm	1973–2001	25	0.8
Vienna	1988–2001	24	1.7
Rome	1995–1997	6	3.0
Paris	1975–1993	33	1.7

Source: NEA *et al.*, 2003

system. Appraisal distinguished between Major Schemes (requiring capital support of over £5 million) and other elements in the Local Transport Plan, a five-year planning document introduced by the 2000 Transport Act, which sets out local transport strategies and policies, along with an implementation programme. Both the Local Transport Plans and the Major Schemes were subject to a full NATA, with the ASTs indicating that there were no adverse effects for any of the sub-objectives.

Data are also available on the impact of integration measures on public transport demand and some of the evidence from the DGTREN study is summarized in Table 12.3. However, it is almost impossible to disentangle integration measures from other factors and rising public transport usage does not necessarily signify an increase in net social benefit. Nonetheless, it can be seen that cities such as Hamburg, Stockholm and Vienna, with a long history of integrated public transport, have also had long-term increases in public transport volumes of up to 25 per cent, although the rate of increase per annum is often below 1 per cent. This is often against a backdrop of falling demand prior to the introduction of the integration policies. Paris illustrated the success of integrated ticketing (the Carte Orange) which led to an uplift in demand of 33 per cent over an almost 20-year period. The Integrate Project in Greater Manchester and the Metrebus integrated ticketing and fare system in Rome show that in the short term increases in public transport volumes of up to 3 per cent per annum can be achieved. However, in Greater Manchester, it appears that this increase was relatively short-lived, with bus demand reverting to the long-run downwards trend.[8]

3.3 Other studies

Further data on the impact of public transport integration on demand have been collated in work on *Smarter Choices* (Cairns *et al.*, 2004). For example, Quality Bus Partnerships and Community Rail Partnerships may be seen as a form of integration by bringing together operators, authorities and the local community. A review of Quality Bus Partnerships in over 20 locations in Britain indicates short-term patronage increases of 18 per cent (15 months or less) and medium-term increases of 36 per cent (18 months or more) (Sloman, 2003). A study of Quality Bus Partnerships in Winchester has undertaken some cost-benefit analysis, albeit in a form that is not convertible into BCRs (Wall and McDonald, 2007). However, it was found that a package of improvements on two routes paid for themselves in social terms in four years and in commercial terms in 12 years. However, it was also found that park and ride investments did not have a net social benefit, largely due to the high cost of provisions and the low level of charges levied, although these calculations did not consider the avoided costs of increased parking provision in the central area. When such costs are taken into account, evidence from France suggests that the social case for park and ride is confirmed (Margail and Auzannet, 1996). There is some evidence that Community Rail Partnerships may have a similar demand impacts to Quality Bus Partnerships. For example, the Partnership for the Bittern rail line between Norwich and Sheringham (Norfolk, UK) was reported as increasing patronage by 40 per cent over five years, with similar effects emerging for the Wherry Lines between Norwich, Great Yarmouth and Lowestoft (Cairns *et al.*, 2004, p. 136).

There is also some evidence in the literature about the demand impacts of integrated ticketing. For example, the introduction of travelcards in London was estimated to lead to a 10 per cent increase in Underground trips and a 16 per cent increase in bus trips (Gilbert and Jalilian, 1991), as a result of lower prices for multiple trips. It has been argued that in Zurich (Switzerland) the introduction of a transferable season ticket or 'rainbow card' (Regenbogenkarte) was the main factor in stimulating public transport passenger growth as, following a period of stability, there was a 33 per cent growth in passenger trips between 1985 and 1990 (Fitzroy and Smith, 1994). A similar effect was observed in Freiburg (Germany), where after years of broadly constant demand, the introduction of an environmental travelcard (Umweltschutzkarte) led to increases of around 7.5 per cent per annum. Between 1983 and 1995 public transport ridership increased by 138 per cent, while population of the region only grew by 13 per cent (Fitzroy and Smith, 1998).

The *Smarter Choices* study has also collated information on the effect of marketing and information on public transport demand in the UK and this is summarized in Table 12.4. It can be seen that marketing can lead to substantial increases in public transport usage.

Case studies of Brighton and Nottingham suggest that around half the increase in demand is related to 'hard' measures (such as service improvements) but the other half is due to 'soft' measures such as promotion and marketing. These

Table 12.4 The effect of marketing and information improvements on public transport patronage

Location	*Measure*	*Demand impact*
South Yorkshire	TOPS (Travel Options Planning Service)	Where applied, bus use up 18%, train use up 10% and tram use up 12%
Nottingham	Re-branding, marketing and better information	Bus patronage up 1.8% pa when previously declining by 1% pa
Brighton	Quality Bus Partnership	Bus usage up 52% between 1993 and 2003
Perth	Service improvements with direct marketing	Passenger growth of 63% over three years
Aylesbury	Service improvements followed by marketing	Initially no increase but marketing led to 28% increase after two months and 42% after 10
Bristol	Showcase bus corridors	Around a 22% increase; where individualized marking, 44% increase
Cambridge	Simplification of network and ticketing and improved information	25% patronage increase over four-month period
London	Service improvements plus flat fare	31% bus patronage increase over four years

Source: Cairns *et al.*, 2004, pp. 149–150

studies also indicate that around 30 per cent of new bus trips would otherwise have been made by car. Analysis suggests that the public sector costs for marketing of city-wide bus services are about two pence per car kilometre saved. Given that official practice calculates the benefits of reduced traffic congestion to be on average about 15 pence per kilometre (Cairns *et al.*, 2004, p. vi), this implies a BCR of around 7.5. It should be noted that this is a partial measure and takes no account of the impact on commercial bus operators, nor of bus user benefits.

In addition to public transport marketing (in its broadest sense), other measures that might promote integrated transport include workplace and school travel plans; personalized travel planning and travel awareness campaigns; car clubs and car sharing schemes; and teleworking; teleconferencing and home shopping. It has been estimated that a major roll out of these 'soft' measures (referred to as a high intensity scenario) could lead to a 21 per cent reduction in peak period urban traffic (13 per cent off-peak) and a reduction in peak period non-urban traffic of 14 per cent (off-peak 7 per cent), resulting in a nationwide reduction in traffic of 11 per cent. By contrast, a low intensity scenario, in which present practice is largely continued, would only lead to a 2–3 per cent reduction in nationwide traffic, increasing to 5 per cent for peak period urban traffic. It is estimated that the public expenditure costs of achieving reduced car use by soft measures are around 1.5 pence per kilometre, which suggests a BCR of at least 10 (Cairns *et al.*, 2004, p. v). It should however be noted that it is assumed that these benefits of congestion relief are not dissipated by induced traffic, while there is also some concern that the high intensity scenario has failed to take into account the possibility of diminishing marginal returns, although this could be offset by complementarities between measures.

4. Integration and city performance

An important policy question relates to whether integrated transport may impact on city performance. A useful, if controversial, analysis is provided by Leunig and Swaffield (2007), who compared the performance of 16 British towns and cities that have been in receipt of various urban regeneration funds (the urban policy sample) between 1994 and 2007,[9] with six more successful towns and cities (see also Table 12.5).

In terms of Gross Value Added (GVA) per capita, the urban policy towns were 9 per cent behind the UK average in 1997 but by 2004 they were 13 per cent behind. By contrast, the successful towns had a GVA per capita 39 per cent above the national average in 1996, which increased to 46 per cent in 2004. Similar trends were detected with respect to personal income, while differentials with respect to house prices were largely unchanged, although unemployment differentials had narrowed. Leunig and Swaffield concluded from this data that urban regeneration policy has been ineffectual. However, there is the problem of the counterfactual. It could have been that the performance of the urban policy sample would have been even worse without the interventions. Moreover, it is difficult to infer causation (or indeed non-causation) from a sample which is in receipt of urban regeneration funding because of structural weaknesses in their economies. Nonetheless, Leunig and Swaffield believed that the urban policy towns are affected by locational rather structural weaknesses. It is noticeable that four of the six successful towns are within (long-range) commuting distance of London[10] but only three of the urban policy towns are in this category. This may provide a case for investment in high speed inter-urban links. Another feature is that seven of the 16 towns and cities in the urban policy sample are in Passenger Transport Authorities, whereas none of the successful towns are, although again it would not be sensible to infer causation from this statistic.

European comparisons are possible from the UITP Millennium Cities database (for 2001) and related sources (Atkins, 2006). Some results are summarized in Table 12.6. In four of the cities (Brussels, London, Munich, Paris), public transport is regulated and centrally planned, while in three UK conurbations (Glasgow, Manchester and Newcastle) it is largely deregulated. It can be seen that in the

Table 12.5 Successful and less successful towns and cities

Successful sample	*Urban policy sample*
Bristol, Edinburgh, Milton Keynes, Peterborough, Swindon, Windsor, Maidenhead	Blackburn, Blackpool, Bradford, Coventry, Hastings, Hull, Leicester, Liverpool, Sheffield, Southampton, Stockton, Stoke, Sunderland, Walsall, Warrington, Wigan

Source: Leunig and Swaffield, 2007

Table 12.6 Some comparisons of public transport (PT)

City	*GDP per inhabitant*	*PT veh. kms per inhabitant*	*PT investment per capita*	*Cost of monthly PT pass*	*PT usage – kms per person*
London	40,068	157.0	231.0	78.9	2,520
Paris	67,564	84.1	121.0	40.1	2,170
Manchester	23,059	58.5	64.9	N/A	561
Glasgow	32,898	99.7	44.4	N/A	978
Munich	47,600	121.0	215.0	N/A	2,910
Newcastle	22,603	84.0	65.0	N/A	976
Brussels	49,645	91.1	140.0	30.8	1,400

Source: Atkins, 2006 (based on 2001 data).

Note: All monetary items expressed in Euros, adjusted for Purchasing Power Parity rates. N/A = Not Available.

regulated (and arguably integrated) cities, public transport investment and usage are substantially higher than in the deregulated cities. There is less difference in terms of public transport supply per capita, although the regulated cities do tend to have higher levels, although they also tend to use larger vehicles (e.g., trains rather than buses). It is also noticeable that the regulated cities have the highest levels of per capita GDP, although again causation should not be inferred.

Public transport fares are generally much higher in the UK than in continental Europe. Some indication of this is given by the cost of a system pass in London being broadly double that of a similar pass in Brussels or Paris. As a result, the mean cost recovery ratio of bus systems in the four UK cities is 87 per cent while the mean cost recovery ratio for all public transport modes in the three continental European cities is 44 per cent.[11] This is suggestive of the higher levels of public transport subsidy in Western Europe.

5. Conclusions

This chapter has shown that an empirical database is emerging concerning some elements of integrated transport policy. In particular, improvements to public transport can be shown to be beneficial in terms of a narrow definition of costs and benefits and the case for such improvements may be strengthened where wider social, economic and environmental effects are taken into account. Furthermore, schemes which have a particular focus on the integration of public and private transport may have particularly good returns. Although the precise numbers of much of the evidence presented can be disputed, a relatively uniform picture emerges that public transport integration measures can represent good value for money in cost-benefit terms and can lead to substantial increases in the use of public transport and to reductions in car travel. This in turn suggests that current levels of integration are sub-optimal and that the implementation of integration measures would generally, but not necessarily always, be beneficial.

However, most of the empirical evidence to date is on the first rung of the integration ladder – namely the integration of public transport. In addition, some attention has been paid to quantifying the synergy between various transport policy instruments – where, for example, the benefits of two policy instruments, such as instrument A (road pricing) and instrument B (improved public transport) are greater than the sum of the individual components. This might be written as Benefit(A+B) > Benefit(A) + Benefit(B) (Mayeres *et al.*, 2003). Those empirical studies that have addressed this issue suggest that this rarely happens in practice but what is more common is either additionality (Benefit(A+B) = Benefit(A)+Benefit(B)) or complementarity (Benefit(A+B) > Benefit(A), Benefit(A+B) > Benefit(B)) (May *et al.*, 2005a). The benefits of one measure can thus be reinforced by the benefits of another. Furthermore, one measure may reduce the barriers related to another. For example, on its own, road pricing may not be politically acceptable because of concern over the adverse impact on low income motorists. However, if combined with public transport improvements particularly targeted to attract low income motorists, these political and distributional constraints may be overcome. Conversely, public transport improvements may not be affordable on their own but may be funded if the revenue from road pricing is hypothecated. May *et al.* (2005a) develop an integration matrix that summarizes the complementarities between infrastructure, management, pricing, land use, information and marketing measures. Optimization studies undertaken in Edinburgh suggest that integrated transport might involve road pricing in the peak, public transport service increases, public transport fare reductions and some expansion in road capacity. Similar results have been found for other cities both in the UK (e.g. Leeds) and in continental Europe (e.g. Oslo, Vienna) (May *et al.*, 2005b). Hence although integration between public and private transport has been considered, such studies have not normally been extended to consider the integration of transport with other policy domains.

Overall, although empirical evidence is available on the benefits of some elements of integration this is largely based on ex-ante appraisals of measures to improve public transport. What are needed are ex-post evaluations, especially those that examine integration in its broadest sense, including the links between transport and other sectors of the economy, and that consider institutional factors. This might include both quantitative evidence of impact benefits using tools such as BCR and more qualitative evidence of process benefits. In the absence of such evidence, it is hardly surprising that the delivery of integrated transport policies has been problematic.

Notes

1 The organization responsible for national rail infrastructure, privatized by a stock market flotation in 1996.
2 See also Chapter 7 in this volume.
3 See www.webtag.org.uk.
4 The shadow price of carbon is based on the social cost of carbon for a particular stabilization goal, adjusted to take into account the marginal abatement costs to achieve that goal and political willingness to pay.
5 Poor, Low, Medium or High – see also Section 2 of this chapter.
6 This was treated as increasing the PVC of a scheme in the Eddington BCR calculations. NATA

Refresh suggests that this should alternatively be treated as a reduction in the PVB of a scheme. For an illustrative public transport scheme examined by Nash and Preston (1991), the former method gives a BCR of 1.25 and the latter a BCR of 1.34, suggesting the change will benefit public transport schemes.

7 Revenue divided by costs.

8 Bus demand in Greater Manchester was back to its 1999 levels by 2005. See NERA (2006).

9 It should be noted that the vast majority of these funds were devoted to initiatives related to employment, housing, education and training and healthcare, with relatively little emphasis on transport.

10 Defined in terms of rail commuting time of 90 minutes or under. Note Bristol calculations based on Parkway.

11 From Jane's *Urban Transport Systems 2005–6* and cited in Atkins (2006).

References

Atkins (2006) *European Best Practice 2006. Update 2.* Report for the Commission for Integrated Transport, London.

Banister, D. (2007) *Quantification of the Non-Transport Benefits Resulting from Rail Investment.* Working Paper 1029, Transport Studies Unit, Oxford University Centre for the Environment.

Cairns, S., Sloman, L., Newson, C., Anable, J., Kirkbride, A. and Goodwin, P. (2004) *Smarter Choices – Changing the Way We Travel.* Chapter 6: Public transport information and marketing. HMSO, London.

Cottham, G.W. (1985) *The Cost Effectiveness of Integrated Public Transport: West Yorkshire – A Case Study.* Metro. West Yorkshire Passenger Transport Executive, Wakefield.

Cresswell, R. (ed.) (1979) *Urban Planning and Public Transport.* Construction Press, Lancaster.

DETR (Department of Environment, Transport and the Regions) (1998) *A New Deal for Transport: Better for Everyone.* White Paper. HMSO, London.

—— (2000). *Transport 2010 – The Ten Year Plan.* HMSO, London.

DfT (Department for Transport) (2004) *The Future of Transport: A Network for 2030.* HMSO, London.

—— (2005), The appraisal process. *Introduction to Transport Analysis TAG Unit 1.1*, London: Integrated Transport Economics and Appraisal (ITEA) Division. Available online at: http://www.dft.gov.uk/webtag/documents/overview/pdf/unit1.1.pdf (accessed 13 April 2010).

—— (2006) *Data on Investment Returns from Transport Schemes Considered by the Eddington Study.* HMSO, London.

—— (2009) *NATA Refresh. Appraisal for a Sustainable Transport System.* HMSO, London.

Docherty, I. and Shaw, J. (eds) (2003) *A New Deal for Transport? The UK's Struggle with the Sustainable Transport Agenda.* Blackwell, Oxford.

—— (eds) (2008) *Traffic Jam: 10 Years of 'Sustainable' Transport in the UK.* The Policy Press, Bristol.

Dodgson, J. (2009) *Rates of Return on Public Spending on Transport.* RAC Foundation Report 09/103. RAC Foundation, London.

DTO (Dublin Transportation Office) (2001) *Platform for Change*, Stationery Office: Dublin.

Eddington, R. (2006). *The Eddington Transport Study. Main Report: Transport's Role in Sustaining the UK's Productivity and Competitiveness.* HMSO, London.

Evans, A. (1987) A theoretical comparison of competition with other economic regimes. *Journal of Transport Economics and Policy*, 21, 1, 7–36.

Fitzroy, F. and Smith, I. (1994) The demand for public transport: some estimates from Zurich. *International Journal of Transport Economics*, 21, 2, 197–207.

—— (1998) Public transport demand in Freiburg: why did patronage double in a decade? *Transport Policy*, 5, 3, 163–173.

Gilbert, C.L. and Jalilian, H. (1991) The demand for travel and for travelcards on London Regional Transport. *Journal of Transport Economics and Policy*, 25, 3, 3–29.

Glaister, S. (1999) Observations on the new approach to the appraisal of road projects. *Journal of Transport Economic and Policy*, 33, 2, 227–234

Hibbs, J. (2000) *Transport Policy: The Myth of Integrated Planning.* Institute of Economic Affairs, London.

Hine, J. (2008) Transport and social justice. In Knowles, R., Shaw, J. and Docherty, I. (eds) *Transport Geographies: Mobilities, Flows and Spaces.* Blackwell, Oxford.

Hull, A. (2005) Integrated transport planning in the UK: from concept to reality. *Journal of Transport Geography*, 13, 4, 318–328.

ICE – (Institution of Civil Engineers) (2008) *The State of the Nation: Transport.* ICE, London.

Luenig, T. and Swaffield, J. (2007) *Cities Limited.* Policy Exchange, London.

Margail, F. and Auzannet, P. (1996) *Evaluation of the Economic and Social Effectiveness of Park-and-Ride Facilities.* Proceedings of the European Transport Conference. Strasbourg, France.

May, A.D., Kelly, C. and Shepherd, S. (2005a) Integrated transport strategies. In Button, K.J. and Hensher, D.A. (eds) *Handbook of Transport Strategy, Policy and Institutions.* Elsevier, Oxford. pp. 237–254.

May, A.D., Shepherd, S. and Emberger, G. (2005b) Optimization of transport strategies. In Button, K.J. and Hensher, D.A. (eds) *Handbook of Transport Strategy, Policy and Institutions.* Elsevier, Oxford. pp. 665–684.

Mayeres, I., Proost, S., Emberger, G., Grant-Muller, S., Kelly, C. and May, A.D. (2003) *Synergies and Conflicts of Transport Policies.* SPECTRUM Deliverable D4. Institute for Transport Studies, University of Leeds.

Mohring, H. (1972) Optimization and scale economies in urban bus transportation. *American Economic Review*, 62, 4, 591–604.

Nash, C.A. (1988) Integration of public transport: an assessment. In Dodgson, J. and Topham, N. (eds) *Bus Deregulation and Privatisation.* Gower, Aldershot.

Nash, C.A. and Preston, J. (1991) Appraisal of rail investment projects: recent British experience. *Transport Reviews*, 11, 4, 295–309.

NEA, OGM and TSU (2003) *Integration and Regulatory Structures in Public Transport. Final Report.* DGTREN, Brussels.

Nellthorp, J. and Mackie, P. (2000) The UK Roads Review – a hedonic model of decision making. *Transport Policy*, 7, 2, 127–138.

NERA (2006) *The Decline in Bus Services in English PTE Areas: The Quest for a Solution.* NERA Economic Consulting, London.

Newell, G.F. and Vuchic, V.R. (1968) Rapid transit interstation spacings for minimum journey time. *Transportation Science*, 2, 4, 303–309.

Potter, S. and Skinner, M.J. (2000). On transport integration: a contribution to better understanding. *Futures*, 32, 3, 275–287.

Preston, J., Marshall, A. and Tochterman, L. (2008) *On the Move: Delivering Integrated Transport in Britain's Cities.* Centre for Cities, London.

Preston, J. and Rajé, F. (2007) Accessibility, mobility and transport-related social exclusion. *Journal of Transport Geography*,15, 3, 151–160.

Preston, J., Whelan, G. and Wardman, M. (1999) An analysis of the potential for on-track competition in the British passenger rail industry. *Journal of Transport Economics and Policy*, 33, 1, 77–94.

Rogers, R. (1999) *Towards an Urban Renaissance.* E&FN Spon, London.

Roumboutsos, A. and Kapros, S. (2008). A game theory approach to urban public transport integration. *Transport Policy*, 15, 4, 209–215.

Sloman, L. (2003). *Less Traffic Where People Live: How Local Transport Schemes Can Help Cut Traffic.* Transport for Quality of Life and Transport 2000, London.

Stern, N. (2006). *The Stern Review on the Economics of Climate Change.* Cambridge University Press.

van Reeven, P. (2003) Competition in scheduled transport. PhD Thesis. Erasmus University, Rotterdam.

Wall, G. and McDonald, M. (2007) Improving bus service quality and information in Winchester. *Transport Policy*, 14, 2, 165–179.

Chapter 13

A decision analysis framework for intermodal transport

Evaluating different policy measures to stimulate the market

Cathy Macharis, Ethem Pekin and Tom van Lier

1. Introduction

Intermodal transport is defined as the combination of at least two modes of transport in a single transport chain, without a change of the loading unit for the goods, with most of the route travelled by rail, inland waterways or an ocean-going vessel, and with the shortest possible final journey by road (ECMT, 2003). The movement from one mode of transport to another usually takes place at an intermodal terminal. In Figure 13.1, a typical maritime-based intermodal chain is shown. Containers are handled in the port and are further transported by barge or rail wagons towards an inland terminal, from which the containers are transported by road to the end destination.

The political and scientific interest in intermodalism has grown significantly in the last three decades. It has been strongly advocated because of environmental concerns, reasons of overall efficiency, and the benefits of coordination of modes to cope with growing transport flows (Bontekoning *et al.*, 2004).

The environmental concerns are mostly examined in the broader context of the externalities within the transport sector. Externalities in the transport sector arise when transport consumers/producers impose additional costs on society without having to bear those costs themselves. Current decision-making agenda centres around concepts of sustainability and environmental concerns not only for their societal importance but also for their interaction with the economy. From a transport

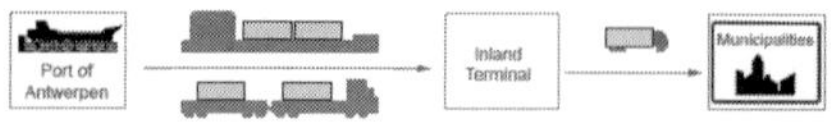

Figure 13.1
The intermodal transport chain
Source: Macharis, Pekin and van Lier, 2009

policy perspective, sustainability concerns are expressed frequently in the ongoing discussions on the internalization of external costs. Indeed, in order to develop more sustainable transport solutions, an internalization of the external costs towards fair and efficient prices between transport means is considered essential. By internalizing these external effects they are made part of the decision-making process of transport users, leading to a more efficient use of transport infrastructure. Therefore externalities are required to be expressed in monetary terms through external costs (on the measurement and use of external costs, and on the pros and cons of the cost-benefit analysis, see Chapter 7 in this volume).

The most important external effects of transport are accidents, noise, air pollution, climate change and congestion.[1] In an attempt to internalize the external costs of transport, taxation of road transport is already applied at a European level (e.g. Eurovignette[2]). Additionally, the member states decide on imposing excises and VAT on fuel, traffic taxes, taxes on insurance premiums and on maintenance of vehicles, vehicle purchase taxes and registration taxes. It is necessary to take into account the effects of this taxation of road transport in order to determine which part of the external costs is already internalized.

In this chapter we focus on the impact of the internalization of external costs on the market share of intermodal transport. The simulation will be done with the LAMBIT-model (Location Analysis Model for Belgian Intermodal Terminals), which makes it possible to simulate different price changes. Also the current subsidy schemes for intermodal transport in Belgium will be simulated and compared to the internalization scenario. In Belgium, subsidies are granted on the federal level for some of the road/rail services and on a regional level for barge container transport, which makes it an interesting country to analyse. Similar conclusions can be drawn on a European level where often several national decisions are made without taking the measures of the neighbouring countries into account.

In Section 2 of this chapter the methodology of the model is briefly described. Section 3 introduces the internalization of external costs. Section 4 discusses the results of the simulation of different policy options. Finally, Section 5 draws conclusions from this chapter.

2. Methodology

LAMBIT is based on three main inputs: transportation networks, transport prices and container flows from/to the municipalities to and from the port of Antwerp. In this section the set-up of the model is briefly described.

2.1 Construction of the model

LAMBIT is a Geographic Information Systems (GIS)-based model, consisting of the different GIS network layers (for each transport mode), the location of the intermodal terminals, and the port of Antwerp (as nodes in the network) (Macharis, 2000, 2004).

The model has been built by connecting the geographic locations of the intermodal terminals (transhipment points) and the municipality centres (end destinations) to the network layers by their corresponding nodes. During the set-up process, possible locations for future terminals can also be included in the model, according to their coordinates on the network.

The GIS network model has two tasks. First of all, it visualizes the real transportation network, including the terminals. The second and vital task of the network is to provide a tool to calculate the transport prices. Figure 13.2 depicts the four layers of the network in the GIS.

The GIS networks for Belgium were built by merging the following digital databases:

- Road layers and municipalities were obtained from the MultiNet database of Tele Atlas.
- Rail and inland waterways layers were extracted from the Environmental Systems Research Institute (ESRI) dataset for Europe.

The reliability of the results of the LAMBIT analysis is dependent on the reliability of the data used in the model. The underlying data are the market prices for each of the transportation modes and handling costs at the terminals and the container volumes from the Belgian municipalities. These data are incorporated into the network layers and nodes.

The LAMBIT methodology is based on the intermodal cost structure and the break-even distance. Considering the total transport prices and the distance

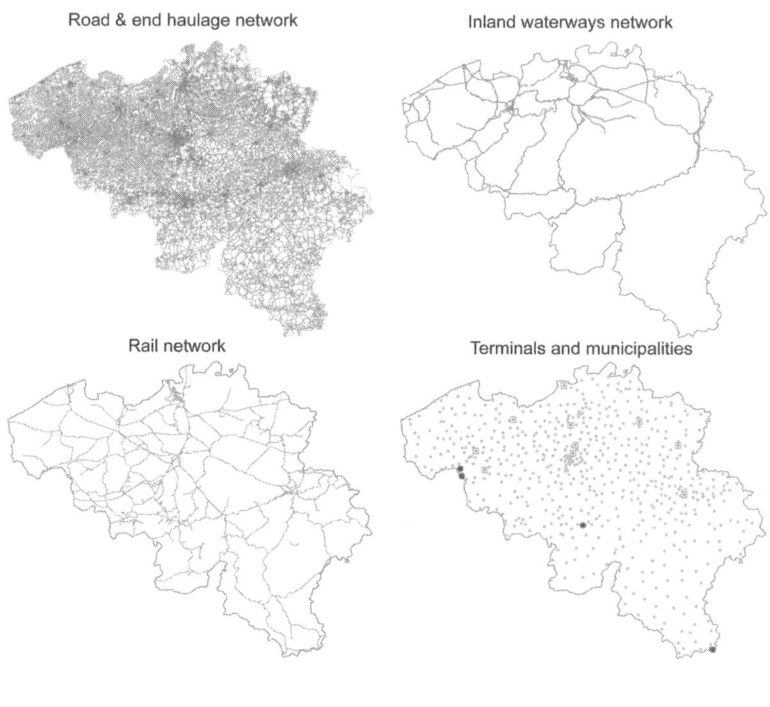

Figure 13.2 Network layers and nodes
Source: Macharis, Pekin and van Lier, 2009

travelled, unimodal road transport is cheaper in the short distances but once the break-even distance is achieved, intermodal transport offers a competitive alternative (Macharis, 2004). The total price of intermodal transport is composed of the transhipment cost in the port of Antwerp to a barge or a wagon; the cost of the intermodal main haul (barge or rail); the transhipment cost in the inland terminal to a truck; and the cost of final haulage by truck. The total intermodal transport cost is obtained by adding all of these fixed and variable costs.

In order to have reliable transport costs, many transport companies were contacted and average market prices were calculated in 2008. For the inland waterways and unimodal road transport, average prices are calculated from the current market prices. On the other hand, the rail prices are based on the market prices of the rail operators and they differ for each inland terminal, as the operators are adapting their prices to the local prices of road transport companies.

2.2 Operating the model

Using a shortest-path algorithm in ArcInfo, various simulations are conducted in order to find the shortest path and the attached transport costs from the port of Antwerp to each Belgian municipality via intermodal terminals and via road only. For each destination, the total transport costs for unimodal road, inland waterways/road and rail/road transport are compared and the cheapest option is selected. The market areas of each inland terminal are then highlighted in the map of the model. These visualizations help us to see how large the market area of each intermodal terminal is. As a further step, the container flows data can be used to show the number of containers that are currently transported by road to the municipalities within the market area, which gives an indication of the existing potential volume that can still be shifted. This is particularly useful when a location of a new terminal needs to be analysed.

It is necessary to indicate that certain assumptions are made in the model. Apart from the transport costs, other modal choice criteria are also important, such as reliability, speed, frequency, safety and customer satisfaction. These other modal choice variables are not yet incorporated in the model. In addition, the transport costs are only one part of the total logistics costs. Warehousing costs and inventory, administration and other processing costs also affect the total logistics costs.

3. Internalization of external costs

3.1 Theoretic background of external costs

Externalities are changes in welfare caused by economic activities without these changes being reflected in market prices (Weinreich *et al.*, 1998). In the field of transport these externalities arise when transport consumers/producers impose additional costs on society without having to bear those costs themselves. External costs are externalities expressed in monetary terms.

In the literature, the most important external costs of transport are the following (Infras/IWW, 2004):

- accidents
- noise
- air pollution
- climate change
- congestion.

In this list, congestion is a somewhat special external cost category. When an additional transport user makes a transport decision, he will only take his own time loss into account, not the time losses he causes for the existing transport users. All other cost categories reflect external costs imposed by transport on the whole of society (including individuals not participating in transport), congestion is thus a phenomenon within the transport sector where users essentially disturb each other mutually, but impose only limited extra costs on the rest of society. Therefore, congestion costs are often considered separately. Given the degree of urbanization in Belgium (congestion is more severe in urban areas), congestion costs are a very important factor in our comparison of intermodal and road transport.

Concerning environmental costs (air pollution, climate change, noise, nature), two aspects are involved: measurement of the environmental effects and monetization of these effects. Measurement of environmental effects falls in the domain of environmental technology (e.g. assessment of dose-response functions), while the conversion to monetary values falls within the field of economics (e.g. use of economic valuation methods). Regarding monetization of external costs, four evaluation methods can be identified: stated preference, revealed preference, shadow prices, and productivity effects (Blauwens *et al.*, 2001).

Calculation of the relevant external costs in this specific case is based on the best practices in the field of marginal external cost (MEC) assessment currently available in scientific literature.[3] Although there is growing consensus on the main methodological issues (CE Delft, 2008), there remain many uncertainties when performing an external cost assessment in practice.[4] Numerous studies have shown that MECs of transport activities depend strongly on parameters such as fuel type, location (urban, interurban, rural), driving conditions (peak, off-peak, night), and vehicle characteristics (EURO standards) (Panis and Mayeres, 2006). As a result, the external cost of one truck-kilometre in urban areas during peak traffic can be up to five times higher than the cost of an off-peak inter-urban kilometre of the same vehicle (CE Delft, 2008).

Distinction should be made between short- and long-run marginal costs. Short-run marginal costs are related to an additional vehicle entering the (existing) system and consider only variable costs (i.e. costs depending on traffic volume), neglecting fixed costs to run the system, or additional costs for possible network improvements in the longer run. Long-run marginal costs consider future system enlargements due to increased traffic volume. Since, in the context considered here, calculation of the societal cost of transport to and from the intermodal terminals on existing transport infrastructure (roads, rail and canals) is required, focus will be on short-run MECs, including the MEC of damage to the road caused by an additional vehicle. However, if long-term policy measures included building new transport infrastructure to connect intermodal terminals, long-term MECs would have to be added as well to account for the external costs caused by these new roads, rails and/or canals. External costs on the intermodal terminals itself (e.g. the external cost caused during transhipment) are not yet included in the model.

In Europe, various transport policies aim to initiate a shift of freight from unimodal road transport to modes that are environmentally more efficient. Intermodal transport, incorporating more environmental modes such as barge, rail and short sea shipping, has lower external costs in most of the trajectories (see Kreutzberger *et al.*, 2006 for an overview of studies). In 2007, the European Commission (EC) announced a European freight transport action plan. One of the concepts introduced is that of 'green transport corridors'. Green transport corridors include short sea shipping, rail, inland waterways and road transport combinations to enable environmentally friendly transport solutions for the European industry. The EC also proposes to revise the directive on the charging of road transport for infrastructure use (Eurovignette). These measures are planned to come into effect before 2011.

The transport policy in Europe, which is driven more by public policy concerns than by business dynamics, suggests a co-modal approach through 'using different modes on their own and in combination' (EC, 2006, p. 4) with the aim to achieve 'a high level of both mobility and of environmental protection' (EC, 2006, p. 3). Since public policy concerns vary in intensity between different member states and regions in Europe, the development of continent-wide solutions is very difficult to implement. European intermodal transport policy and national intermodal policies of the member states should complement each other. This situation highlights the involvement of the EC in modal shift through the formulation of various programmes such as PACT[5] and Marco Polo,[6] and its support of decisions concerning state aid cases of national initiatives to promote intermodal transport.

3.2 Relevance

From a social point of view it is important to calculate the amount of potential external cost savings from barge transport and rail transport compared to road transport; these calculations are also relevant from a transport policy perspective. The importance of external costs and the ever-growing attention that they receive in the transport sector is explained by the fact that non-internalization of external costs gives the wrong market signals and thus leads to significant inefficiencies, such as congestion and environmental nuisances. In order to develop more sustainable transport solutions, an internalization of the external costs towards fair and efficient prices between transport means is considered essential. By internalizing these external effects they are made part of the decision-making process of transport users, leading to a more efficient use of transport infrastructure. To some extent the introduction of market-based instruments for internalization of external costs has already been substantiated in EU directives, particularly the Eurovignette Directive on road charges. The EC has recommended this policy of internalization in several strategy papers such as the Green Paper on fair and efficient pricing (EC, 1995); the White Paper on the overall transport strategy, *Time to Decide* (EC, 2001); its midterm review, *Keep Europe Moving* (EC, 2006); the *Greening Transport Package*

(EC, 2008); and most recently in the document, *A Sustainable Future for Transport: Towards an Integrated, Technology-led and User-friendly System* (EC, 2009), which was published in June 2009 in preparation for the next White Paper of 2010. The EC proposes a stepwise strategy for the internalization of external costs in all transport modes, which contemplates, among other measures, the inclusion of aviation in the EU emission trading scheme from 2012 and the introduction of internalization charges for heavy goods vehicles.

3.3 External costs of transport: the Belgian case

Since the external costs of road transport depend highly on the location, time and vehicle type, there are significant country-related differences for most external cost categories. Therefore, use was made of the external cost figures for road transport from De Ceuster (2004), for four main reasons:

- The study calculates values on a Belgian level (more specifically the Flemish part of Belgium).
- The study takes into account the five most important short-term MEC categories: air pollution, climate change, accidents, noise and congestion.[7] In addition, the short-term MEC of damage to the road, caused by additional trucks on the road, is taken into account ('MEC ROAD' in Figure 13.3).
- The study differentiates between different vehicle types, giving figures for diesel trucks.
- The study takes into account the effects of taxation of road transport (including excises and VAT on fuel; traffic taxes; taxes on insurance premiums and on maintenance of vehicles; Eurovignette; and vehicle purchase taxes and registration taxes) in order to determine which part of the external costs is already internalized.[8]

This last point is especially useful, since it allows taking only the part of non-internalized external costs into account. Figure 13.3 shows values for the MECs and taxes for a heavy-duty diesel truck for Flanders over the period 1991–2002. In 2002 total MECs amounted to €0.52/km. Note the high and increasing proportion of congestion costs over the years, accounting for 74 per cent of total short-term MECs in 2002, whereas the other external cost categories remained stable or gradually decreased. As can also be seen from this figure (the shaded area), the existing taxation system on heavy-duty diesel trucks compensated for 26 per cent of short-term MECs in 2002, leaving 74 per cent of external costs non-internalized. This equals the proportion of congestion costs, so it could be said that this taxation system internalizes all the external cost categories, except the largest category, namely congestion.[9] Since congestion is very time and location dependent, this implies that a full internalization of external costs requires the introduction of some form of differentiated congestion charging.

In a recent publication of the Belgian Federal Planning Office, Hertveldt *et al.* (2009) calculated the proportion of congestion costs in the total of congestion and direct environmental external costs for Belgium for 2005 and 2030. The results showed that in 2008, during peak traffic the proportion of congestion was well above 80 per cent and still more than 50 per cent during off-peak (above 95 per cent in peak and about 80 per cent off-peak in 2030), but since important external cost categories such as accidents and noise are not considered in Hertveldt *et al.* (2009), comparison with the results from De Ceuster (2004) is not straightforward. However, the study seems to confirm the trends visible in Figure 13.3: an increase in marginal congestion costs and a decrease in the other MEC categories (due to advances in the field of car technology, traffic safety, legislation, etc.) between 2002 and 2009. It is expected that this trend will continue in the future.

For rail and barge, figures are provided by De Vlieger *et al.* (2004) in Table 13.1. Here comparison is made between key figures for the different MECs of transport for three transport modes, provided by three different studies (from VITO, EC and Planco). The differences between the three studies show that there is no such thing as standard key figures for this type of costs. This is explained by the fact that, as mentioned, MECs of transport activities depend strongly on parameters such as fuel type, location, driving conditions and vehicle characteristics. For road transport especially, parameters such as network type and driving conditions can vary strongly. However, as can be seen in Table 13.1, MECs for trucks are consistently and significantly higher than for the other two modes. For air pollution and climate change, MECs of barge and train are comparable; for the other categories barge has lower MECs than train (for infrastructure the data is inconclusive). In this analysis we will use the figures from the Belgian VITO study for barge and rail, since these best reflect local conditions. Truck figures in this table are just for comparison and will not be used in this analysis since the above figures from De Ceuster (2004) are partly based on VITO data for environmental aspects and better account for congestion costs in the Belgian situation.

4. Results of simulations

Figure 13.4 presents the existing intermodal inland terminals with their market area. In this reference scenario, nine barge/road (inland waterways) terminals and four rail/road terminals are included. Current market prices are used to show the market area

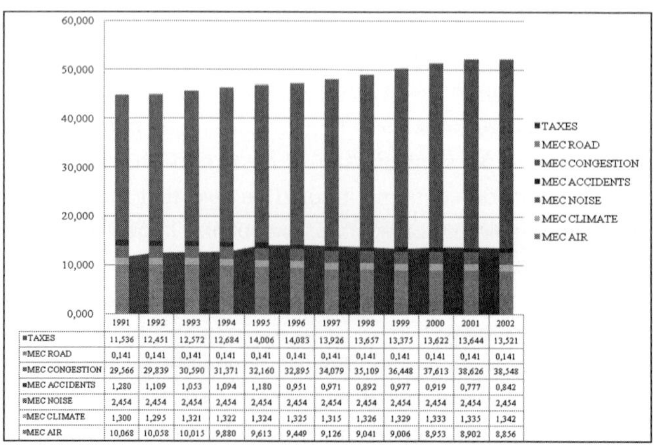

	1991	1992	1993	1994	1995	1996	1997	1998	1999	2000	2001	2002
TAXES	11,536	12,451	12,572	12,684	14,006	14,083	13,926	13,657	13,375	13,622	13,644	13,521
MEC ROAD	0,141	0,141	0,141	0,141	0,141	0,141	0,141	0,141	0,141	0,141	0,141	0,141
MEC CONGESTION	29,566	29,839	30,590	31,371	32,160	32,895	34,079	35,109	36,448	37,613	38,626	38,548
MEC ACCIDENTS	1,280	1,109	1,053	1,094	1,180	0,951	0,971	0,892	0,977	0,919	0,777	0,842
MEC NOISE	2,454	2,454	2,454	2,454	2,454	2,454	2,454	2,454	2,454	2,454	2,454	2,454
MEC CLIMATE	1,300	1,295	1,321	1,322	1,324	1,325	1,315	1,326	1,329	1,333	1,335	1,342
MEC AIR	10,068	10,058	10,015	9,880	9,613	9,449	9,126	9,041	9,006	8,953	8,902	8,856

Figure 13.3 Marginal external costs versus taxes – heavy truck diesel, Flanders, 1991–2002 (2002 prices)
Source: Translated from De Ceuster, 2004. MEC climate and MEC air are derived from VITO (Flemish Institute for Technological Research), 2003.
Note: MEC = marginal external cost.

Table 13.1 Marginal average external costs per transport mode from various sources

Externalities	*Truck*			*Barge*			*Train*		
	VITO	*EC*	*Planco*	*VITO*	*EC*	*Planco*	*VITO*	*EC*	*Planco*
Accidents	22.8	5.4	37.8	0.01	0.0	0.3	1.6	1.5	2.3
Noise	4.4	2.1	7.4	<0.1	0.0	0.0	2.8	3.5	12.7
Air	6.8	7.9	29.1	4.8	3.0	4.2	0.0–8.2	3.8	3.5
Climate	2.3	0.8	0.0	0.6	ne	0.0	0.4–1.26	0.5	0.0
Congestion	5.4	5.5	1.2		ne	0.0	ne	0.2	0.0
Infrastructure	1.9	2.5	0.0	0.7	1.0	0.0	0.2	2.9	0.0
Space availability			1.3	–		0.0	–		0.4
Soil and water-pollution	–		8.6	–		0.0	–		0.0
Total	**43.5**	**24.1**	**85.4**	**6.1**	**5.0**	**4.5**	**7.1**	**12.3**	**19.0**
difference with truck	–	–	–	**37.4**	**19.1**	**80.8**	**36.4**	**11.8**	**66.3**

Source: Translated from De Vlieger *et al.*, 2004.

Note: ne = negligible.

of the current terminals. No subsidies for rail and inland waterways transport are taken into account in the reference scenario. The municipalities are highlighted in Figures 13.4–13.8, when intermodal transport has a more attractive transport price compared to unimodal road transport based on the current market prices. Different shades are used for each terminal in order to distinguish their market area.

4.1 Subsidy scenario

An option that can be used to stimulate intermodal transport is to decrease the costs linked to the transport operations at the terminals by means of subsidies. On a regional scale, subsidies are granted for each container handled by an intermodal barge terminal. This kind of measure is applied for intermodal barge transport in the Wallonia, Brussels and Flemish regions but at different tariffs for each. As rail transport is a federal matter, on a Belgian/federal level another subsidy measure has been taken for road/rail transport. Figure 13.5 presents the regional and federal subsidy schemes in Belgium.

The subsidy is specifically designed to promote national intermodal rail transport. The Belgian government grants an annual budget of €30 million to the intermodal operators, which offer transport services within Belgium over a minimum of 51 kilometres. The subsidy applies to each transport of a container from the Port of Antwerp to an intermodal rail terminal a fixed part (€20) and a variable part (maximum €0.40 per kilometre). In order to assess the impact of this subsidy scheme, LAMBIT was used to show the impact on the market areas of the terminals.

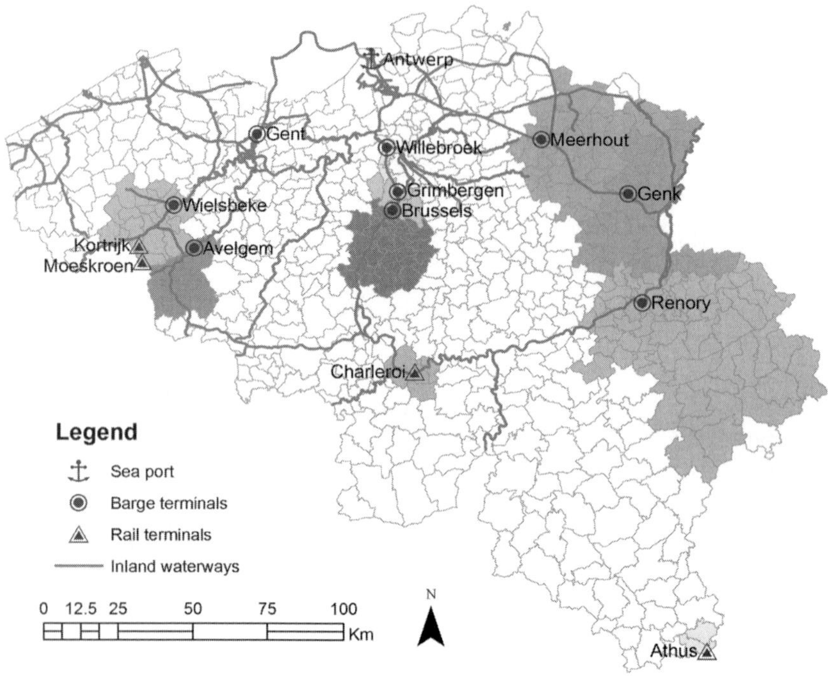

Figure 13.4 The reference scenario
Source: Macharis, Pekin and van Lier, 2009

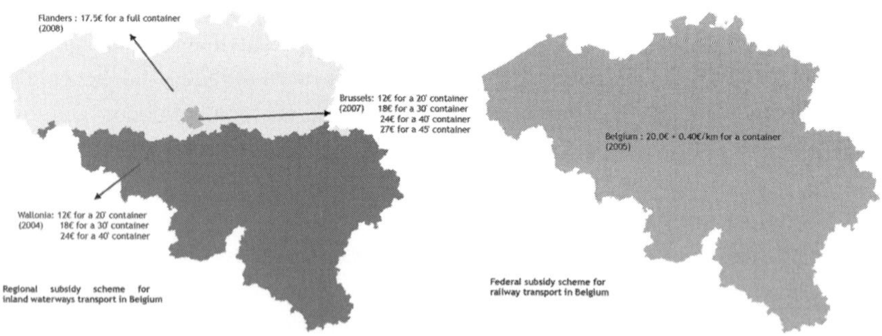

Figure 13.5 Regional and federal subsidy schemes
Source: Macharis, Pekin and van Lier, 2009

In Figure 13.6 the result of this simulation is shown. Compared to the reference scenario, the terminal in Charleroi increases its market area by attracting various municipalities due to the cheaper transport prices of rail transport. Furthermore, the rail terminals in Kortrijk and Moeskroen also gain market area, but this at the expense of the barge terminals in Avelgem and Wielsbeke, which is not the aim of the intermodal transport policy. Overall, the rail subsidy decreases the market area of unimodal road transport by 134 municipalities. Changes in the market areas of terminals (in terms of the number of municipalities served) with respect to their reference scenario are provided in Table 13.3.

The current regional subsidy schemes on inland navigation are analysed in Figure 13.7. The simulations of these subsidies show that intermodal transport

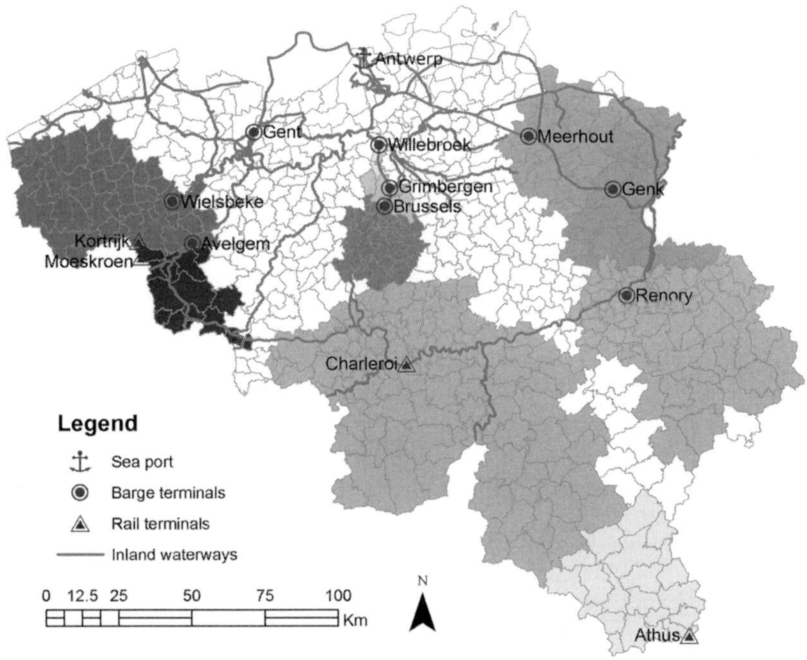

Figure 13.6 Subsidy scenario for the rail terminals *Source: Macharis, Pekin and van Lier, 2009*

grows enormously thanks to the subsidies, despite a decrease in the market areas of the rail terminals compared to the reference scenario. An interesting finding from this scenario is that the rail terminals in Kortrijk and Moeskroen lose some of their market area in terms of the number of municipalities to the barge terminals in Avelgem and Wielsbeke. The effect of the inland navigation subsidy on unimodal road transport is similar to the rail subsidy. Compared to the reference scenario, the market area of the intermodal terminals increases by 219 municipalities when subsidies on inland navigation and rail are applied. In Table 13.3 the market areas for all of the subsidy scenarios are summarized.

Both the federal rail and regional inland navigation subsidies have a clear impact in favour of intermodal transport compared to the reference scenario. The rail subsidy scenario developed in this chapter shows that the market areas of rail terminals would not be able to survive without government subsidy. When the regional subsidy schemes are directed to inland waterway transport, a further growth in the market share for intermodal transport is achieved. However, a drawback of uncoordinated policy measures on different governmental levels is that a modal shift is induced from barge to rail due to the stimulation of intermodal rail transport in regions where barge transport is possible. The global aim should be a modal shift from road to the more environmental intermodal options. Acknowledging the potential benefits of subsidies on promoting a further growth in intermodal transport, one also has to mention their costs to society. Federal subsidy on rail transport, which has recently been extended for four more years, foresees an annual budget of €25 million for the

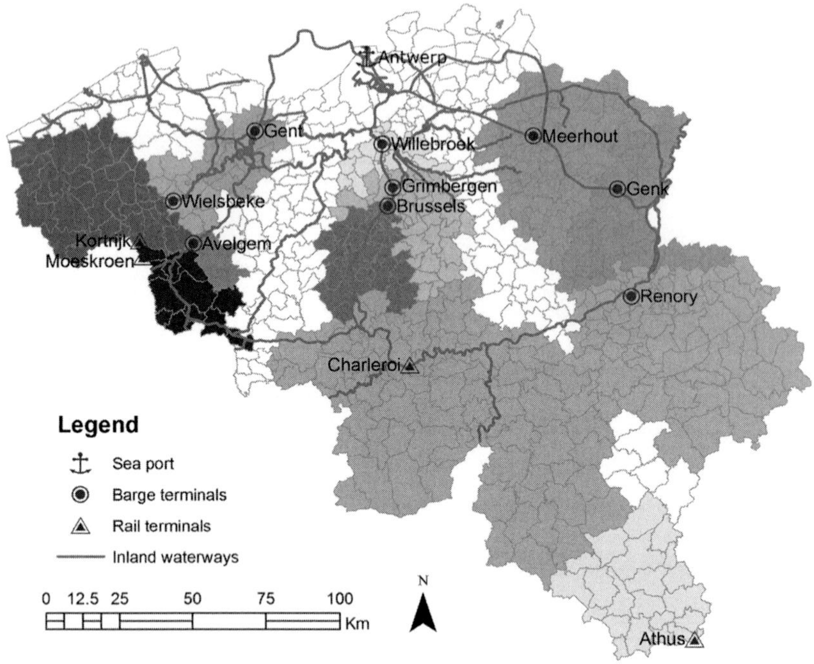

Figure 13.7 Subsidy scenario for the inland waterway terminals
Source: Macharis, Pekin and van Lier, 2009

intermodal rail operations within Belgium. Budgets of regional subsidies on inland navigation are modest compared to the rail subsidy. In Flanders, an annual budget of €3.75 million is foreseen for a period of three years from 2007–09. In 2008, the subsidy in Wallonia has been extended for five more years, which represents an annual budget of €3.5 million for the development of intermodal transport in this region. Finally, the subsidy in Brussels accounts for an annual budget of €150,000 for the period 2007–09.

4.2 Internalization of external costs scenario

In order to simulate the internalization of external costs, first the cost functions of the LAMBIT model were adapted. As described in Section 3, taxes on road transport do not cover all of the external costs of this mode. However, not taking them into account would bias the analysis. Therefore we have subtracted the current road taxes from the total external costs for road. It should be mentioned that no taxes are applied to the inland waterways and rail transport. The next step was to calculate external costs per TEU.[10] Table 13.2 shows the MECs for each transport mode that are used in the LAMBIT simulations. These costs include all costs related to usage of transport infrastructure, such as accident costs, noise, air pollution, climate change and congestion (See Section 3).

Figure 13.8 shows the internalization of external costs for all modes of transport. The costs of transport by different modes are internalized based on current market prices and no subsidies for rail and inland waterways transport are taken into account. The simulation shows an increase in the market areas of intermodal

Table 13.2 The marginal external costs in €/TEU

Year	*Road*	*Rail*	*Barge*
2002	0.39	0.07	0.06

Source: Own calculations based on De Ceuster, 2004, and De Vlieger *et al.*, 2004.

terminals but the impact differs for rail and barge terminals individually. This is explained by the difference in the external costs for each transport mode, which shows that barge transport has a very small advantage in terms of external costs.

The major change occurs in the market area of barge terminals. While the terminals in Genk, Willebroek and Grimbergen face a moderate growth in their market areas, the terminals in Brussels, Renory and Wielsbeke experience a considerable growth. Considering the rail terminals, the terminals in Athus and Charleroi increase their market areas. Both types of terminals now reach municipalities which were formerly in the market area of the unimodal road transport. Nevertheless, an internalization of the external costs does not lead to any market area for the rail terminals in Kortijk and Moeskroen. These two terminals can only compete with the barge terminals in Wielsbeke and Avelgem when the rail subsidy is introduced.

Internalization of the external costs maximizes the market area of the barge terminals compared to both subsidy scenarios. This situation can be considered as the ideal situation although the decrease in the market area of unimodal road transport is less than in the case where inland waterway transport as well as

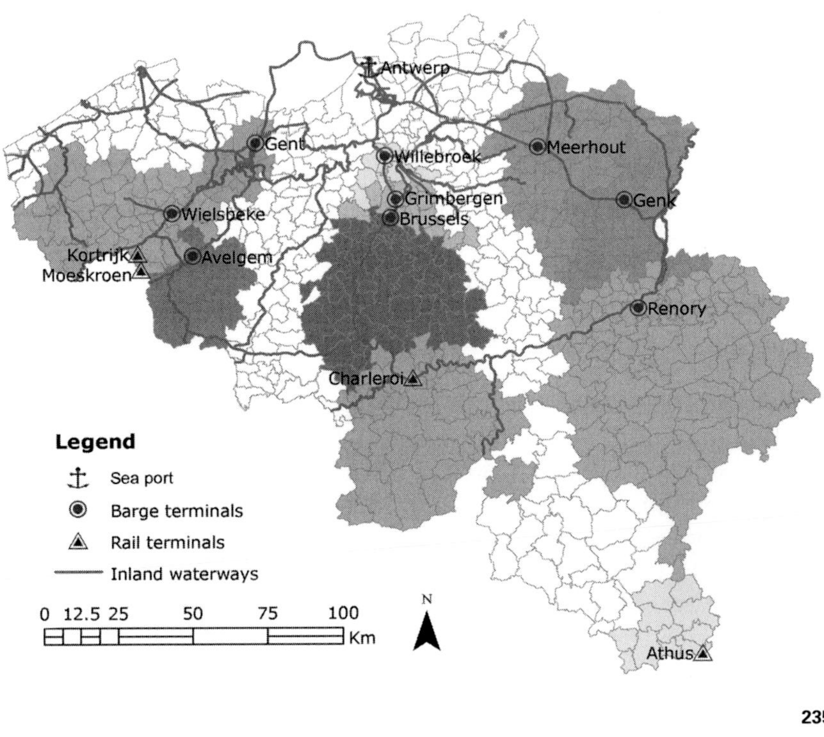

Figure 13.8 Internalization of the external costs: all modes of transport
Source: Macharis, Pekin and van Lier, 2009

Table 13.3 Market areas of terminals in the different scenarios in terms of the number of municipalities

Terminal type	*Intermodal terminals*	*Reference scenario*	*Internalization of external costs*	*Rail subsidy*[1]	*Barge subsidy*[2]
BARGE TERMINALS	Meerhout	24	35	24	41
	Gent	0	10	0	14
	Wielsbeke	19	42	0	11
	Willebroek	3	10	3	18
	Avelgem	8	18	0	5
	Renory	60	86	60	64
	Genk	30	33	30	37
	Grimbergen	6	14	6	33
	Brussels	39	74	39	39
RAIL TERMINALS	Athus	2	11	20	20
	Charleroi	5	38	90	86
	Kortijk	0	0	45	35
	Moeskoren	0	0	13	12
Total barge terminals		189	322	162	262
Total rail terminals		7	49	168	153
Total road transport		393	218	259	174
Total		**589**	**589**	**589**	**589**

Source: Macharis, Pekin and van Lier, 2009

Notes:

1 Rail subsidy: Athus €120; Charleroi €70; Kortrijk €71; Moeskroen €73.

2 Barge subsidy: Meerhout, Gent, Wielsbeke, Willebroek, Avelgem, Genk and Grimbergen €17.5; Renory and Brussels €12.

rail transport is subsidized. Market areas of terminals (in terms of the number of municipalities) compared with the reference scenario are provided in Table 13.3.

5. Conclusions

Intermodal transport can help in reducing the environmental harm and congestion problems caused by our transportation system. By shifting freight, especially containers to barge and rail transport, and only using road transport for the pre- and final haulage, congestion and exhaust emissions can be reduced. Also, in economical terms, intermodal transport can be a cost-efficient option compared to unimodal road transport as the cost per unit decreases because in rail and barge transport economies of scale are experienced. These lower unit costs on the long haul should compensate for the extra handling costs at the inland terminal. This is possible if the distance between origin and destination is long enough (the so-called critical distance).

In order to stimulate intermodal transport, several subsidy schemes have been introduced to lower the price of intermodal transport and to avoid the cost of the extra handling at the terminal. In this chapter, a GIS-based simulation model called LAMBIT was used to analyse the impact of these subsidy schemes. The case of Belgium was used, as a federal rail subsidy scheme is foreseen for road/rail transport and several barge subsidy schemes were introduced on a regional level for the road/barge combination. The simulations show that the subsidies are indeed increasing the market area of intermodal transport, but also that a shift from barge to rail is induced in areas where both types of terminals are available. The same is true on a European level, where the different regions/countries are introducing different funding schemes to support the intermodal transport market. Attention should be paid, however, to possible negative side effects.

In a second round of simulations, the possibility of internalizing the external costs was examined. The external costs were internalized into the market prices of all transport modes, keeping in mind which taxes were already paid by these transportation modes. The simulation shows that an internalization of external costs would also stimulate intermodal transport in a substantial way and with a better balance between the two types of intermodal transport.

Notes

1 In addition, some other, harder to measure and quantify external cost categories of transport have been identified, such as the effect of transport on nature (disturbance of ecosystems, loss of biotopes) and landscape (visual disturbances caused both by transport infrastructure and vehicles); additional costs in urban areas (such as separation effects for pedestrians caused by traffic crossing through cities and loss of available space for cyclists due to car traffic on roads); and external costs caused during up- and downstream processes related to transport (external effects that occur during the manufacturing, maintenance and demolition of both transport infrastructure and vehicles, and external effects that occur during the so-called pre-combustion processes (the manufacturing of fuel)) (INFRAS/IWW, 2000).

2 The Eurovignette Directive was adopted in May 2006 in order to create a harmonized EU framework for charging heavy goods vehicles on European motorways. The directive's main novelty was to introduce the possibility for individual states to integrate the 'external costs' of road transport into toll prices.

3 Note that we calculate the impact of additional units of transported goods via road or intermodal transport, which means that we are interested in marginal rather than average external costs.

4 For a recent summary of the different external cost categories, the most relevant studies in a European context, and recently recommended key figures as proposed by the European Commission, see *Handbook on Estimation of External Cost in the Transport Sector* (CE Delft, 2008), produced within the IMPACT study (Internalization Measures and Policies for All External Cost of Transport).

5 The PACT (Pilot Actions for Combined Transport) programme is a mechanism launched by the EC in 1992 to provide start-up assistance for innovative projects which are likely to increase the use of intermodal transport by improving its competitiveness. For the period 1996–2001, the programme financed 92 projects that developed innovative modal shift concepts.

6 As a successor of the PACT programme, the Marco Polo programme was set up in 2003 with the aim of shifting road freight transport to short sea shipping, inland waterway and rail. The current programme (Marco Polo II) runs from 2007 to 2013, with a total budget of €400 million. Between 2003 and 2006, 55 projects were financed.

7 In De Ceuster (2004), the short-term component of up- and downstream processes which consists mainly of pre-combustion processes (= external costs due to energy production) is not taken into account as a short-run marginal external transport cost.

8 Abstraction is made of the fact that, ideally, transport users that cause higher external costs should be taxed higher in order to give the correct price signal to transport users. Therefore, taxes should vary according to place (urban, non-urban), time (peak, off-peak) and vehicle characteristics (EURO-norm). This differentiation however is achieved only very partially at present.

9 Recently, Hertveldt *et al.* (2009) calculated that the taxation level for trucks in Belgium was €0.14/km for 2005 and would decrease to €0.12€/km in 2030 with unchanged taxation policy (in year 2000 Euro values). The marginal external congestion cost for trucks during peak traffic was calculated to be €0.80/km in 2005 and would rise to €2.21/km in 2030 with fixed road infrastructure capacity (in year 2000 Euro values). During off-peak, values were €0.19/km and €0.34/km for 2005 and 2030 respectively. Since only air pollution, climate change and congestion were considered in Hertveldt *et al.* (2009) (no accidents and noise), figures from De Ceuster (2004) were used for truck transport in our research.

10 TEU = twenty-foot equivalent unit.

References

Blauwens, G., De Baere, P. and Van de Voorde, E. (2001) *Vervoerseconomie.* Standaard Uitgeverij, Antwerp, Belgium.

Bontekoning, Y. M., Macharis, C. and Trip, J. J. (2004) Is a new applied transportation research field emerging? A review of intermodal rail-truck freight transport literature. *Transportation Research Part A* 38(1), 1–34.

CE Delft (2008) *Handbook on Estimation of External Cost in the Transport Sector (Version 1.1). Produced Within the Study Internalization Measures and Policies for All External Cost of Transport (IMPACT).* Delft, the Netherlands.

De Ceuster, G. (2004) *Internalisering van externe kosten van wegverkeer in Vlaanderen.* Study performed for Vlaamse Milieumaatschappij (Flemish Environmental Agency), MIRA, Transport and Mobility Leuven, Belgium.

De Vlieger, I., Cornelis, E., Joul, H. and Int Panis, L. (2004) *Milieuprestaties van de binnenvaart in Vlaanderen; Eindrapport.* Study performed for Promotie Binnenvaart Vlaanderen, Vito, Mol, Belgium.

European Commission (EC) (1995) *Green Paper: Fair and Efficient Pricing in Transport.* Luxembourg: Publications Office of the European Union.

—— (2001) *White Paper: European Transport Policy for 2010: Time to Decide*, Luxembourg: Publications Office of the European Union.

—— (2006) *Keep Europe Moving. Mid-term Review of the European Commission's 2001 Transport White Paper.* Luxembourg: Publications Office of the European Union.

—— (2008) *Greening Transport Package.* DG Energy and Transport, Luxembourg: Publications Office of the European Union.

—— (2009) *A Sustainable Future for Transport: Towards an Integrated, Technology-led and User-friendly System*, DG Energy and Transport, Luxembourg: Publications Office of the European Union.

European Conference of Ministers of Transport, United Nations Economic Commission for Europe Statistical Division and European Union Eurostat (ECMT) (2003) *Glossary for Transport Statistics*, 3rd edn, Luxembourg.

Hertveldt, B., Hoornaert, B. and Mayeres, I. (2009) *Langetermijnvooruitzichten voor transport in België: referentiescenario*, Federaal Planbureau, Planning Paper 107.

INFRAS/IWW (2000) *External Costs of Transport: Accident, Environmental and Congestions Costs in Western Europe*, Karlsruhe/Zürich/Paris: The International Union of Railways (UIC).

INFRAS/IWW (2004) *External Costs of Transport: Update Study*, Karlsruhe/Zürich/Paris: The International Union of Railways (UIC).

Kreutzberger, E., Macharis, C. and Woxenius, J. (2006) Intermodal versus unimodal road freight transport – a review of comparisons of the external costs. In: Jourquin, B., Rietveld, P. and Westin, L. (eds), *Towards Better Performing Transport Systems*, London: Taylor & Francis, pp. 17–42.

Macharis, C. (2000) Strategische modellering voor intermodale terminals. *Socio-economische evaluatie van de locatie van binnenvaart/weg terminals in Vlaanderen*, PhD Thesis, Vrije Universiteit Brussel, Brussel, Belgium.

—— (2004) A methodology to evaluate potential locations for intermodal barge terminals: a policy decision support tool. In: Beuthe, M., Himanen, V., Reggiani, A. and Zamparini, L. (eds), *Transport Developments and Innovations in an Evolving World*, Berlin: Springer-Verlag, pp. 211–234.

Macharis, C., Pekin, E., van Lier, T. (2009) *The Internalisation of External Cost: Opportunity or Threat for Intermodal Transport?* NECTAR Conference, 18–20 June, Washington, DC.

Panis, L. I. and Mayeres, I. (2006) Externe kosten van personenvervoer. In: Despontin, M. and Macharis, C. (eds), *Mobiliteit en (groot)stedenbeleid. 27ste Vlaams Wetenschappelijk Economisch congres*, 19–20 October, Brussels, Belgium, pp. 417–446.

Weinreich, S., Rennings, K., Geβner, C., Schlomann, B. and Engel, T. (1998) *External Costs of Road, Rail and Air Transport – A Bottom-up Approach*. Paper presented at the 8th WCTR, Antwerp, Belgium.

Chapter 14

Integrating the railways

Key assessment issues

Torben Holvad

1. Introduction

Integration is a prerequisite for quality in public transport provision and hence a necessary condition for maintaining or improving the position of public transport in the passenger transport market, principally in terms of increased modal share for public transport modes (including heavy rail, urban rail, metro, tramway and bus) relative to the private car. Therefore, public transport integration becomes important in achieving one of the key ambitions for passenger transport proposed in the 2001 European Commission (EC) White Paper *Time to Decide* (European Commission, 2001) to increase the modal share of public transport. The initial objective is that the share of the market held by each mode is returned to 1998 levels by 2010 for the EU15 and to keep the share of rail freight traffic at 35 per cent for the future new member states in Central and Eastern Europe. For passenger rail this would imply a modal share of 5.8 per cent, while for bus and coach the relevant share would be 9.0 per cent (both largely achieved). However, the need for further increases is anticipated to ensure the sustainability of the transport sector. In the White Paper rail is given a pivotal role as it is mentioned that 'rail transport is literally the strategic sector, on which the success of the efforts to shift the balance between the modes will depend' (European Commission, 2001, p. 13). The policy aims embedded in this White Paper were confirmed in the midterm review undertaken in 2006 where it was emphasized that 'Europe needs all modes of transportation – road, rail, air and waterborne. However, these need to be efficient, well integrated and complement each other in order to ensure seamless transport routes and well integrated transport networks, in full co-modality' (European Commission, 2006, p. 5).

Public transport integration covers a wide range of elements, including information, tickets and fares, network planning and wider integration. The aim of this chapter is to examine an optimization framework for integration of railway services, considering both the costs and benefits of such integration. In this context, rail integration would concern both intra-rail measures (e.g. timetable coordination between rail services provided by different operators) and initiatives enhancing the interfaces between rail and other modes (e.g. access to bus stops within the rail station).

At the practical level, transport integration is often identified as implying that 'whatever modes or types of transport are involved they all operate as one seamless entity for the benefit of the customer' (City Transport Info, 2001–10). A recent EC study 'Integration and regulatory structures in public transport' used a more abstract definition of transport integration (NEA *et al.*, 2003), which is given at the beginning of Chapter 12 in this volume. This definition highlights that transport integration involves multifaceted elements across the different modes and even policy areas. Key integration elements would in accordance with this definition concern (listed in ascending order of complexity):

- *Integrated information on routes (such as common maps, etc.), timetables and fares*
  For railways only, this would, for example, involve availability of maps, timetables and fares information not only for the whole range of services by an individual operator but also for competing operators and operators providing connecting services. The recently created 'Railteam' allows access to integrated information about the high-speed rail services of all the railway operators that participate in this collaboration (see Section 4 for further details about the Railteam activities).
- *Integrated ticketing (total or partial), availability of tickets (points of sales) and integrated fares (partial or integral)*
  Integrated ticketing, in the context of railways, refers to the case(s) where the same ticket will be accepted by more than one operator (in principle both intra- and intermodal journeys) and will allow passengers to use a single ticket for all journeys. It also addresses the physical appearance of the ticket. Integrated ticketing should be considered distinct from integrated fares, as the latter refers to a system in which a journey involving several trips, regardless of mode or operator, is 'integrated' into a single fare for the journey without any transfer penalties. Integrated ticketing and fares systems are often introduced jointly as part of the same package.
- *Network integration, both at the planning and operational stages and in terms of coordination of infrastructure planning*
  In the case of integration of heavy rail and other public transport modes an important example of network integration relates to the interchange facilities at railway stations to facilitate intra- and intermodal transfers.
- *Wider integration with the rest of the transport system (mainly car, taxis, buses and bicycles)*
  An example of wider integration from a railway (transport) perspective could be the provision of bicycle parking at stations or the possibility for passengers to take bicycles onboard the train.
- *Wider integration with other policy domains*
  This form of integration covers a broad range of initiatives that would influence the links between transport and other policy areas such as environment,

education and health care. It could also encompass steps towards improved coordination of railway schemes with land use planning (see also Chapter 3 in this volume).

In principle, any change of the parameters listed above would have an impact on the level of integration. These could be introduced either as standalone measures or combined in bigger transport packages. The aim of the optimization framework is to assess measures that change the level of integration.

The rest of the chapter is structured as follows. Section 2 outlines a framework for Cost-Benefit Analysis (CBA) by identifying the key principles, impact categories and stakeholders to be considered in assessments of integration measures from the perspective of integration initiatives of railways. In Section 3 findings regarding customer valuations of journey characteristics linked to integration with focus on rail are outlined. Section 4 includes a rail-specific illustrative example regarding integration within this mode in the context of the Railteam partnership activities. In Section 5 final remarks are presented.

2. A cost-benefit analysis framework

A given level of rail integration generates various benefits but also involves costs. As such it is likely that up to a certain level of integration benefits will increase more than costs, such that net benefits are increasing. However, there could be an optimum level of integration (where overall net benefits are maximized) and increasing integration beyond this level would result in lower net benefits compared to the optimum (see Figure 14.1). This point coincides with the marginal social benefits of integration being equal to the marginal social costs of integration. In accordance with this it would be possible to look at existing levels of integration for different rail transport systems to determine whether integration is sub-optimal (too little integration) or super-optimal (too much integration).

However, this approach is not directly applicable in practice because it would require estimation of the current integration level, as well as an optimum level of integration (in terms of net benefits), in order to be able to recommend whether the integration level should be increased or decreased. A practical way forward consistent with the core message of Figure 14.1 would be to undertake CBA of specific integration measures to examine whether benefits outweigh costs.

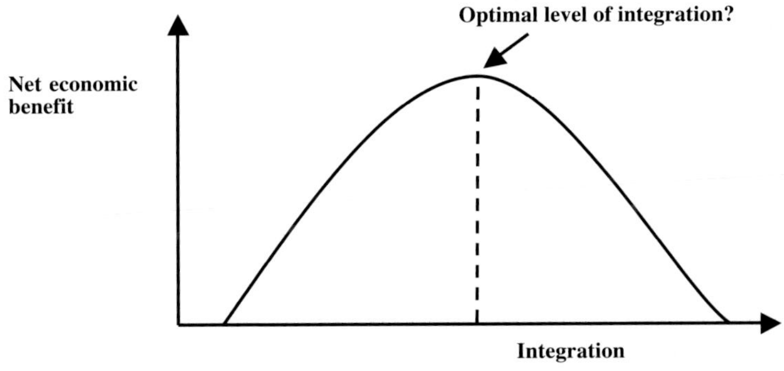

Figure 14.1 A curve of net economic benefits from public transport integration *Source: NEA* et al., *2003*

Following the standard principles of CBA, if the decision-making entity is not facing a budget constraint then all integration measures with positive Net Present Value (NPV) should be implemented, as such schemes would imply that the integration level is moving towards the optimal level. However, as most entities deciding on integration measures for rail transport[1] have budget constraints, the Benefit/Cost Ratio (BCR) should be used to rank the different schemes and select those with the highest ratio until the budget is exhausted. Again this approach would ensure a movement towards the optimal integration level. In the following section a CBA framework is outlined in broad terms considering benefits, costs and relevant incidence groups (see also Chapters 7 and 12 in this volume).

A CBA framework seeks to set out the main benefits and costs with reference to the groups affected by rail integration measures. This framework can be used both in the context of ex-ante and ex-post analysis, where the former in fact not only informs about the benefits of introducing a given integration measure but can also be used to indicate the loss of not introducing that integration measure.

The central characteristic of the proposed framework is that the assessment of benefits and costs could be undertaken within a before-after approach relying solely on information for the area in which the integration measure is being introduced. This framework can be used to demonstrate whether quantifiable benefits outweigh quantifiable costs. Often the practical application of the framework will involve using simple calculations to determine the benefits obtained by new and existing rail transport users compared to the implementation and operating cost associated with the integration measure. It is likely that these benefits will represent the main categories (outlined below). However, there will be benefit/cost elements which cannot be incorporated in the CBA analysis. These impacts should, where relevant, be considered in other ways in order to ensure a comprehensive analysis of the implications of rail transport integration measures and policies (see also Chapters 7 and 12 in this volume on Multicriteria Analysis – MCA).

The starting point for the CBA framework is that a given rail integration measure will generate impacts (consequences). These impacts will represent benefits and costs according to the perspective of the various incidence groups. Below, the impact categories included in the CBA framework are outlined (i.e. those impacts which in principle can be quantified and monetized). In addition, the various impacts will be allocated to different incidence groups. This will accommodate two levels of analysis: (1) comparison of total benefits and costs (across incidence groups) and (2) examination of how each incidence group is influenced (i.e. whether the benefits for a specific incidence group are larger than the costs incurred for that group). Some of the benefits (costs) for one incidence group will be costs (benefits) for another group and hence cancel out when the total net benefits are calculated – one of the disadvantages of standard CBA if equity issues are not explicitly addressed. It is important to emphasize that the impacts would not necessarily be limited to the rail mode alone. For example, rail integration measures could lead to changes in bus patronage (increases and/or decreases). A possible positive effect could arise because the integration measure facilitates bus use (e.g. when the improvements concern interchange facilities between bus and rail). On the other hand, negative effects on bus use could happen in those cases where rail and bus services compete along a route, so that a rail integration measure would result in transfer of passengers from bus to rail.

The outlined framework is consistent with other appraisal frameworks for rail infrastructure investment projects (e.g. RAILPAG[2] developed by the European Investment Bank and the European Commission; see European Investment Bank, 2006).

Impacts considered include:

- changes in capital costs;
- changes in operating costs (including management expenses and maintenance costs);
- changes in revenue for rail and other public transport operators;
- changes in generalized cost for existing rail transport users;
- changes in generalized cost for new rail transport users;
- changes in generalized cost for other public transport users.

In addition, a rail integration measure could have impacts on:

- congestion costs (experienced by non-public transport users), provided (some) car users transfer to the improved public transport mode and freed capacity is not used for new (generated) car journeys;
- accident levels for the transport system as a whole;
- environmental costs taking into account both rail transport, other public transport and other modes;
- economic and social development benefits.

Possible effects on markets with (indirect) taxes and subsidies should be taken into account as required, as well as consideration of the marginal cost of public funds (due to the distortionary effects of taxation). Quantification of this set of impacts is challenging and it is not expected that it can be achieved in every application. This challenge is not unique to the assessment of integration measures.

The generalized cost (Bruzelius, 1979) for rail and other public transport users consists of fares and journey times (converted into money units using values of time). Attributes related to service quality (e.g. service frequency, interchange period, service reliability/punctuality, comfort, overcrowding, vehicle quality and information) can also in principle be included in the generalized cost concept (provided these can be converted into money units). This is important in the context of railway integration measures where service quality attributes could be closely associated. For example, provision of improved information at stations would mainly influence time elements (e.g. wait time) and service quality attributes.

The following incidence groups should be considered:

- rail infrastructure authorities and companies (where relevant);
- rail operators;

- other public transport operators (where relevant);
- rail users;
- other public transport users (where relevant);
- car users (where relevant);
- local and/or central government;
- society (referring to external effects, e.g. environmental consequences).

It is important to emphasize that the specific groups present in a given area are dependent on the institutional structure of rail and other public transport and could therefore differ from the ones listed above. The main source of differences concerns whether or not infrastructure authorities/companies are present. In some cases, rail infrastructure management and service operations are organized within one integrated entity while in others there are separate entities.

This framework allows for assessment of the total benefits and costs as well as how the aggregated incidence groups are influenced. The analysis of benefits and costs across the aggregated incidence groups should be supplemented by assessment of impacts for individual entities, in particular individual operators and authorities. This would contribute to the understanding of why schemes that are socially profitable (positive net benefits) are not being implemented. For example, it would be important to know that the introduction of an integrated ticketing system would lead to increased revenues for rail transport operators as a whole, but one or more operators may experience lower revenues. The latter group of operators would therefore be reluctant to participate in integrated ticketing arrangements.

Furthermore, a comprehensive CBA of discrete integration measures/packages needs to be supplemented with consideration of spatial, temporal and social aspects.

3. Customer valuation of integration measures

As part of a rail-focused CBA framework one critical issue would be to determine how specific rail integration measures influence rail users (both existing and new users). A common characteristic of integration measures is that they are designed to make rail transport easier by addressing the barriers preventing or reducing the use of rail. Such barriers include:

- Logistical barriers: lack of ticket integration; uncoordinated timetables.
- Financial barriers: cost differential between public and private (car) transport.
- Psychological barriers: poor perception of travel time and image; lack of control over journey, poor perception of true cost of travel.
- Institutional barriers: impact of competition between operators; impact of deregulation.
- Information barriers: lack of appropriate information; lack of coordinated information.
- Physical barriers: physical accessibility; comfort; travel time differential.
- Social barriers: personal safety and security.

In this context, rail integration measures will generate benefits for existing rail users and new users attracted to rail transport due to the improvements. Existing users get benefits directly according to the value they put on service quality improvements, as well as the possible consequences on journey time (e.g. wait and walk times) and fares (if these are reduced as a consequence). As noted earlier these consequences for users can be examined within a generalized cost framework.[3] Reduced generalized cost for rail transport would provide both benefits to existing users as well as generating new traffic. New traffic may be demand transferred from other modes (e.g. other public transport; car and other motorized modes; more frequent use of rail by current passengers; and/or from people who did not travel beforehand).

The relative importance of quality-of-service characteristics is often expressed by assigning weights to each quality-of-service attribute. The weight can be represented in travel time units; for example, a real-time information system may lead to X minutes reduction of in-vehicle time per trip. Alternatively, and based on service attributes, it may be expressed in money terms, such as one minute of waiting time being equivalent to Y cents in fare. Information on fare elasticities can then be used to forecast the implications of service quality improvements on rail transport demand. Obviously, the validity of these calculations depends on the robustness of the attribute weightings collected from mainly Stated Preference (SP) and Revealed Preference (RP) studies.

The key issue for determining the impact of integration measures (in cost-benefit terms) is the empirical evidence required to adopt a generalized cost framework. Many elements are relatively well understood, such as:

- fares elasticities;
- service level elasticities;
- values of in-vehicle, walk, wait and interchange times;
- value to rail travellers of new rolling stock, terminal facilities, information provision, etc.

For a detailed review of available evidence regarding all these elements for public transport see TRL (2004). Litman (2007) includes a survey of recent studies regarding public transport fare elasticities and cross-elasticities. In Litman (2009) an overview of elasticities for out-of-pocket expenses, service level and travel time is presented covering public transport as well as car modes. Zamparini and Reggiani (2007) undertake a meta-analysis of some 90 Value of Travel Time Savings (VTTS) studies in North America, Europe and Australia. Evidence regarding values for walk, wait and transfer times is examined in Litman (2008). Jia *et al.* (2008) reviews available evidence regarding the benefits to users and transport agencies of introducing public transport management systems (including real-time information systems). Evidence on the impact of access/egress times is provided from 183 British studies in a meta-analysis in Wardman (2001), where the estimated mean walk time weightings (in units of in-vehicle time) range from 1:37–2:02, depending on mode, journey purpose and urban/inter-urban travel.

A recent study,[4] however, highlighted the lack of information on how out-of-vehicle travel conditions influence public transport patronage, including the possible role played by transfer facilities (Iseki *et al.*, 2006). For transfer facilities the authors conclude that 'we know little about which attributes are most important under which circumstances and in what combinations with other factors' (Iseki *et al.*, 2006, p. 3). Reasons mentioned by the authors for the limited information regarding the linkage between transfer facilities and public transport patronage include:

- More attention has been given towards quantity and quality of in-vehicle travel with an intuitively clearer link to demand.
- There is difficulty in quantitatively evaluating transfer facility attributes compared to factors such as fare, service level and headway.

A number of other elements are not yet well understood in terms of impact on the patronage for rail and other public transport modes:

- regular ('clock face') timetables (e.g. departure at 13:00 as opposed to 13:12);
- simplified fare systems including integrated ticketing and fares;
- enhanced connectivity arising from changes in transport network structures (e.g. hub and spoke);
- synergistic effects of a package of measures;
- dynamics of rail and other public transport demand impacts (e.g. there may be substantial difference between the level of modal shift in the short term compared to a longer term).

Therefore, it appears that lack of available empirical evidence regarding customer valuations for rail and other public transport modes relates particularly to those elements that are of importance for integration; for example, integrated ticketing, transfer facilities and public transport network structures of one or several modes.

Below, we will consider in more detail one of the abovementioned group of service characteristics which are of importance for integration of railways as part of the public transport network: railway stations as transfer facilities. Indeed, the significance of transfer facilities for rail use is linked to the context where 'a railway journey is almost always part of a journey "chain" that includes a journey to, and later from, the railway station by different modes of transport' (Givoni and Rietveld, 2007, p. 357).[5] This characteristic is compounded by available information showing the penalty assigned to out-of-vehicle travel time by travellers. Travellers value the out-of-vehicle parts of a journey significantly higher than other (in-vehicle) journey elements because they are viewed as substantially more inconvenient. The importance of the cost of transferring (transfer penalties) during the course of a typical public transport journey as part of the generalized cost is considered in Iseki *et al.* (2006); Table 14.1.

The journey consists of: (1) walking to a bus stop; (2) waiting at a bus stop; (3) a bus ride; (4) walking from a bus stop to a rail station; (5) waiting at a rail station; (6) travel by train, (7) walking from a rail station to a final destination. Information about journey characteristics for this typical journey is included in the second column

Table 14.1 Typical public transport trip and its associated time and costs

	Time (min)	*Cost*
Access by walk from trip origin to a bus stop	8	$1.66
Wait at a bus stop	4	$0.74
Bus fare ($1.35)	–	$1.35
Travel in vehicle from a bus stop to a rail station	20	$2.50
Transfer penalties		
Transfer from a bus stop to a rail station: walking	6	$1.25
waiting	10	$1.84
Other transfer penalties (a)	–	$1.32
Travel in vehicle from rail station to another	30	$3.75
Train fare ($1.35)	–	$1.35
Egress time from a rail station to a trip destination	6	$1.25
Total	84	$16.99
% of transfer penalties in total generalized costs		26%

	Weight	*Value per*	
		Hour	*Minute*
Wage	2.00	$15	$0.25
In-vehicle travel	1.00	$7.50	$0.13
Walking (b)	1.66	$12.45	$0.21
Waiting (b)	1.47	$11.03	$0.18
Other transfer penalties (b)	17.61	$132.08	–

Source: Adapted from Iseki *et al*., 2006.

Notes:

a Other transfer penalties is further weighted by 0.01 to make its costs comparable to other costs.

b The ratio relative to in-vehicle time is taken from Wardman, 2001.

in terms of minutes for the different stages. The (generalized) costs are listed in the third column by converting the time spent into costs using appropriate values of times and adding the out-of-pocket expenses (fares) and other transfer penalties. The bottom part of Table 14.1 illustrates the importance of removing transfer penalties: (a) in terms of waiting time; (b) in terms of walking time; and (c) in terms of both waiting and walking time.

In this context, the term 'transfer penalty' is used to refer to all monetary costs, time, inconvenience and stress experienced by public transport users when making transfers during a door-to-door journey. Transfers can involve one (e.g. change from one rail service to another at a railway station) or several modes of transport (e.g. change from rail to bus).

Table 14.1 shows that transfer penalties for a typical journey would account for 26 per cent of the total generalized costs. The main components are composed of walking time, waiting time and other transfer penalties. It is important to highlight the distinction between actual time used for walking/waiting in the transfer situation and perceived time spent. Walking and waiting time may be perceived to be more difficult (therefore getting a higher value within a generalized cost calculation) if the transfer facility is perceived insecure (e.g. is without access to an escalator or lifts, involving long walks, etc). The example also highlights that substantial reductions in overall generalized costs can be facilitated by optimizing the time schedule for public transport arrivals and departures at transfer points by reducing the wait time. Furthermore, there is scope for reducing transfer penalties through changes in the physical attributes of transfer facilities (e.g. comfortable and secure waiting room facilities) (Guo and Wilson, 2004). Cost-effective measures would require comprehensive evidence regarding values of transfer penalties and how they vary according to transfer facility characteristics and conditions.

An overview of possible ranges of values for transfer penalties involving different types of transfer (e.g. subway-to-subway or bus-to-rail) according to different empirical studies is provided in Table 14.2. Values for transfer penalty in Table 14.2 (fourth column) are all expressed in minutes of in-vehicle travel time.

As noted in Iseki *et al.* (2006), this type of information provides a general indication of the variation in transfer penalties according to the modes involved in the transfer and the local circumstances. However, additional evidence is required on how the conditions of the transfer facilities influence transfer penalties, since the values may vary substantially between different locations. Thus, the need for more empirical research on the relative importance of different attributes at stations and other public transport transfer points is emphasized; this is in addition to evidence required on the role of higher frequencies and improved service reliability in reducing transfer penalties (see Chapter 6 in this volume for estimates of the value of reliability). This also highlights the possible problems regarding value transfer (i.e. the extent to which values derived in one specific set-up can be used in CBA elsewhere). Generally, values used should, as much as possible, reflect the context in which the appraisal is undertaken. For example, if transfer penalties are estimated for cases where stations had poor customer facilities, such values are likely to be inappropriate to use in assessing improvements to stations with significantly better amenities.

This is taken forward by Taylor *et al.* (2007) who recommend using importance-satisfaction analysis to provide empirical information about improvements to transfer facilities. This approach is used in Chapter 11 in this volume.

Diagrams of the importance-satisfaction matrix provide guidance as to what areas would be effective in terms of increasing the overall satisfaction of rail-based journeys, taking into account both the individual satisfaction scores and the assigned importance of the different dimensions. Particular focus should then be on those dimensions with relatively low satisfaction scores but relatively high importance rankings, such as travel time reliability. However, Brons and Rietveld (2009) conclude that due to the potential costs involved in travel time reliability improvements, more cost-effective options may be represented by improving access-to-station facilities, especially among the group of infrequent passengers (which may face relative substantial transfer penalties at railway stations compared to frequent users).

Table 14.2 Valuation of transfer penalties

Studies	*Variables in the utility function*	*Transfer types (modal structure)*	*Transfer penalty equivalence (in minutes of in-vehicle time)*
Han (1987) Taipei, Taiwan	Initial waiting time Walking time to stop In-vehicle time Bus fare Transfer constant	Bus-to-bus (path choice)	30
Hunt (1990) Edmonton, Canada	Transfer constant Walking distance Total in-vehicle time Waiting time Number of transfers	Bus-to-light rail	17.9
Liu, Pendyala and Polzin (1997) New Jersey, NJ	Transfer constant In-vehicle time Out-of-vehicle time One-way cost Number of transfers	Auto-to-rail Rail-to-rail (modal choice)	15 5
CPTS (1997) Boston, MA	Transfer constant In-vehicle time Walking time Initial waiting time Transfer waiting time Out-of-vehicle time Transit fare	All modes combined (path and mode choice)	12 to 15
Wardman, Hine and Stradling (2001) Edinburgh, Glasgow, UK	Utility function not specified	Bus-to-bus Auto-to-bus Rail-to-rail	4.5 8.3 8
Guo and Wilson (2003)	Transfer constant Walking time In-vehicle time Transfer walking time Transfer waiting time Assisted level change Station dummies Pedestrian environment dummies	Subway-to-subway	1.6 to 31.8

Source: Adapted from Guo and Wilson, 2004.

Similar analyses in other countries would be worthwhile in order to establish more comprehensive empirical information on rail travellers' experience as customers and their perceptions regarding the elements of the service which should be improved.

4. Illustrative example: integration of high-speed rail services

The creation of Railteam, an alliance between seven European high-speed rail operators,to promote seamless and integrated high-speed train travel in Western Europe, demonstrates the evidence given above about potential benefits of integration.

Railteam, launched in July 2007, currently involves the following operators: DB (Germany), SNCF (France), Eurostar (UK), NS Hispeed (Netherlands), ÖBB (Austria), SBB (Switzerland) and SNCB (Belgium), along with their subsidiaries Thalys, Lyria and the DB/SNCF cooperation between France and Germany. Other operators may join in the future.

In particular, Railteam is taking a range of initiatives such as:

- facilitating tickets and platform changes at main hubs in the Railteam network;
- harmonizing rates, on-board services and in-station services to increase awareness of high-speed services;
- access to dedicated services (e.g. waiting rooms, internet access);
- combining passenger loyalty schemes and determining passenger rights;
- combining marketing and communication activities.

The initiatives are expected to make high-speed rail travel easier, especially when trips involve more than one operator in terms of service quality and comfort. For example, access to information such as timetables and fares should be enhanced and there could be improved possibilities for interchange (Fried, 2008).

In the spring of 2008 (before the onset of the economic crisis), it was expected that there would be 23.7 million passengers on the Railteam network in 2010 compared to 13.5 million in 2005 (Fried, 2008). Of these, 1 million would be the result of the creation of Railteam and the implementation of the activities towards intra-modal integration. This would mean a 4 per cent increase in rail travel due to Railteam. A similar demand effect from reduced fares would require an 8 per cent reduction, assuming a fare elasticity of –0.5 (TRL, 2004).

There is only limited information about the costs of the Railteam integration activities, although European Agenda (2007) mentions a figure of €30 million as the costs associated with setting up the distribution system, which will allow international travellers to find an attractive combination of prices, make a reservation, pay and receive their ticket from any distributor of European rail in one transaction. There may though be additional start-up costs as well as on-going costs not considered in the €30 million figure. While there is uncertainty regarding the demand and cost data it is likely that from the participating companies' perspective Railteam will be beneficial.

Further analysis of the implications of Railteam on customers' experiences and value is required and could possibly follow the satisfaction-importance framework described in Chapter 11 in this volume. This could contribute to determining areas of priority for Railteam operators to enhance overall customer satisfaction.

5. Summary

Current rail transport policy initiatives across Europe often include elements involving better integration. These measures are likely to bring benefits to various stakeholders, including existing and new rail transport users. However, there may also be substantial costs associated with improved integration. Informed decision making regarding integration initiatives to be taken forward by public authorities will require appropriate impact assessment, including appraisal of costs and benefits.

A key element in assessing the benefits of integration measures concerns how (existing and new) users will be affected and the valuation of the different service characteristics. The standard approach is to undertake this assessment within a generalized cost framework. However, as there is a lack of evidence concerning the user valuation of important service characteristics, further research is required to close such gaps. In particular, there is a need for more empirical evidence regarding transfer facilities and their influence on transfer penalties and public transport patronage.

Overall, there is a need to undertake empirical studies to establish rail user experiences in order to ensure sufficient knowledge about the relative importance assigned by different user groups to the journey characteristics. This information is essential for authorities and transport operators to ensure that integration initiatives are cost-effective and fulfil their objectives. The information requirements in this area are linked to the multifaceted aspects of integration, whereby a range of different measures in different combinations are available in order to promote integration. In this context, the problem is that not all measures are equally effective, and there are variations in the resources required for implementation and operation. Hence, an assessment of integration measures is crucial before proceeding to implementation.

Notes

1 In most cases the decision-making entities for integration measures involving rail would be government authorities or agencies, as well as railway and other public transport operators.
2 RAILPAG = Railway Project Appraisal Guidelines.
3 All components considered in the assessment of generalized cost changes should be transformed into monetary units (see TRL, 2004, and IER, 2006).
4 The study, *Tool Development to Evaluate the Performance of Intermodal Connectivity (EPIC)* to improve public transportation, was undertaken by the Institute of Transportation Studies (ITS, 2008).
5 The main research focus in Givoni and Rietveld (2007) concerns how access to railway stations can contribute to developing the railway network and to increased rail use. This aspect is part of the role that transfer facilities may have in encouraging greater use of public transport.

References

Brons, M. and Rietveld, P. (2009) 'Improving the quality of the door-to-door rail journey: a customer-oriented approach', *Built Environment* 35, 1: 30–43.

Bruzelius, N. (1979) *The Value of Travel Time: Theory and Measurement*. London: Croom Helm.

Central Transportation Planning Staff (CTPS) (1997) *Transfer Penalties in Urban Mode Choice Modeling*, Travel Model Improvement Program, US Department of Transportation and US Environmental Protection Agency.

City Transport Info (2001–2010) 'Integrated transport', Available online at: http://www.citytransport.info/Integ.htm

European Agenda (2007) 'Revitalization of train transport in Europe – European railways companies launch a new initiative', European Agenda Event Report, 2 July 2007, Brussels: European Agenda. Available online at: http://www.european-agenda.com/_files/eventreport/173.pdf (accessed May 2009).

European Commission (2001) *European Transport Policy for 2010: Time to Decide*, White Paper, COM(2001) 370, Brussels.

—— (2006) *Keep Europe Moving: Sustainable Mobility for our Continent*, Mid-term Review of the 2001 Transport White Paper, DG TREN Graphical Memo, European Commission Directorate-General for Transport and Energy, Brussels. Available online at: http://ec.europa.eu/transport/strategies/doc/2006_keep_europe_moving.doc/2006_transport_policy_review_memo.pdf (accessed May 2009).

European Investment Bank (2006) *RAILPAG – Railway Project Appraisal Guidelines*, Luxembourg. Available online at: http://www.eib.org/attachments/pj/railpag_en.pdf (accessed October 2009).

Fried, J. (2008) 'Competition "on the tracks" in Germany', 6th World Congress on High Speed Rail, Amsterdam. Available online at: http://www.uic.org/apps/presentation/fried.pdf (accessed May 2009).

Givoni, M. and Rietveld, P. (2007) 'The access journey to the railway station and its role in passengers' satisfaction with rail travel', *Transport Policy* 14, 5: 357–365.

Guo, Z. and Wilson, N. H. M. (2004) 'Assessment of the transfer penalty for transit trips: Geographic Information System-based disaggregate modeling approach', *Transportation Research Record* 1872: 10–18.

Han, A. F. (1987) 'Assessment of transfer penalty to bus riders in Taipei: a disaggregate demand modeling approach', *Transportation Research Record* 1139: 8–14.

Hunt, J. D. (1990) 'A logit model of public transportation route choice', *ITE Journal* 60, 2: 26–30.

IER (2006) 'Proposal for harmonised guidelines', HEATCO (Developing Harmonised European Approaches for Transport Costing and Project Assessment) Project, Deliverable D5. Available online at: http://heatco.ier.uni-stuttgart.de/ (accessed May 2009).

ITS (2008) *Tool Development to Evaluate the Performance of Intermodal Connectivity (EPIC)*, Institute of Transportation Studies, University of California, Los Angeles. Available online at: http://www.its.ucla.edu/research/EPIC/

Iseki, H., Taylor, B. D. and Miller, M. (2006) 'The effects of out-of-vehicle time on travel behaviour: implications for transit transfers', Report submitted by Institute of Transportation Studies, University of California to California Department of Transportation. Available online at: http://www.its.ucla.edu/research/EPIC/Appendix%20A.pdf (accessed May 2009).

Jia, X., Sullivan, E., Nuworsoo, C. and Hockaday, N. (2008) 'Literature review report on benefit/cost studies and evaluations of transit management systems', California PATH Working Paper, Institute of Transportation Studies, University of California, Berkeley. Available online at: http://www.its.berkeley.edu/publications/UCB/2008/PWP/UCB-ITS-PWP-2008-6.pdf (accessed May 2009).

Litman, T. (2007) 'Transit price elasticities and cross-elasticities', Victoria Transport Policy Institute. Available online at: http://www.vtpi.org/tranelas.pdf (accessed May 2009).

—— (2008) 'Valuing transit service quality improvements', *Journal of Public Transportation* 11, 2: 43–63.

—— (2009) 'Transportation elasticities: how prices and other factors affect travel behavior', Victoria Transport Policy Institute. Available online at: http://www.vtpi.org/elasticities.pdf (accessed September 2009).

Liu, R., Pendyala, R. M. and Polzin, S. (1997) 'Assessment of intermodal transfer penalties using stated preference data', *Transportation Research Record* 1607: 74–80.

NEA, OGM, University of Oxford, Erasmus University, TIS.PT and ISIS (2003) 'Integration and regulatory structures in public transport', Final Report to the European Commission, DG TREN. Available online at: http://www.eltis.org/docs/INTEGRATION.pdf (accessed May 2009).

Taylor, B. D., Iseki, H., Miller, M. A. and Smart, M. (2007) 'Thinking outside the bus: understanding user perceptions of waiting and transferring in order to increase transit use', Final Report submitted by Institute of Transportation Studies, University of California to California Department of Transportation. Available online at: http://www.its.ucla.edu/research/EPIC/EPIC%20Final%20Report.pdf (accessed May 2009).

TRL (2004) 'The demand for public transport: a practical guide', Transport Research Laboratory Report, Crowthorne, Berkshire: TRL Limited.

Wardman, M. (2001) 'A review of British evidence on time and service quality valuations', *Transportation Research Part E – Logistics and Transportation Review* 37, 2–3: 107–128.

—— Hine, J. and Stradling, S. (2001) 'Interchange and travel choice', Report for the Scottish Executive, Scottish Executive Central Research Unit, Vol. 1. Available online at: http://www.scotland.gov.uk/Resource/Doc/156530/0042033.pdf (accessed May 2009).

Zamparini, L. and Reggiani, A. (2007) 'Meta-analysis and the value of travel time savings: a transatlantic perspective in passenger transport', *Networks and Spatial Economics* 7, 4: 377–396.

Chapter 15

Assessing intermodal re-balance and integration in urban transportation planning

An illustration on the basis of a sub-lagoon tube plan for Venice

Vincenzo Punzo, Vincenzo Torrieri, Maria Teresa Borzacchiello, Biagio Ciuffo and Peter Nijkamp

1. Introduction

The aim of this chapter is to present the results of a study carried out in the city of Venice in which a modal re-balance and intermodal integration have been experimented to increase the sustainability of the existing transportation system. It is clear that the pathway to a more sustainable transportation system has to take into account a strong reduction in private car transport, but it is not as clear how to make this feasible. A universal recipe still does not exist, even though several policies have been tested in the world. In a recent study (Garling and Schuitema, 2007), the authors analysed the impact of several travel demand management measures and concluded that the only way to make such measures effective is to combine together coercive and non-coercive ones (i.e. introducing obstacles to car usage but at the same time increasing the attractiveness of other transportation modes). In this light an important role in achieving a sustainable mobility is played by a mix of policies of intermodal integration and modal re-balance.

This idea of modal reconfiguration is also supported elsewhere. In particular, in Potter and Skinner (2000), an interesting perspective is proposed, that transport planning should follow a *holistic objective-led approach*. The basic idea is that an effective planning process has to take into account all the dimensions of travel choices: the choice of travelling, the choice of destination, the choice of mode, and finally the choice of a path. Only by regulating all these dimensions it is possible to create a well-organized system. This idea is not very easy to implement, above all because personal choices are very difficult to identify and influence. However, the idea of integrated planning is becoming fashionable and can be used on a case-by-case approach in order for it to be adapted in the best possible way.

Another interesting analysis that was carried out concerns the Greater Toronto Area (Kennedy, 2002) and calls for a new transportation system organization in which a mixture of modes have to be managed at once. Also, in other studies (see Musso *et al.*, 2008), the steps followed in the Italian cities of Rome and Napoli to improve the cities' environmental conditions and liveability, and the role played by the integration of all transportation modes of the city are described.

This chapter presents an integrated approach that has been adopted to design a feasible solution to the congestion problems in the city of Venice. As will be detailed below, the proposal for an integrated policy intervention in Venice is based on the construction of a sub-lagoon tube connection between Venice and the mainland. The analyses are focused on the impacts on the modal split of a new access infrastructure to Venice Island (i.e. the Historic Centre), and of the new intermodal terminal on the mainland access point, where the terminal, connected to the airport station, will also integrate the Regional Metropolitan Rail System and the highway system. The study represents an illustrative case of the possibility of incisively impacting on the integrated land use/transport system by means of sustainable mobility measures, aimed at modal re-balance and intermodal integration. Such a study is magnified by the peculiarity of Venice, which represents a unique laboratory for experimenting with sustainable mobility policies, due to its morphological, urban, social and land use characteristics.

Simulations carried out in the study allowed the identification of the most suitable layout of the sub-lagoon tube and the best localization of the Metro stations. The three dimensions of sustainability, namely the social, the economic and the environmental (Button and Nijkamp, 1997), are reflected in the main evaluation criteria used in the remainder of the chapter, which are respectively (i) the reduction of the total 'running' time; (ii) the increase of revenues for the areas impacted by the project; and (iii) the reduction of pollution due to a reduction in private car traffic flows. The outcomes are then used to evaluate the integration of the sub-lagoon tube within the wider transportation system of the area, performing Cost-Benefit Analysis (CBA) and multicriteria analysis (MCA) by means of the regime method of the scenarios.

2. The island of Venice: need for an integrated transport solution

The uniqueness of the city of Venice for experimenting with sustainable mobility policies is highlighted by different factors. First of all, the city of Venice is composed of 118 small islands connected by means of 354 bridges and divided by 177 channels. For these unique geographic features the transport supply of the city is, by nature, an intermodal supply. The access from the mainland to the Historic Centre (HC) is by lagoon boat services, or alternatively by the only available land connection, the Ponte della Libertà ('Freedom Bridge'), which accommodates rail and road transport, both for private use and public (transit) services. Once in the HC, the transport alternatives reduce to canal boat services and to pedestrian mobility. Hence, movements towards internal destinations of the HC may be realized only as trip chains involving at least two transport modes.

Second, the transportation demand pattern in Venice is peculiar as well. Very few inhabitants still live on the islands. Actually, the city is dominated by a lot of economic activities and touristic attractions. For this reason there exist two main demand categories for trip purposes: work and tourism. They have roughly the same weight but, of course, are very different in terms of both daily and seasonal trends and in the underlying travel behaviours and travel choices.

Moreover, the very high numbers of daily commuters who reach Venice from Venice-Mestre (the mainland) and the singular land connection to the HC leads to heavy congestion, which involves to a large extent the road network on the mainland. Even the pedestrian network of the HC is affected by significant congestion problems.

Finally, the environmental constraints in the area strongly influence the possibility of suitably managing and modifying the transportation system. In particular, the potential transport capacity of the bay and the various channels within the HC cannot be realized since boat movements must be limited in order to preserve the vulnerable microcosm of the lagoon and to limit damage to the ancient buildings caused by wave motion.

For all these reasons, to study the impact of an important infrastructural intervention, like the construction of a sub-lagoon Metro line, a detailed and careful scenario analysis is necessary. The methodology used in this study follows the typical path of a transportation study. In particular, after the analysis of the current transportation system and the definition of the scenarios of intervention, a simulation of the transportation system will be performed in the different scenarios, in order to generate the data for the evaluation phase, carried out by means of CBA and MCA.

2.1 The current transportation system

As described above, the transportation system which connects Venice to the mainland is composed of road infrastructure as well as rail infrastructure: the Ponte della Libertà, a group of parking areas for cars and buses concentrated at the end of this bridge, and a set of docks for lagoon boats. Apart from the lagoon boat services, all the connections to Venice are focused towards the Ponte della Libertà (Figure 15.1), with a heavy concentration of vehicular and pedestrian traffic on the extreme points of the system (bridges, station, harbour). Lagoon transportation services are rather slow (the maximum allowed speed in the lagoon is 6 knots), and not frequent; in particular, the system is influenced by the lack of a direct connection between the island and the airport.

Moreover, the current configuration of the transportation system has led to an excessive localization of activities next to the Ponte della Libertà, while further

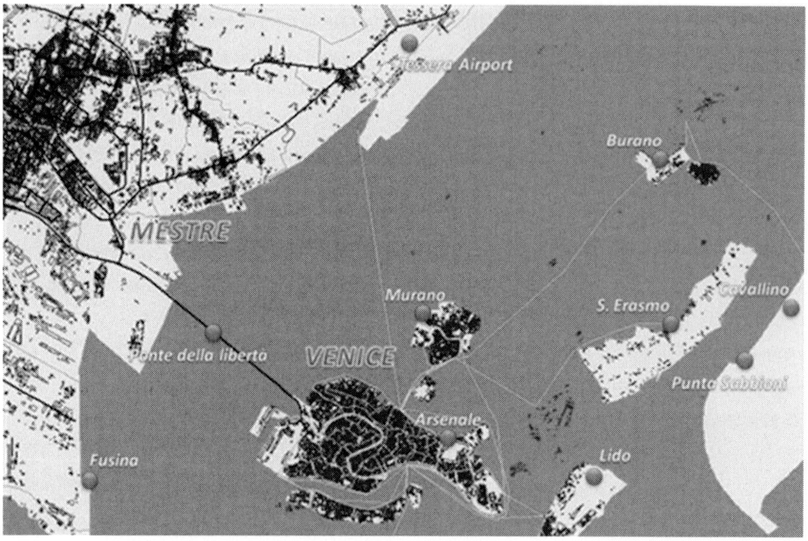

Figure 15.1 Current connections between Venice Island and the mainland

zones, in particular the area of Arsenale, have been progressively abandoned, and thus suffer nowadays from urban decay (Figure 15.1).

A lively debate about the urban recovery model of the HC of Venice has led to the formulation of the so-called PRUST (plan for the urban requalification and the sustainable development of the land use) for the lagoon area, which, among other things, requires a new high-frequency transport connection between Tessera Airport and the Arsenale area that is non-polluting and not too invasive, and has a twofold aim: first, to reduce congestion on the Ponte della Libertà and the surrounding areas; second, to contribute to the revitalization of the Arsenale area.

In order to evaluate the technical and financial feasibility of the initiative, the municipality and the Chamber of Commerce, Industry, Crafts and Agriculture (CCIAA) of Venice have promoted a set of planning studies, of a technical, environmental, financial and economic nature. In particular, a specific study (Torrieri *et al.*, 2002) has been conducted on the system of connections between Venice island and the mainland in order to forecast traffic volumes on the new links, and to evaluate the land use benefits together with rate of return on the investment.

2.2 The planned scenarios

Currently the connections of the HC (Figure 15.1) with the mainland (Mestre, and the coast of Cavallino) and with the other islands of the lagoon (Murano, Lido) are assured by the following transportation services (Figure 15.2):

- national and regional rail services, based at the station of S. Lucia;
- tourist bus services, based in Tronchetto, a small island connected to the Ponte della Libertà;
- public road transport services, based in Piazzale Roma;
- private transport (cars and motorcycles) with parking areas located next to Piazzale Roma and the island of Tronchetto;

- lagoon boat services, located at the southern side (Fusina) and northern side (Tessera) of Mestre; and at the eastern side of the lagoon (Punta Sabbioni-Lido).

The planned sub-lagoon tube described below is included in a more comprehensive programme to strengthen the transportation system of the area.

In this framework, the realization of the sub-lagoon tube, with its new terminal point in Tessera, aims at strengthening public transport in the Venice area by offering an alternative for both current individual users of private transport and tourists who reach the Venice Island by bus, thus providing a real alternative to the terminal of Piazzale Roma.

In particular, the project includes the construction of an underground Metro line in the lagoon, which will support the transport of vehicles with an average capacity of 150 to 200 seats and will be comprised of three lines, as described in Figure 15.3:

- Line 1 (L1): with two stations on the mainland (Tessera Airport and Terminal Tessera), a stop on Murano island and three stops on Venice Island (Cannaregio, Fondamenta Nuove, Arsenale);
- Line 2 (L2): continuing from Line 1 towards the Island of Lido with stops in Arsenale, Giardini di S Elena, S.M. Elisabetta and Nicelli Airport (Lido);
- Line 3 (L3): adding the following stops to Line 2: S. Lucia, Dorsoduro and Fusina.

Plans for such an ambitious underground system for integrated transport require a systematic assessment and evaluation. We will present here now the various possibilities in the form of a policy scenario analysis. For the evaluation of these scenarios the following configurations have been considered:

Figure 15.2 Location of the main transportation services

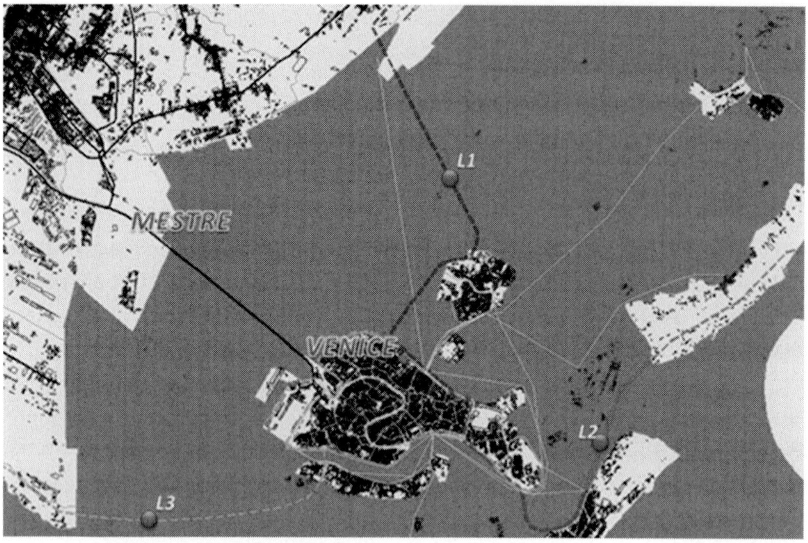

Figure 15.3 Scheme of the designed lines of the sub-lagoon project

- *Scenario 0*: the system in the current situation is simulated and integrated with the works in progress on the existing transport infrastructures (Regional Metropolitan Service, highway link of Mestre, Tram line).
- *Scenario 1 (L1)*: the designed system is simulated, with the construction of line L1: Tessera-Murano-Arsenale, and the parking facility of Tessera; the start year for the running of this service is set at T1 = 6 years from commencement.
- *Scenario 2 (L1 + L2)*: includes the addition of a new stretch (line L2): Arsenale-Nicelli Airport (Lido); the start year of the service is set at T2 = T1 + 4.
- *Scenario 3 (L1 + L2 + L3)*: the construction of line L3, Arsenale-S. Lucia-Dorsoduro-Fusina, is modelled and planned; the start year of the service is set at T3 = T1 + 14.

It is worth noting that the scenarios analysed are to be considered integrative and sequential.

3. The project appraisal

In order to evaluate the different scenarios considered in the study, two approaches were adopted: CBA (which includes a financial analysis and an economic analysis) and MCA.

While CBA was explicitly required by Venetian public authorities – first to investigate the project financial feasibility, and then to justify the investment by means of the derived socio-economic benefits, MCA will be performed here to compare the results. In the following sections various details on both approaches are provided (for a detailed description, see Chapter 7 in this volume).

3.1 Cost-benefit analysis (CBA)

CBA is an appraisal technique based upon the assumption that all relevant costs and benefits of an investment can be valued in terms of money. Usually the project evaluation is made against a reference scenario (i.e. the 'do nothing' scenario). Once evaluated, costs (with negative signs) and benefits (with positive signs) occurring during all the years of the investment are summed, actualized and compared by means of different indicators in which a discount rate, depending on the general economic conditions, is included. In the CBA analyses described below the indicator used is the net present value (NPV), which, in this context, represents 'the equivalent value, in year 0, of the time stream of annual project costs and benefits' (Cascetta, 2009). If NPV is positive, then the project is worth doing; when comparing two or more projects, the most preferable project is the one with the higher NPV.

While in the financial CBA the benefits are the revenues of the investment in terms of cash inflows and the costs are represented by the investment costs (in terms of construction, service production, maintenance, and so on), in the economic CBA, which has more social meaning, costs and benefits are considered in terms of direct and indirect impacts on society and environment, alongside the financial balance. The impacts on which the present study will focus are assessed in terms of reduction of the total travel time, increase in income, and reduction in air pollution.

3.2 Multicriteria analysis (MCA)

MCA has over the past decades been applied to a wide array of evaluation problems; for example, in the fields of environmental management, energy policy, industrial planning, land use planning, public facilities management, and so forth. It has also found a wide application in the field of transportation planning.

MCA is an established decision support tool in decision and choice problems that are characterized by (i) a set of distinct choice alternatives; (ii) a set of mutually incompatible (conflicting) decision criteria; (iii) the presence of different priorities (weights) attached to the various decision criteria; and (iv) the need to identify 'the best possible' choice alternative, compatible with the context needs. The information necessary to perform a MCA is normally included in a so-called effect (impact) matrix which contains the expected outcomes of each alternative to be considered for all relevant choice criteria. In addition, it is necessary to have insight into the decision makers' preferences for each of the decision criteria (see Munda, 2008).

A wide variety of software packages have been developed to offer an operational decision support system to planners and decision makers. In the following section, the hierarchical regime method (Hinloopen and Nijkamp, 1990) will be applied to evaluate the scenarios discussed so far.

The hierarchical regime analysis is able to deal with either qualitative or quantitative plan or project effects. In our scenarios, the cost-benefit indicators are numerical in nature, but as there are some missing data, we have preferred to transform such cardinal information into rank order information, in order to consistently compare the outcomes.

4. The model for an integrated transportation system

4.1 The study area and its traffic zones

The study area includes the entire territory of the province of Venice. In order to use data already available from other studies, the same zoning adopted by the Venice Province and municipality has been maintained. As a result, the activities are supposed to be localized in 398 zones, whose dimensions are bigger as the distance from the Venice HC increases.

4.2 The transportation supply model

The transportation network of the Province of Venice is clearly multimodal. Despite the difficulty of modelling all transportation modes together, it was vital to achieve a realistic estimation of the effects of the new transportation system. For this reason the supply model used here is composed of the following infrastructures and services:

- *The private transport network (road network, parking areas).* This comprises the main road infrastructures of the Province of Venice (obviously in the HC there are no roads). For each infrastructure the main physical (length, lane number, etc.) and service (capacity, speed, delays, etc. in both peak and off-peak hours) features have been defined.
- *The transit road network (buses).* A specific local transit network does not exist, but was assumed to do so. This was necessary since the future transit lines' layout will be different from the existing one, being influenced by the new sub-lagoon system. In this light, simulation results can be used to design the most attractive future transit lines' layout.
- *The pedestrian network.* The pedestrian network was only modelled for the HC of Venice. Even if not realistic (considering the large number of people that daily walk on the narrow streets of the city), the pedestrian system was considered to be uncongested for the sake of simplicity and this assumption also holds for the existing pedestrian flow models.
- *The lagoon boat services network.* All the routes covered by the official boats (those of the ACTV agency) moving in the lagoon were considered. The main service features (frequency, speed, stop delays, and so on) were also taken into account.

Notice that the current *railway network* has not been considered in the simulations, due to the scarce traffic demand that can be potentially attracted (the railway system is mainly used by commuters and visitors reaching Venice directly from outer parts of the province or even from other provinces). On the other hand, the pedestrian network has been included, because the traffic in the future scenarios can be significantly influenced by the connections between the sub-lagoon stations and the pedestrian roads on the island.

In our future scenarios, the network model is integrated with the *sub-lagoon tube services*. The specification of the new transportation services offered by the tube system depends on the particular technological choice. However, the system performances were hypothesized to be equal on average to those of other similar light rail transit systems. For further details on the transportation network characteristics and the systems model, see Torrieri (2007).

All network models have been represented by means of a geo-referenced oriented graph.

4.3 The demand model

On the basis of detailed information from previous studies and surveys, the mobility demand among lagoon islands and the mainland, as well as its temporal, spatial and motivational characteristics are known. Also the modal split among the existing transportation services is known.

It is worth noting that the present study uses data from several years. By means of the data available from various sources, it was possible to estimate the complete O-D matrix for the Venice Province, as well as to retrieve the temporal, spatial and other characteristics of the traffic demand in the area. In the following figures and tables, the main results are summarized. In Table 15.1, the matrix dimensions and the aggregated daily volumes are reported. These data are shown here to give the reader an idea of the entire traffic pattern in the region.

In Figure 15.4 the spatial characteristics of the traffic demand in the HC are shown. In particular, the traffic zones attracting the highest demand in the HC appear to be S. Marco, S. Croce, Dorsoduro, Castello and S. Paolo. Among these, S. Marco and Castello particularly suffer from the distance to the only current access point of the city and these zones are supposed to be, together with S. Paolo, the traffic zones that will benefit most from the introduction of the new transit system.

The variability in demand due to the seasonal cycle is shown in Figure 15.5. On an average day (there are about 310 'average' days per year in Venice), average

Table 15.1 Pedestrian and vehicular daily average volumes directed to the lagoon islands and passing through the Ponte della Libertà (2002)

Demand volumes towards the Historical Centre (HC) of Venice	
Access mode to the HC	*Daily volume (pax/day)*
Private road transport	19,855
Road transit	37,559
Rail transit	23,681
Tourist buses	12,443
Lagoon navigation Lido-P.ta Sabbioni	12,728
Lagoon navigation from Tessera	600
Total daily volume of passengers entering HC	106,866
Demand volumes in the study area	
Access mode to the HC	*Daily volume (veh/day)*
Private road transport	980,657 light vehicles 106,581 heavy vehicles

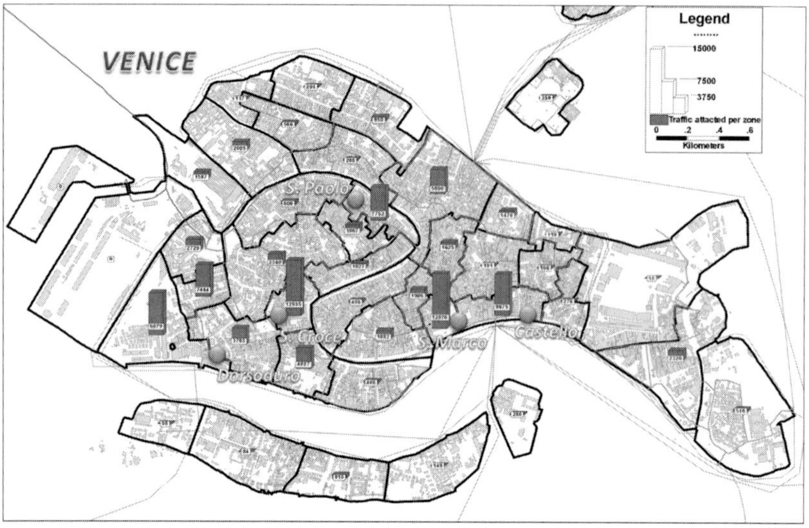

Figure 15.4 Mobility distribution subdivided by destination in the lagoon

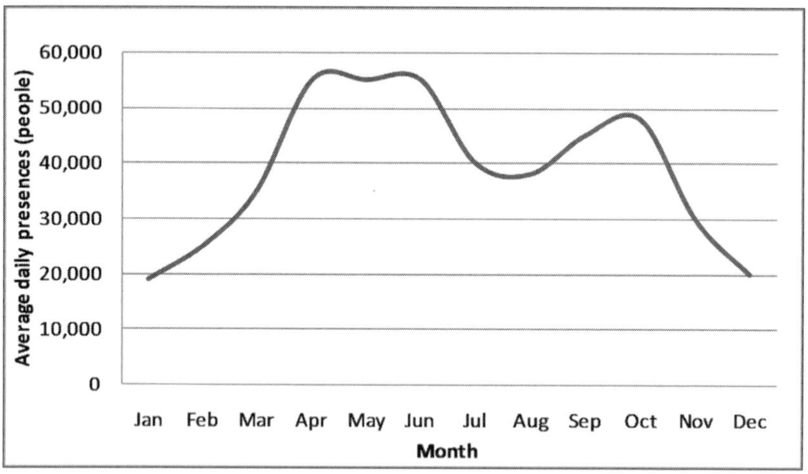

Figure 15.5 Yearly mobility trend (Venice – HC 2000–01)

annual daily traffic amounts to 16,181 vehicles per day. This volume is divided between two periods: the peak period lasting 6 hours, in which about 70 per cent of the daily average traffic is registered, and the off-peak period, lasting 8 hours, in which about 30 per cent of the daily average traffic is registered. The remaining 10 hours, occurring during the night, are not considered in the analysis, as these are not very relevant for the purpose of the study.

The other characteristics concern almost entirely the drivers of commuting trips (or otherwise trips by local residents) and the trips of tourists visiting the region. For both groups the demand elasticity with respect to the supplied transportation network performances (in particular those related to the travel time) can be assumed to be about nil.

5. The traffic forecasting model

5.1 Expected growth

On the basis of surveyed historical trends and existing previous studies about the mobility patterns in the study area, the trend curve has been estimated by means of a logistic model. This indicates that the growth in 20 years is approximately equal to 15 per cent of the registered volume of 2003.

5.2 Simulation experiments

Once the set-up of the transportation supply and demand model were operationalized, the different scenarios described above were simulated using TransCAD commercial software (Caliper Corporation, 2005). For each scenario, two simulations have been performed; one for the peak period and one for the off-peak period.

The traffic implications of the new transportation system were assessed by comparing the results of the different scenarios. Below, several interesting outcomes are reported. In Table 15.2 the percentage of passengers transferring from the existing transport modes to the sub-lagoon tube in Scenario 1 is highlighted. The distribution of the traffic volume attracted for each traffic zone per different access mode, as simulated for Scenario 1, shows, as expected, that as the distance of the zone from the Ponte della Libertà increases, so does the percentage of passengers using the sub-lagoon tube connection.

Next, in Table 15.3 the estimated modal split of passenger transport towards the HC in all three future scenarios is reported. It is evident from these results that in Scenario 2 the sub-lagoon system is already the most used access mode to the HC of Venice.

The estimated data also presents the possibility of designing the supplied service capacity in the peak hour period, proportional to the maximum expected value in the 20-years projection (in particular, the maximum estimated value is about 2,326 passengers per hour per direction).

Figure 15.6 further clarifies that in Scenario 3 the integration of all transportation modes and the implementation of the three lines of the sub-lagoon transit system allows the realization of a balanced distribution of the traffic among the available transportation modes. In particular, it can be seen that only a few places

Table 15.2 Percentage of passengers transferring from the existing transport modes to the sub-lagoon tube in Scenario 1

Sub-lagoon tube traffic volume	*Scenario 1*
From tourist buses	5%
From cars and motorcycles	7%
From transit system	4%
From lagoon boat services from P.ta Sabbioni to Lido	–
From lagoon boat services Tessera	1%
Total sub-lagoon tube (L1)	17%

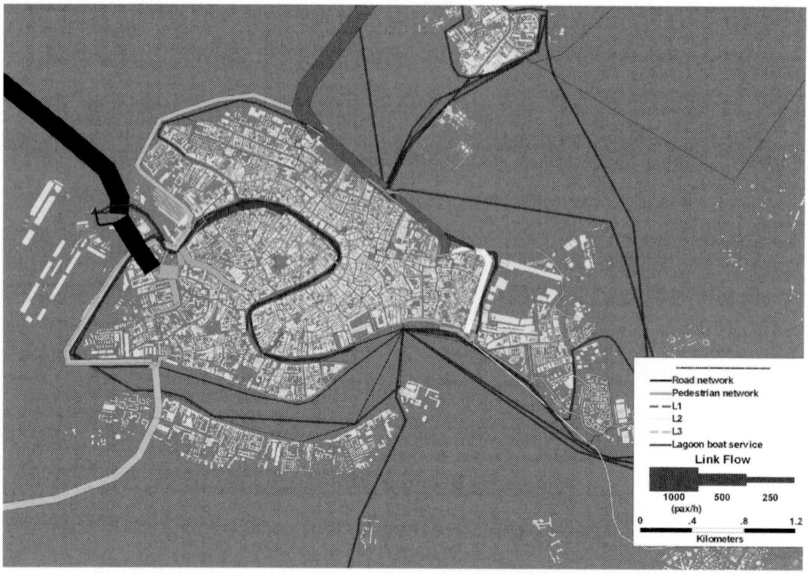

Figure 15.6 Traffic flows entering the historic centre of Venice with the sub-lagoon lines L1, L2 and L3

are a long-distance walk from the (Metro and boat) transport networks. Moreover, the zones of the islands closer to the Arsenale now appear to be very close to the airport and this will surely influence their future development (McDonald and Osuji, 1995).

Regarding the expected impacts of the project, Figure 15.7 shows how the travel time distribution between the HC zones (by foot) changes in Scenario 1 to significantly lower values. Moreover, the resulting average reduction in travel time for each trip in Scenario 3 is 21.4 minutes and 14.0 minutes for commuting/local trips by car and transit respectively. The reduction for tourist trips is 25.4 minutes, an

Table 15.3 Modal split of passenger transport towards the Historic Centre

Projected start years for each scenario	*T0*	*T1*	*T2*	*T3*
	0	*6*	*10*	*20*
Total sub-lagoon tube	–	17%	29%	43%
Cars and motorcycles (P.le Roma and Tronchetto)	19%	12%	12%	8%
Tourist buses (P.le Roma and Tronchetto)	12%	7%	7%	0%
Extra-urban road transit (P.le Roma and Tronchetto)	21%	21%	21%	21%
Urban road transit (P.le Roma and Tronchetto)	13%	9%	9%	0%
Tram (P.le Roma and Tronchetto)	–	–	–	6%
Lagoon transport services from P.ta Sabbioni to Lido	12%	12%	0%	0%
Lagoon transport services (Tessera)	1%	0%	0%	0%
Rail transport	22%	22%	22%	22%
Total	100%	100%	100%	100%

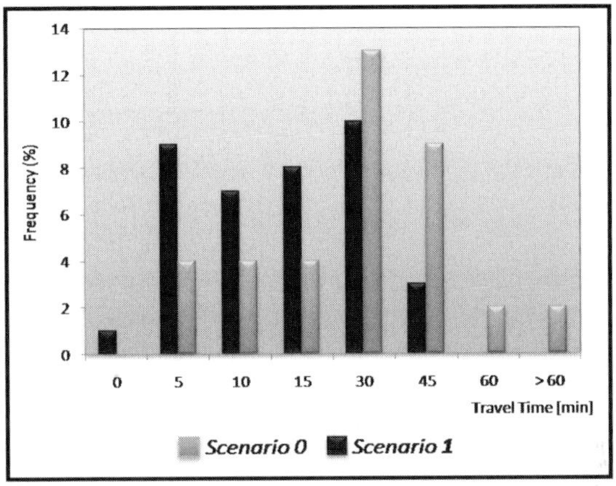

Figure 15.7 Pedestrian travel time distribution within the city centre, with and without the sub-lagoon line L1

unexpected result which shows how it is possible to positively change the existing conditions by better harmonizing the transportation system.

6. The Venice sub-lagoon tube appraisal

The transportation study results were the basis for the subsequent appraisal phase. In fact, the social benefits of the project, in terms of travel time reduction, can be directly derived from these findings. The environmental benefits are estimated from the reduction of pollutant emissions which are connected with the changes in the traffic flows in the area, while the economic benefits, due to the new activities which will result from the new infrastructure, can only be estimated once the mobility patterns are known.

The first concern of the public administration was to check the financial feasibility of the study, namely to what extent the income generated by the potential traffic attracted by the new infrastructure would have justified the cost of its implementation. As mentioned, CBA was chosen for this purpose.

The revenues from the new transportation system have been evaluated on the basis of the following assumptions on fares: €2/trip for a commuter/local passenger and €6/trip for occasional users (e.g. tourists). The maintenance costs were estimated at €0.033/seat-km. The financial analysis for Scenario 3 based on these assumptions results in a balance sheet of €29 million/year revenues against €5 million/year management costs (all estimates in NPV) and total investments of €330 million.

Using a 5 per cent interest rate (due to the economic context at the time of the study), the NPV indicator appears to be negative in all scenarios, so that from a financial perspective the investment appeared to not be justified, at least not without any public investment. However, with about 30 per cent public finance the NPV becomes positive and the internal rate of return (IRR) takes a value of about 7 per cent for Scenario 3.

This situation led the public administration to request an economic analysis in order to understand if the use of public money would be justified. As already pointed out, the main benefits considered in the analysis are:

- a reduction of the total travel on journeys to Venice Island;
- an increase in income for the areas affected by the project;
- a reduction in pollution due to lower private car traffic flows.

For this purpose CBA was used. Considering a value of time equal to €0.065/min, i.e. a value of around €4/hour; and between 7,000 and 8,000 Italian Lira, which was commonly used in transportation studies before the introduction of the new currency in Europe[1], and assuming a zero economic value to the saved time by occasional users, then the economic value of the total time savings is at least €10.2 million/year.

The environmental benefit is tied to the reduction of pollution and greenhouse gases. In particular, they derive from the reduction of private transport and from the increase of the average speed on the network (thus reducing stop-and-go traffic conditions that heavily affect the emissions). In order to quantitatively estimate pollutant and greenhouse gases emissions, the Tartaglia model has been applied (Tartaglia, 1999). The estimation of the total emission reduction has led to their economic evaluation. The methodology used has followed the indicators proposed in Freeman (1993), leading to an economic value ranging between €50 million and €100 million/year for the three scenarios.

The increase in income from the proposed sub-lagoon tube has been evaluated. This is the most uncertain element in the evaluation, but its inclusion is crucial in order to understand the total benefits carried by the investment. For this purpose it was considered that the areas of the HC of Venice connected by the sub-lagoon system would have the same density of activities as the areas that are currently more accessible from the mainland, which could lead to the establishment of about 100 new firms in the HC and 30 new firms in Murano in Scenario 3. This leads to an added value of about €34.4 million in 20 years. Table 15.4 shows the NPV of benefits which are expected as a result of the sub-lagoon tube. Assuming a 5 per cent interest rate, the NPV is positive and the IRR is estimated to be around 14 per cent for Scenario 3.

These results led the public administration to recognize the public benefits of the project and to promote its construction, based on 30 per cent public finance. However, the project never started, probably because the results of the economic CBA are not always easily understood by the stakeholders concerned. In this case, the use of MCA in which stakeholders are involved in giving weights to different criteria/effects of the project would have likely been more useful. For this

Table 15.4 Estimated net present value of the direct and indirect benefits generated by the sub-lagoon tube in 20 years

Category	*Actual benefits*	*%*
Time	€138 million	21%
Income impacts	€463 million	69%
Pollution reduction	€67 million	10%
Total	€669 million	100%

reason, a basic application of a regime MCA is considered, using as an input to the analysis the impacts estimated in the economic CBA. Since it was not possible to undertake surveys in order to obtain stakeholders' weights for the different criteria, equal weights were assumed.

The total effect matrix related to our project assessment model is presented in Table 15.5, where all entries have a positive meaning (i.e. the higher, the better).

Our hierarchical MCA using the regime method was implemented in a stepwise (hierarchical) way by grouping together first the effects regarding total kilometres (effects 2, 3 and 4); then the effects regarding the total travel time (5, 6, 7 and 8); then the effects related to congestion (9, 10 and 11); and then effects regarding traffic volume (12 and 13). Thus, a regime analysis was applied to each of these sub-sets. The analysis was carried out without assigning weights to the effects. Next, the overall regime method was applied to the remaining aggregate effect matrix. The outcome of the MCA clearly shows that Scenario 3 is the preferable option with an index of 1, compared to an index of 0.667 and 0.333 for Scenarios 2

Table 15.5 Effect table for regime analysis

No.	*Effects*	*Scenario*			
		0	*1*	*2*	*3*
1	Average network speed	1	3	3	4
2	Total kilometres made by cars directed towards HC	1	3	3	4
3	Total kilometres made by tourist buses directed towards HC	1	3	3	4
4	Total kilometres made by heavy vehicles directed towards HC	1	3	3	4
5	Total travel time – cars	1	3	3	4
6	Total travel time – tourist buses	1	3	3	4
7	Total travel time – lagoon boats – Tessera	1	2	3	4
8	Total travel time – lagoon boats – Lido	1	1	3	4
9	Average value of congestion (VOC) (peak hour)	1	1	1	1
10	Percentage of saturated links	1	3	3	4
11	Total vehicle kilometres travelled (VKT) – peak hour	1	3	3	4
12	Percentage of passengers reaching HC with the sub-lagoon tube	1	2	3	4
13	Percentage of traffic volume using sub-lagoon tube	1	2	3	4
14	Investment cost	4	3	2	1
15	Time savings	1	3	3	4
16	Income impacts	1	2	3	4
17	Pollution reduction	1	3	3	4

and 1 respectively (the index for the 'do nothing' scenario is 0). This result is in line with previous findings reported above.

7. Conclusions

The main outcomes of a feasibility study regarding an integrated new transportation infrastructure for Venice have been presented, with particular attention paid to the current and future organization of the intermodal transportation system of the area. After a brief overview in which the integration of different transport modes was considered as a possible significant contribution to the challenge of sustainability, the case of Venice has been presented as an illustrative case, even though it is full of peculiarities due to the physical morphology, the demand pattern and the multimodal structure of the study area.

The case study presented here, although still lacking a comprehensive methodology for the impact evaluation, is interesting for different reasons. First, it has brought to light some of the usual contradictions in the decision-making process. Supported by earlier decisions stated in official urban plans, the project has been the object of a complex and broad feasibility study, requiring research efforts from several panels of experts in the field of transportation and economic evaluation, as well as the use of time and financial resources. Such studies, as explained in our study, have predicted the success of the application of a multimodal approach from different points of view. Nevertheless, despite the conclusions of these studies and especially Torrieri *et al.* (2002), the project is still not ready for implementation, due to the difficulties encountered in the public debate with several private and public stakeholders involved, as well as due to lack of both private and public funds.

Second, the abovementioned general urban plan, promoting multimodal and integrated approaches to the transportation system management, has allowed and encouraged the fulfilment of an in-depth analysis whose main outcomes, presented in this chapter, have highlighted the necessity of the planned approach in order to solve effectively the problems of congestion, pollution and progressive urban decay of some areas of Venice.

Third, the study confirms the need for a clear future perspective for the area concerned, in order to assess the feasibility as well as the sustainability of the foreseen investments. Indeed, looking only at the financial indicators, there is no evidence of significant and sufficient financial benefits: the project is worth realizing only if we consider the economic translation of social and environmental benefits. This emphasizes, moreover, the importance of developing more reliable and precise methods for the evaluation of the economic impact of new transportation infrastructures.

It is evident that further steps are still needed to complete the study, including the use of fine-tuned indicators to perform a solid and comparative methodology for the evaluation of the impacts of a new public transport investment plan.

Note

1 It is interesting to notice that nowadays the time is roughly estimated to have a value of €7–€8 (e.g. see HEATCO, 2007) which corresponds to the late-1990s cost of 7,000–8,000 Italian Lira (i.e. the cost has doubled since then).

References

Button, K.J. and Nijkamp, P. (1997) Social change and sustainable transport. *Journal of Transport Geography*, vol. 5, pp. 215–218.

Caliper Corporation (2005) TransCad Overview. Available online at: http://www.caliper.com/tcovu.htm (accessed 4 February 2010).

Cascetta, E. (2009) *Transportation Systems Analysis: Models and Applications*, 2nd edn. Springer: New York.

Freeman, M. (1993) *The Measurement of Environmental and Resource Values: Theory and Methods*. Resources for the Future Press: Washington, DC.

Garling, T. and Schuitema, G. (2007) Travel demand management targeting reduced private car use: effectiveness, public acceptability and political feasibility. *Journal of Social Issues*, vol. 63, no. 1, pp. 139–153.

HEATCO (2007) Developing harmonised european approaches for transport costing and project assessment. Available online at: http://heatco.ier.uni-stuttgart.de/ (accessed 4 February 2010).

Hinloopen, E. and Nijkamp, P. (1990). Qualitative multiple criteria choice analysis: the dominant regime method. *Quality and Quantity*, vol. 24, pp. 37–56.

Kennedy, C.A. (2002) A comparison of the sustainability of public and private transportation systems: study of the Greater Toronto Area. *Transportation*, vol. 29, pp. 459–493.

McDonald, J.F. and Osuji, C.I. (1995) The effect of anticipated transportation improvement on residential land values. *Regional Science and Urban Economics*, vol. 25, no. 3, pp. 261–278.

Munda, G. (2008) *Social Multi-criteria Evaluation for a Sustainable Economy*, Springer-Verlag, Berlin.

Musso, A., Vuchic, V.R. and Corazza, M.V. (2008) Toward an intermodal balanced transport system in Rome. In *Proceedings of the 87th TRB Annual Meeting*. Washington, DC.

Potter, S. and Skinner, M.J. (2000) On transport integration: a contribution to better understanding. *Futures*, vol. 32, pp. 275–287.

Tartaglia, M. (1999) *L'inquinamento dell'aria da traffico stradale*. BIOS, Consenza, Italy.

Torrieri, V. (2007) *Tecnica ed Economia dei Trasporti – Manuale introduttivo all'ingegneria dei trasporti*. Edizioni Scientifiche Italiane.

Torrieri, V., Punzo, V. and Sica, M. (2002) Progetto preliminare Filovenezia, studio trasportistico di sistema. Unpublished.

Chapter 16

Weather and travel time of public transport trips

An empirical study for the Netherlands

Muhammad Sabir, Jos van Ommeren, Mark Koetse and Piet Rietveld

1. Introduction

The main modes of transportation in the Netherlands are car, bicycle and walking. They cover about 90 per cent of all trips. About 50 per cent of trips are made by car, and the main alternative for the car is the bicycle, with a share of about 25 per cent. Distance appears to be an important moderator, as people prefer to not use the bicycle for longer distances. Three-quarters of all bicycle journeys to work are less than 5 kilometres, one-fifth are between 5 and 10 kilometres, and 5 per cent are longer than 10 kilometres (Statistics Netherlands, 2008). Longer commuting trips are mostly made either by car or by public transportation. Since using the car is not always a viable option (e.g. because no driving licence is available or because of parking restrictions, especially in the Randstad region[1]), public transport in the Netherlands is frequently a good alternative for trips over longer distances. Public transport will be classified in two main categories: trips made by bus/tram/metro on the one hand, and trips made by train on the other.[2]

Public transport in the Netherlands has about an 8 per cent share of trips. When travelling by public transport, an individual has to go to an access point (bus stop, train station, etc.). The more time spent getting to an access point, the larger the total time required to reach the final destination. Transferring between trains or buses during a trip has a similar effect. Furthermore, waiting time, delays and adverse weather may also influence the speed of a trip and hence total travel time. Thus, the need for and importance of integrating the different segments of the journey is clear.

Compared to car trips, the speed of public transport is often a weak element of multimodal transport chains. In the large majority of the cases, car trips are considerably faster than public transport trips, and it is therefore not surprising that the modal share of the car is much higher than that of public transport. There are two exceptions to this, leading to market segments where public transport tends to perform relatively well: in congested areas; and for long-distance trips where the high speed of trains is a large advantage. The first sub-market demonstrates that push factors (congestion on the road, parking problems in cities) are an important factor contributing to the success of public transport. In all cases it is clear that a key success factor for integrated public transport is that it achieves speeds that are competitive to those offered by car transport.

A related consideration is that reliability of transport services is an important determinant of the quality of multimodal trips: when delays occur in part of a trip, travellers may miss their connection, leading to extra waiting time at transfer points and possibly high scheduling costs (when one is late for an important appointment – see Chapter 6 in this volume on the value of reliability and its relevance in transport networks). Thus variations in speed in a certain part of a trip chain may lead to substantial extra waiting or travel time in the rest of the trip.

The potential success of integrated transport therefore depends considerably on the degree of its reliability. To make integrated public transport successful it is therefore important that transport organizations can cope with these uncertainties in an adequate way (i.e. so that the effects for the travellers are minimized). An example of such policies would be that timetables are made in such a way that there is some slack at transfer points. This may make the average speed slightly slower, but it would reduce the negative impact of disturbances (Rietveld *et al.*, 2001). Another element would be a high level of integration between services at the operational level, for example, the bus driver waits a few minutes when it is known that the train is some minutes late.

This underlines the importance of various 'soft' dimensions to achieve integrated public transport of high quality. One is the dimension of human resource management: drivers should be motivated to serve passengers properly. Timetables should be obeyed where possible, but in case of disturbances flexibility is needed. Another important aspect of integrated public transport is the institutional dimension: when one integrated company is responsible for the overall quality of a multimodal chain this creates favourable conditions for high quality, reliability and flexibility. On the other hand, with one integrated (bus-rail) company the potential benefits of competition may be lost. This leads to challenging questions about how to combine the better elements of both worlds.

In the present chapter we carry out an analysis of speed fluctuations as a determinant of the quality of public transport. We do this by focusing on a special cause of unreliability: variations in weather conditions. We find that this is a relatively under-explored theme of research. We will focus on commuting trips. This is an important part of the trips made by public transport. The advantage of focusing on commuting is that weather variations will most probably not affect the decision of travellers to stay at home, which would lead to selectivity effects. This makes it easier to analyse commuting trips compared to, for example, recreational trips. Please note that by travel time of a trip we mean travel time of a complete door-to-door trip; this not only includes the in-vehicle time but also includes time spent on access and/or egress modes, waiting times and delays.

During the past decades comparatively little attention has been paid to the effects of weather on transportation in general, and on public transport in particular. An overview of the empirical literature on weather and transport can be found in Koetse and Rietveld (2009). Most of the available empirical studies on weather and transport focus on road transport and report a reduction in speed during adverse weather conditions (see e.g. Martin *et al.*, 2000; Hranac *et al.*, 2006; Maze *et al.*, 2006). The major reduction in speed of road transport is due to precipitation and snow. Martin *et al.* (2000) report a 10 per cent speed reduction in wet conditions and a 25 per cent reduction in slushy and wet conditions. These results are confirmed by Hranac *et al.* (2006) using detailed traffic and weather data from 2002 to 2004 for the Baltimore, Seattle and Minneapolis-St. Paul metropolitan areas. They find that light rain reduces the free flow speed by 3 per cent and the speed at full capacity by 9 per cent. Reduction in speed increases with rain intensity, with maximum reductions of around 6–9 per cent for free flow speed and 8–14 for speed at capacity (see also Ibrahim and Hall, 1994; Hall and Barrow, 1988; Maze *et al.*, 2006). Finally, Sabir *et al.* (2008) report negative but small effects of adverse weather on the speed of car commuter trips. However, the effect of snow is substantial, with speed reductions of around 7 per cent. Furthermore, although the effects of rain on speed are small in general, rain causes a speed reduction of 10–15 per cent for trips made during rush hours on congested routes.

Interestingly, these studies mainly focus on road transport (but see Hranac *et al.*, 2006, for an exception). Public transport is largely ignored. Whereas trip speed reductions for car transport are mainly caused by congestion, public transport delays are also and perhaps mainly caused by technical failures. Therefore, the current study contributes to the limited available empirical evidence by providing a closer insight into the effects of weather conditions on the speed of public transport commuting trips and the welfare effects associated with the changes in travel time.

The rest of the chapter is organized as follows. Section 2 discusses the data as well as the descriptive statistics of the variables. Section 3 explains the econometric methodology used to analyse the effects of weather and individual characteristics on the speed of commuting trips made by public transport. Section 3 also discusses the explanatory variables included in the model. Section 4 provides the empirical results and discusses the welfare effects of weather conditions for the Netherlands. Section 5 concludes.

2. Data

The data used in this paper are taken from two main sources. We make use of a transportation survey (MON) for the years 2004 and 2005. The MON transportation survey contains information about 130,000 persons who reported the trips they made on one particular day during these two years, leading to about 450,000 reported trips.[3] The data also contain information about important individual and household characteristics. The second data source is a weather database of the Royal Netherlands Meteorological Institute (KNMI) for the years 2004 and 2005.

It contains weather conditions on an hourly basis recorded by 32 weather stations spread over the entire Netherlands. We use the weather conditions from the weather stations which are nearest to places of departure (in almost all cases). The average distance to a weather station is about 12 to 13 kilometres, which means that our measurement of weather conditions is fairly local.[4] The weather conditions refer to temperature, wind speed, visibility, rain and snow. By combining these two data sources we are able to analyse for each trip the local weather conditions of the hour in which the trip took place.

We select only commuting trips for a number of economic and statistical reasons. First, because the demand for commuting is derived from the demand for workers, which does not directly depend on weather, while the derived demand for other trips (in particular, leisure trips) is affected by weather conditions. Hence, for commuting trips, the interpretation of the welfare effect of weather is more straightforward. Second, commuters are a relatively homogeneous group of travellers for which assumptions on the value of time are likely to be more accurate. Third, we select public transport trips because for other trips, in particular bicycle trips, the welfare of commuting is directly affected by weather (e.g. it is unpleasant to take the bicycle on a rainy day because one gets wet, and not so much because of reductions in speed). Fourth, the selection of a sample may generate biased estimates of the coefficients of variables (Wooldridge, 2003). Fifth, adverse weather may increase the risk of travel speed for car use due to accidents, but this is less likely for public transport users. Finally, commuters generally take at least two trips per day, and panel estimation techniques can be employed to deal with this issue.

Given these selections, our sample contains 13,618 public transport trips made by 2,225 commuters. Average trip distance is 32.9 km, average trip speed is 31.7 km/hour, and average travel time per trip is 57.5 minutes. The descriptive statistics for bus/tram/metro, and train are given along with other explanatory variables in Appendix A.

It is important to realize that the speed, as we measure it, is based on the travel time of the *whole* trip rather than only the in-vehicle travel time. This implies that this travel time also includes the time to reach the access point, waiting time, in-vehicle time, and time to get to the final destination from the arrival. The average in-vehicle travel time for public transport is 24.6 minutes. Average in-vehicle travel time for bus/tram/metro is 21.6 minutes, while the average in-vehicle travel time for train is 30 minutes. It appears that the share of in-vehicle time in the total trip time of bus/tram/metro trips and train trips is about 50 per cent and 45 per cent, respectively. This implies that public transport travellers spend a significant part of their total travel time on access/eggress modes or on waiting time. Additionally, this also explains why the average speed of public transport trips is low.

3. Model specification and estimation procedure

Similar to Sabir *et al.* (2008) the interpretation of our empirical analysis is based on standard micro-economic theory, such as that used in Van Ommeren and Dargay (2006), who derived a structural model for commuting speed and then used that model for Great Britain, as well as Fosgerau (2005), who applied it to Denmark.[5]

Van Ommeren and Dargay (2006) assume that commuters optimally chose their speed given a specified cost function (the only restriction is that the cost function is a power function of speed) and the travel time costs are proportional to the wage. Furthermore, they show that the marginal effect of an exogenous environmental characteristic, such as weather, on the logarithm of speed can be interpreted as the marginal effect of this characteristic on the logarithm of the commuter's total commuting costs (the sum of the travel time and other costs which vary with speed, such as accident costs and fuel costs). Given an estimate of average worker value of time, it is meaningful to estimate the effect on the welfare of commuters through a loss in travel time only. We will use a log specification in line with the theoretical considerations discussed in Van Ommeren and Dargay (2006). It takes the form:

$$\ln(S_{itd}) = \beta_0 + \beta_1 W_{itd} + \beta_2 \ln(D_{itd}) + \beta_3 \ln(y_i) + \beta_4 X_i + \beta_5 F_{td} + \xi_{itd}, \qquad (1)$$

where the β's are parameters to be estimated and subscripts i represent individuals; t represents hour of departure and d represents day of the year. S is speed; W is a vector of individual-specific time-varying variables (including weather variables); D denotes the distance travelled and y is the income of individuals. X is a vector of individual variables (including gender and age); F refers to time-specific characteristics such as urbanization, hour of travel, and seasonal variations; and ξ denotes an unobserved error term.

Using OLS for analysing the impact of weather on the speed of commuting trips is not ideal since it assumes that the residuals are uncorrelated. We face two shortcomings if we employ OLS estimation for equation (1). First, OLS does not control for differences in unobserved preferences of individuals and differences in other unobserved features (such as the exact location of the individual). We therefore exploit the panel structure inherent in the data to control for these issues. Specifically, we include individual fixed effects in order to control for selection and unobserved heterogeneity.[6] Second, using OLS for analysing the impact of weather on the speed of commuting trips is not ideal since it assumes that the residuals are uncorrelated. This implies that if a person makes two trips on the same day the residuals from the model of both trips are assumed to be uncorrelated. This obviously does not hold in the current case.[7] As a result, OLS is inefficient (Wooldridge 2003). Therefore, a random effects panel data model that controls for the correlation between errors is employed.

Most of the variables included in the model are self-explanatory. However, some other variables need additional explanation. We include personal characteristics, because they may effect the optimally chosen travelling speed to and from access point. Furthermore, in this way we control for selection effects. The travel time of a trip may be different during peak and off-peak hours because of the difference in frequency of the service, difference in number of people using the public transport, etc. Therefore, a dummy variable is included for rush hours.[8] In order to control for congestion effects on roads for bus and tram trips we distinguish between trips on congested and non-congested roads. Specifically, we distinguish those trips that originate in non-congested areas during the morning peak hours and that are directed towards congested areas. Similarly, we distinguish those trips during evening peak hours that are directed from congested areas to non-congested areas (see Sabir *et al.*, 2008). Ultimately, we include a congestion dummy variable that controls for these specific trips.

In order to analyse the effects of weather on travel time we use hourly measured wind strength, temperature, precipitation, snow and visibility. Dummy variables are used to measure the effects of most weather variables. Wind strength is measured by a dummy variable that represents wind strengths larger than 6 Beaufort (BFT).[9] We define three temperature categories: a dummy for temperatures less than 0 °C; a dummy for temperatures between 0 °C and 25 °C; and a dummy for temperatures higher than 25 °C. Precipitation effects are captured by using a dummy variable that is equal to 1 for the presence of precipitation during the hour in which the trip took place. The visibility variable measures the horizontal visible distance; a dummy is used to indicate a visibility distance less than 300 metres. We do not have an exact measure for snow. However, we use interaction effects of temperature lower than or equal to 0 °C and the presence of rain as a proxy for measuring snow. However, it should be noted that this is a crude measure of snow and will only capture the effects of *falling* snow.

4. Results and discussion

4.1 Speed of bus/tram/metro trips

The estimation results are presented in Table 16.1. The results are robust and most of the variables have plausible signs. Observe that although temperature does not have a strong impact, snow, limited visibility and rain on congested routes all substantially reduce the speed of bus/tram/metro trips. Remember that trip speeds are computed on the basis of the sum of in-vehicle time and other time components, including access and egress times, waiting times and delays. Apparently, these three specific circumstances (snow, limited visibility and rain) cause an increase in travel time by affecting one or more of these components.

The fixed-effects model shows that the speed of bus/tram/metro commuting trips is reduced by 12 per cent in snow. A potential reason, at least for bus trips and maybe partly for tram trips, is that it is more risky to drive in snow and that the capacity of roads is reduced because of an increased distance between vehicles. Another reason may be that snow causes people to switch from biking and walking to public transport, thereby increasing the demand for public transport. This may mean that more and longer stopping and waiting times are required for passengers to enter and leave the vehicles. Although rain in general has no effect on trip speeds, it does have a substantial impact on trip speeds on already congested routes (meaning that the results are mainly driven by the effect of rain on the speed of bus trips). Specifically, rain reduces trip speed on congested routes by approximately 18 per cent. This result is consistent with the result obtained in Sabir *et al.* (2008), who reported a 10–15 per cent reduction in the speed of car commuting trips under the same circumstances. There also is a 6 per cent reduction in trip speeds when visibility is under 300 metres (also this effect is likely driven by the effects of

Table 16.1 Analysis of logarithm of speed of public transport commuting trips (individual specific effects)[a,b]

	Bus/Tram/Metro				*Train*			
	Fixed effects		*Random effects*		*Fixed effects*		*Random effects*	
	Coeff.	*S.E.*	*Coeff.*	*S.E.*	*Coeff.*	*S.E.*	*Coeff.*	*S.E.*
Strong wind	**–.005**	.002	**–.004**	.001	.001	.001	.0003	.001
Temperature ≤ 0 °C	.02	.02	.01	.02	**.04**	.01	**.04**	.01
Temperature > 25 °C	–.01	.02	–.01	.02	**–.04**	.01	**–.03**	.01
Rain	.01	.01	.01	.01	.003	.00	.005	.005
Snow	**–.12**	.04	**–.08**	.04	**–.06**	.02	**–.05**	.02
Limited visibility	**–.06**	.02	**–.06**	.02	.03	.02	.03	.02
Congestion x Rain	**–.18**	.05	**–.17**	.05	–	–	–	–
Congestion	*–.08*	.04	–.01	.01	–	–	–	–
Rush hours	–.01	.01	*–.06*	.03	**.02**	.01	**.01**	.01
Distance	**.56**	.02	**.57**	.01	**.62**	.02	**.52**	.01
Gender (males)	–	–	.01	.02	–	–	–.02	.01
Age less than 18 years	–	–	.04	.07	–	–	–.02	.09
Age between 30 and 40 years	–	–	.02	.03	–	–	**.05**	.02
Age between 40 and 65 years	–	–	–.01	.02	–	–	.02	.02
Age greater than 65 years	–	–	–.14	.12	–	–	*–.19*	.12
Weekdays	–	–	–.05	.04	–	–	–.01	.04
Very urbanized	–	–	**–.11**	.04	–	–	**–.08**	.03
Urbanized	–	–	–.05	.04	–	–	–.04	.03
Moderately urbanized	–	–	–.05	.04	–	–	.0004	.03
Little urbanized	–	–	.02	.04	–	–	–.04	.04
Summer	–	–	.002	.02	–	–	.003	.01
Autumn	–	–	–.03	.02	–	–	.01	.01
Winter	–	–	**.05**	.02	–	–	–.01	.01
Constant	–	–	**1.66**	.06	–	–	**1.73**	.06
R^2	.95		–		.94			
Number of groups	1124		1124		1441		1441	
Variance of random error	–		.02		–		.01	
Variance of group specific error	–		.09		–		.06	
Correlation between error terms	–		.84		–		.83	

Notes:

a Bold coefficients are statistically significant at 5%; italic coefficients are statistically significant at 10%.

b The reference categories for temperature, urbanization, age, and seasonal variables, are temperature between 0 °C and 25 °C, rural, age between 18 and 30 years, and Spring, respectively.

visibility on the speed of bus trips). This finding is plausible, as one would expect people and vehicle operators to change their behaviour under risky conditions such as limited visibility. The effects of other weather variables are small and, except for strong wind, statistically insignificant.

The effect of log distance on the speed of bus/tram/metro trips is around 0.6: on average, trip speed increases by 0.6 per cent when distance increases by 1 per cent. This makes sense because longer trips are likely to make more use of roads with higher speed limits than shorter trips. The congestion variable shows a reduction of 8 per cent in trip speed, which is comparable to the results achieved by Sabir *et al.* (2008), who reported an 8 per cent reduction in the speed of car commuting trips on congested routes. The effects of other characteristics on the speed of bus/tram/metro commuting trips are generally small and most are statistically insignificant. An exception is commuting trips made in highly urbanized areas, which are on average 11 per cent slower compared to trips made in rural areas. This makes sense, since in these areas public transport is confronted with a larger number of crossings and traffic lights. Also the speed of the access mode (walking or cycling) to the public transport stop will be lower in highly urbanized areas.

Our analysis shows that snow, fog, wind and rain indeed have an impact on the speed of bus/tram/metro trips. Part of the explanation is that the speed of these vehicles themselves will be affected, implying increases in in-vehicle time. Another part of the explanation is that adverse weather leads to longer waiting times at platforms, in particular when people miss a connection, and leads to longer access and egress times. Note also that adverse weather has a doubly negative effect on integrated public transport: not only does it lead to longer and less reliable travel times, but also the comfort at transfer points will be worse. This provides a challenge to operators that aim at offering integrated transport services. Timetables should be made in such a way that they are reasonably robust under conditions of adverse weather, and also the comfort levels at transfer points should be adequate under various weather conditions.

4.2 Speed of train trips

The results of the fixed-effects model on the speed of train commuting trips show that at temperatures below 0 °C train trips are 4 per cent faster than train trips made at temperatures between 0 °C and 25 °C. Similarly, train commuting trips made during temperatures higher than 25 °C are 4 per cent slower compared to normal temperatures. Remember that trip speeds are computed on the basis of the sum of in-vehicle time and other time components, including waiting times, delays, access and egress times to get to the station by foot, bicycle, bus, car, etc. A likely explanation is therefore that people may prefer to walk or bike rather than use public transport to go to or from a train station. Another reason may be that demand for train trips is lower in cold weather, which may result in a smaller number of people on access points, implying lower probabilities of delays. Similarly, if demand for train trips is higher in warm weather, we would observe an increased probability of delays. The results furthermore show that train trips are also slower during snow; the speed reduction is around 5 per cent. Comparing the effects of snow on the speed of bus/

tram/metro trips on the one hand and train trips on the other shows that train trips are less affected by snow. This is not surprising given the technology of the train compared to the bus. Both types of trips share the possible delay during the access and egress mode, but the bus (and up to some extent trams) travel on road networks with other vehicles, whereas the train has a separate network. Trains will therefore suffer less congestion and one may expect a smaller effect of snow on the speed of train trips compared to the effect for other modes.

Again the effects of other characteristics are small and generally insignificant for train commuters. However, the age variable shows some interesting results. The results suggest that trips made by people in the oldest age category are 19 per cent slower compared to trips made by younger people. This likely reflects that older people take more time to reach access points and spend more time transferring between trains. It is also possible that older people have less access to cars, so they have to use public transport even when they live further away from an access point. Another interesting finding is the speed reduction in very urbanized areas compared to rural areas. This probably reflects a difference in access modes: residents of highly urbanized areas typically will not use the car to get to the railway station and other access modes are typically slower than the car.

We find that, compared to bus/tram/metro trips, the impact of weather on rail trips is considerably smaller. The main effect we observe relates to snow, and this most probably is a consequence of the impact of snow on the access and egress modes used, not on the railway trip itself. This robustness makes rail an attractive transport mode compared to bus/tram/metro, and also compared to the car. This does not mean that reliability is not an issue in rail trips, because it certainly is. It does mean that weather is not an important factor here and that the negative effects of certain weather conditions on rail trips are confined to the comfort level at railway stations. Thus, from the perspective of adverse weather, the main challenge to railway operators that aim at high-quality public transport services is to build railway stations that are comfortable under various weather conditions.

4.3 Welfare effects through changes in travel time

An important purpose of the current study is to assess welfare effects of weather through changes in travel time of public transport.[10] For this we use information on the average value of travel time. There is a vast empirical literature on the value of travel time (see e.g. Small and Verhoef, 2007). Based on a meta-analysis of 56 value-of-time estimates from 14 different countries, Waters (1996) finds an average ratio of value-of-time equal to 48 per cent of gross wage rate and a median ratio of 42 per cent for commuting trips made by automobile. In another review, Wardman (1998) finds similar values. In this chapter we follow the standard literature on values of time and use 50 per cent of hourly gross wages as our measure. In the Netherlands the average gross hourly wage rate is about €18, implying a value of time of €9 per hour (Statistics Netherlands).[11] The welfare effects are based on the estimates from the fixed-effects models, and are obtained by taking the product of the percentage effects, the average travel time and the value-of-time. The results are presented in Table 16.2.

Table 16.2 Welfare effects of weather through changes in travel time

	Welfare loss/gain (in €)	
Variables	*Bus/Tram/Metro*	*Train*
Wind strength	–0.03	0
Temperature ≤ 0 °C	0	0.40
Temperature > 25 °C	0	–0.40
Rain	0	0
Rain x Congestion	–1.78	–
Snow	–0.76	–0.50
Visibility	–0.38	0

The highest welfare loss due to adverse weather is observed for bus/tram/metro trips. The welfare loss for these trips due to snow is €0.76 per commuting trip per person.[12] Similarly, bus/tram/metro commuting trips made in rainy conditions and on congested routes experience a welfare loss of €1.78 per commuting trip per person. Furthermore, the welfare loss due to limited visibility is around €0.38 per commuting trip per person. The highest welfare loss for train trips is that of snow, which leads to a loss of €0.50 per commuting trip per person. Additionally, train trips made during high temperatures experience a loss of €0.40 per commuting trip per person. However, there is a gain of €0.40 per commuting trip per person when trips are made during temperatures below 0 °C. Note that these calculations only address the travel time element and disregard the comfort element of adverse weather. No doubt, comfort levels of waiting at platforms and walking to access points will be lower under such circumstances. It is beyond the scope of the present study to provide estimates for this aspect.

5. Conclusion

In this study we analysed the effects of weather on the speed of commuting trips made by public transport in the Netherlands. We use micro data at the trip level obtained from a national transportation survey for the Netherlands. The data cover trips made by bus/tram/metro and train during 2004 and 2005. Hourly measured weather data for this period were obtained from the Royal Netherlands Meteorological Institute. The weather and transport data were matched in such a way that each trip was assigned the weather data for the hour in which that trip took place and from the weather station that was nearest to the place of departure.

We estimated panel data models with individual specific fixed and random effects in order to control for possible selection problems and unobserved heterogeneity. We used a large number of variables in our model to explain the speed of public transportation. Our main interest, however, was in the effect of weather variables on the speed of public transport and the associated welfare effects.

In general, the results are robust and most of the coefficients have plausible signs. The results show that wind strength has only a small negative effect on the speed of bus/tram/metro commuting trips. Snow has a substantial negative effect on the speed of public transport. The associated welfare loss is €0.50 per commuting trip per person made by train and €0.76 per commuting trip per person made by bus/tram/metro. Rain strongly affects the speed of bus/tram/metro commuting trips on congested routes. The associated welfare loss is €1.78 per commuting trip per person.

Effects of other characteristics are generally absent. However, one interesting finding is that train trips made by older people are 19 per cent slower than those made by younger people. This may indicate that older people have less chance of taking the car on their way to the train station. They may also walk slower to their final destination on the egress part of their trips. Of course it may be that they are just in less of a hurry, but one should not forget that in our analysis we focus on commuting trips.

In terms of integrated transport we find that the effects of weather on trip speed are relatively strong in the case of bus/tram/metro trips. They may well lead to changes in in-vehicle time, but most probably also in waiting times at transfer points. This implies a challenge to public transport operators to develop timetables and operating routines that lead to reasonably robust outcomes for travellers. In the case of railway trips the impact of weather on speeds is clearly smaller. For both types of trips a general observation is that the comfort of trips under adverse weather likely depends substantially on the quality of the facilities at transfer points such as bus stops and railway stations. This is one of the fields where efforts to improve the quality of integrated transport should focus.

Acknowledgements

We would like to thank Nuffic, the Higher Education Commission of Pakistan, and the 'Climate Changes and Spatial Planning' and Transumo research programmes for sponsoring this research.

Notes

1 The Randstad consists of a ring of the four largest cities of the Netherlands (Amsterdam, Utrecht, Rotterdam and the Hague) and their surrounding areas. The population of the Randstad is over 7 million inhabitants which is almost 50 per cent of the total Dutch population. The Randstad is the main centre of employment and business activities, so in the morning congestion occurs on roads towards the Randstad, while evening congestion occurs on roads from the Randstad.

2 We combine trips made by bus/tram/metro for two main reasons. First, average speed, distance and travel time of trips made by bus/tram/metro were similar. Second, bus/tram/metro are mostly used for medium-distance trips, unlike the train which is mostly used for long-distance trips.

3 The exact number of individuals in the sample is 130,534. These people reported 453,885 trips out of which 13,618 were made by public transport.

4 We have estimated the average distance as follows: the total land area of the Netherlands is 33,889 km^2. Given the assumption that stations are homogenously spread over the country and each weather station covers a circle, the maximum distance is 18.78 km. The average distance of a location to the centre of the circle is two-thirds of the maximum distance, so the average distance to a station is 12.52 km. Although there may be differences in rain/no rain conditions within this range, especially during the summer, this is the smallest range available for hourly weather conditions in the Netherlands.

5 We improve on the statistical analyses of Fosgerau (2005) and Van Ommeren and Dargay (2006) by explicitly taking the time dimension of the moment of travel (in days and hours) into account, as well as the unobserved heterogeneity of commuters.

6 Note that some commuters have two different distances on the same day, which allows us to identify the effect of distance-using individual fixed effects.

7 The correlation between the two errors is higher than 0.80 for all public transport modes.

8 Morning peak hours are from 06:00 to 10:00, evening peak hours from 16:00 to 18:00, and all other hours are off-peak hours.

9 The Beaufort scale (BFT) measures wind strength on a scale of 1 to 12. On this scale, 6 BFT represents powerful winds with a speed between 39 and 49 km/hr (or 10.8 to 13.8 metres per second) over a period of at least 10 minutes. Similarly, 12 BFT represents a hurricane with wind speeds greater than 117 km/hr (or greater than 32.6 metres per second).

10 The calculations for welfare effects are computed on a per person basis, implying that total welfare loss for a trip by train or bus should be multiplied by the average load factors. This holds for all welfare calculations in this study.

11 Gross wage is €19 for the whole population. It may be noted that the gross wage can be lower for bus commuters and higher for train commuters. Therefore, results will be slightly biased.

12 There is a 12 per cent reduction in speed for bus/tram/metro trips when it snows. This implies an increase of 0.0852 hours in average travel time ($0.12 \times 0.71 = 0.0852$). Given a value of time of €9 per hour, the welfare loss due to snow is $0.0852 \times$ €9 = €0.76.

13 This is the average distance of the entire trip. This implies that it includes not only in-vehicle distance but also distance travelled by access/aggress modes. The average in-vehicle distance for bus/tram/metro trips and train trips is 13.1 km and 36.7 km, respectively.

References

Fosgerau, M. (2005) 'Speed and Income', *Journal of Transport Economics and Policy* 39, 2: 225–240.

Hall, F.L. and Barrow, D. (1988) 'Effect of Weather Conditions on the Relationship Between Flow and Occupancy on Freeways', *Transportation Research Record* 1194: 55–63.

Hranac, R., Sterzin, E., Krechmer, D., Rakha, H. and Farzaneh, M. (2006) 'Empirical Studies on Traffic Flow in Inclement Weather', Publication No. FHWA-HOP-07–073, Washington, DC: Federal Highway Administration.

Ibrahim, A.T. and Hall, F.L. (1994) 'Effect of Adverse Weather Conditions on Speed-Flow-Occupancy Relationships', *Transportation Research Record* 1457: 181–191.

Koetse, M.J. and Rietveld, P. (2009) 'The Impact of Climate Change and Weather on Transport: An Overview of Empirical Findings', *Transportation Research Part D* 14, 3: 205–221.

Martin, P.T., Hansen, J.P. and Quintana, I. (2000) 'Inclement Weather Signal Timings', UTL Research Report MPC01–120, UTL, University of Utah, Salt Lake City.

Maze, T.H., Agarwal, M. and Burchett, G. (2006) 'Whether Weather Matters to Traffic Demand, Traffic Safety, and Traffic Operations and Flow', *Transportation Research Record* 1948: 170–176.

Rietveld, P., Bruinsma, F.R. and Van Vuuren, D. (2001) 'Coping with Unreliability in Public Transport Chains: A Case Study for the Netherlands', *Transportation Research* 35: 539–559.

Sabir, M., Van Ommeren, J., Koetse, M.J. and Rietveld, P. (2008) 'Welfare Effects of Adverse Weather through Speed Changes in Car Commuting Trips', Tinbergen Institute Discussion Paper 08–087/3, Amsterdam: VU University.

Small, A.K. and Verhoef, E.T. (2007) *The Economics of Urban Transportation*, London/New York: Routledge.

Statistics Netherlands (2008) 'Commuter Cyclists Prefer Short Distances', *CBS Web Magazine*, 16 September.

Statistics Netherlands, Statline, http://statline.cbs.nl/statweb/?LA=en.

Van Ommeren, J. and Dargay, J. (2006) 'The Optimal Choice of Commuting Speed: Consequences for Commuting Time, Distance and Costs', *Journal of Transport Economics and Policy* 40, 2: 279–96.

Wardman, M. (1998) 'The Value of Travel time: A Review of British Evidence', *Journal of Transport Economics and Policy* 32: 285–316.

Waters, W.G.I. (1996) 'Values of Travel Time Savings in Road Transport Project Evaluation', *Proceedings of 7th World Conference on Transport Research*, Vol. 3, Oxford: Pergamon.

Wooldridge, J.M. (2003) *Introductory Econometrics: A Modern Approach*: South-Western College Publishing.

Appendix A

Table A1 Descriptive statistics of variables included in the empirical analyses

	Bus/Tram/Metro		*Train*	
	Mean	*S.D*	*Mean*	*S.D*
Speed (km/hr)	21.19	11.29	38.17	14.93
Travel time (hr)	0.71	0.34	1.11	0.48
Travel time (congested areas) (hr)	1.10	0.34	–	–
Distance (km)[13]	15.63	13.15	43.34	29.10
Strong wind	0.02	0.16	0.02	0.16
Temperature ≤ 0 °C	0.05	0.22	0.05	0.23
Temperature > 0 to ≤ 25	0.93	0.26	0.93	0.25
Temperature > 25	0.02	0.15	0.02	0.12
Rain	0.19	0.40	0.19	0.39
Snow	0.01	0.10	0.01	0.09
Visibility	0.02	0.14	0.01	0.11
Morning peak hours	0.45	0.50	0.47	0.50
Evening peak hours	0.37	0.48	0.40	0.49
Non peak hours	0.18	0.39	0.13	0.34
Weekday dummy	0.94	0.24	0.98	0.16
Spring	0.25	0.43	0.22	0.41
Summer	0.21	0.41	0.23	0.42
Autumn	0.32	0.46	0.34	0.47
Winter	0.22	0.42	0.21	0.41
Very urban	0.37	0.48	0.23	0.42
Urbanized	0.30	0.46	0.41	0.49
Moderately urbanized	0.11	0.32	0.21	0.41
Little urbanized	0.13	0.34	0.09	0.29
Rural	0.09	0.28	0.05	0.22
Age less than 18 years	0.02	0.15	0.00	0.07
Age between 18 to 30 years	0.31	0.46	0.25	0.43

(continued)

	Bus/Tram/Metro		*Train*	
	Mean	*S.D*	*Mean*	*S.D*
Age between 30 and 40 years	0.19	0.39	0.27	0.44
Age between 40 and 65 years	0.47	0.50	0.47	0.50
Age greater than 65 years	0.01	0.07	0.00	0.05
Male	0.42	0.49	0.59	0.49
Congestion	0.05	0.21	–	–
Congestion x Rain	0.01	0.08	–	–
Number of observations	5,126		8,492	

Part 4

The challenges in achieving integrated transport at city, regional and national levels

Chapter 17

Impediments to integrative transport policies

Lessons from the new town of Modiin

Eran Feitelson and Josef Gamlieli

1. Introduction

The growing realization that no transport measure can sufficiently address the issues that face transport systems and users, and the rising number and scope of these issues, has generated widespread calls for integration (Geerlings and Stead, 2003; Macario, 2007). However, it is not entirely clear what is meant by integration. Potter and Skinner (2000) advanced a typology of five levels of integration: functional integration (mainly ticketing that enables multimodal trips); modal integration; transport and planning integration (integrating land use and transport planning so as to reduce both trips and trip distances); social integration (the integration of transport policies and other policy arenas); and finally a holistic environmental, economic and transport policy integration. To these issues, the integration between institutions involved in transport planning and management, and integration between the public and private sector, can be added (Viegas, 2005); as well as integration across spatial scales (Keil and Young, 2008).

It is possible, therefore, to determine varying levels of integration. The factors that may affect the type and degree of integration can be hypothesized to be those that determine the benefits of integration, on one hand, and the costs associated with the specific level of integration on the other. As Viegas (2005) argues, the benefits are essentially the reduction in door-to-door transition costs within the transportation system. These are a direct function of the interoperability of the transport system. Yet, the ability to interoperate is affected by the physical layout of the relevant transportation systems, and the relations between these systems and

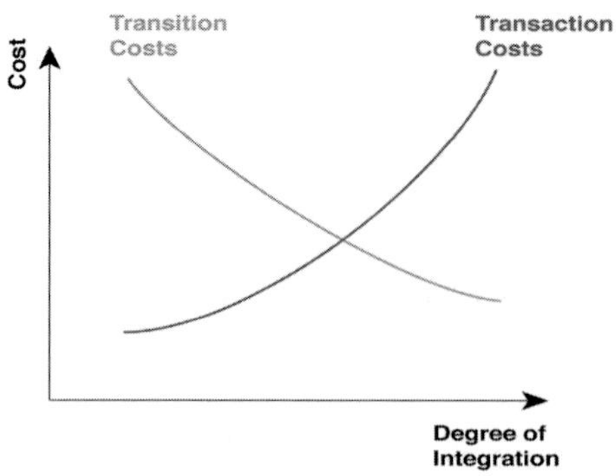

Figure 17.1 Transition and transaction costs of integration

other land uses. These, in turn, are a function of interagency relations and institutional structures. It can be argued, therefore, that all the levels noted above ultimately affect the transition costs.

The costs associated with integration are essentially the dynamic transaction costs associated with meshing together different systems. These can be hypothesized to be a function of the institutional structure, social capital and the physical layout. The physical layout affects the number and nature of players that will need to be involved in improving the interoperability of transportation systems. The greater the number of players involved, the higher the transaction costs are likely to be (Furubotn and Richter, 2000). Thus, if the institutional setting is highly fragmented the number of players that will need to be engaged will be greater, and hence the transaction costs can be expected to be higher. However, transaction costs may vary between cases, even when the number of players remains the same. This is due to the importance of the relationships between the players (Hull, 2008). If there is a history of cooperation between the different players and if the organizational culture of the most pertinent organizations is one of openness, then the transaction costs are likely to be lower in comparison with cases where there is no history of cooperation and the organizational culture of the agencies is a closed one. Similarly, if there is trust among the players (high social capital) the transaction costs will be lower than if there is mistrust (Laws and Hajer, 2006).

If the different types and levels of integration are mapped out as a function of the degree to which they come close to a holistic integration, it is likely that the transition and transaction costs can be depicted as in Figure 17.1. That is, the degree of integration will be determined by the point at which the transaction costs of further integrating will exceed the reduction in transition costs. While this figure provides a conceptual answer to the question, "What determines the level of integration?", the factors that actually determine the degree of integration in practice remain opaque (Geerlings and Stead, 2003).

In order to help identify the factors that actually determine the degree of integration, this chapter scrutinizes one case study which, arguably, provided a particularly propitious opportunity for integration, but failed to deliver on this promise. This is the case of Modiin, a new town built in Israel since the mid-1990s.

In the next section we describe the case study, and make the argument that this is a particularly good case to test the questions posed above. Then we describe the plan of Modiin, focusing on the transportation issues, and how it was implemented. This is followed by a necessarily brief description of the implementation of the plan. Then the factors that impeded integration in Modiin are analysed systematically. Finally, the implications and lessons that can be derived from the Modiin case are spelled out.

2. Modiin: the potential for transport integration

Modiin is a new city in Israel. It was built between the two main metropolitan centers, Tel Aviv and Jerusalem (see Figure 17.2), as a central element in Israel's strategy to accommodate the unexpected immigration wave from the former Soviet Union in the early 1990s. The area in which it was built was formerly an army training ground, and hence it was largely barren, with very few restrictions imposed on it. As a result planners of the city had a relatively free hand in designing the city. Thus, in contrast to most other cases, the urban context was largely malleable, thereby reducing the restrictions that such a context may impose on the integration of transport systems and land use.

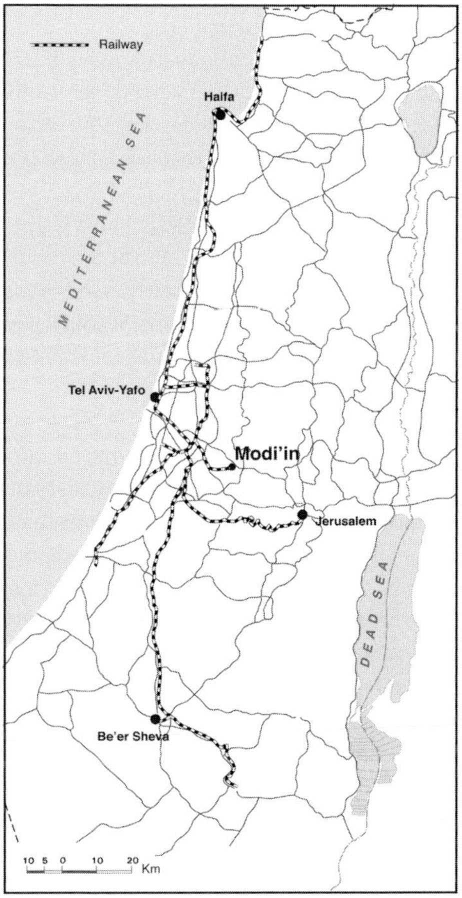

Figure 17.2 Modiin and its environs

The planning of the city was tendered by the Ministry of Housing and Building (MHB) to Moshe Safdie. A well-known architect who has worked worldwide on a wide variety of projects, Safdie was particularly interested in the place of transport in city design. In a book published concurrently with the planning of Modiin, entitled *The City After the Automobile*, Safdie argued that planners should "rethink all facets of urban transportation as a united system" (Safdie, 1997, p. 125). He went on to focus on the interfaces between transportation systems, and particularly on the adverse effects of parking requirements near the main nodes on the urban environment. To address these effects he even suggested a new "utility car" concept, whose purpose was to reduce transition costs to a minimum in terms of the area required for parking. In Safdie the city had thus a planner who was well aware of the importance of transport integration, at least at the modal and land use levels.

Modiin was conceived, planned, financed and built largely by the MHB. Thus, in contrast to the situation in existing cities, the institutional structure that accompanied the early years of Modiin was extremely centralized. While the MHB did need the approval of the planning authorities (in which it was a voting member), it essentially faced very little opposition or constraints in the planning and building of the city. Hence, also from an institutional perspective Modiin seems to offer a particularly favorable opportunity for integrating transport and residential development. Moreover, as the Israeli system of governance is a unitary system where authority in all planning and building issues resides largely at the national level, the MHB could conceivably coordinate the development of Modiin with the regional transport systems quite easily, as it had to coordinate these with a very limited set of national actors.

In summary, it would seem that Modiin had a truly exceptional potential to integrate the transport system both within it and at the regional scale. Moreover, it had the potential to integrate not only the functions and modes, but also the land use with the transport systems and the institutions. It could, conceivably thus come close to a holistic integration with relatively low transaction costs.

3. Modiin: the plan

Modiin is situated at the foot hills of the Judean Mountains, within a hilly terrain. As Safdie describes in his book, he tried to maintain the main features of the terrain in the plan (Safdie, 1997, pp. 105–110). Thus, Modiin was planned as a series of residential neighborhoods on the hills, where all residences, except those at the top, will be at the height of the future vegetation to give it a green image. The transportation system and the centers of activity, both shopping and schooling, were relegated to the wide valleys. These were planned so the services and schools will be at the center of the valley, while two one-way streets will run down the valley on the two sides of these services. It was conceived that school children and residents will walk down to these inter-neighborhood centers.

At the center of the city Safdie planned a large transportation center, set at the edge of the Central Business District (CBD), adjacent to the main shopping mall planned for the city. This transportation center was planned to include a subterranean train station and the central bus station, also below grade. On grade a wide traffic

Figure 17.3 The planned transport center of Modiin
Source: Modiin Municipality, 2008a
Site address: http://www.modiin.muni.il/modiinwebsite/ArticlePage.aspx?PageID=370_266

circle was planned (see Figure 17.3). The transportation center was thus planned to be the multimodal hub of the transport system of Modiin. It was to be the interface between intra-city buses, intercity buses, and the intercity rail, in proximity to the main shopping mall, thereby allowing for trip chaining by commuters.

In the western part of the city a large employment center was planned. A smaller employment center was planned in the southern part. Both centers are located on the main exits from the city toward the Tel Aviv metropolitan area (see Figure 17.4). Hence, the city was to provide a balance of residences and employment, thereby providing residents with opportunities to work locally rather than having to commute to either the Tel Aviv or the Jerusalem metropolitan areas. Overall, the city was planned for a population of 120,000 in the first stage, with a possibility of expanding it to 250,000 at a second stage. The transport system was therefore planned to accommodate this latter population in mind.

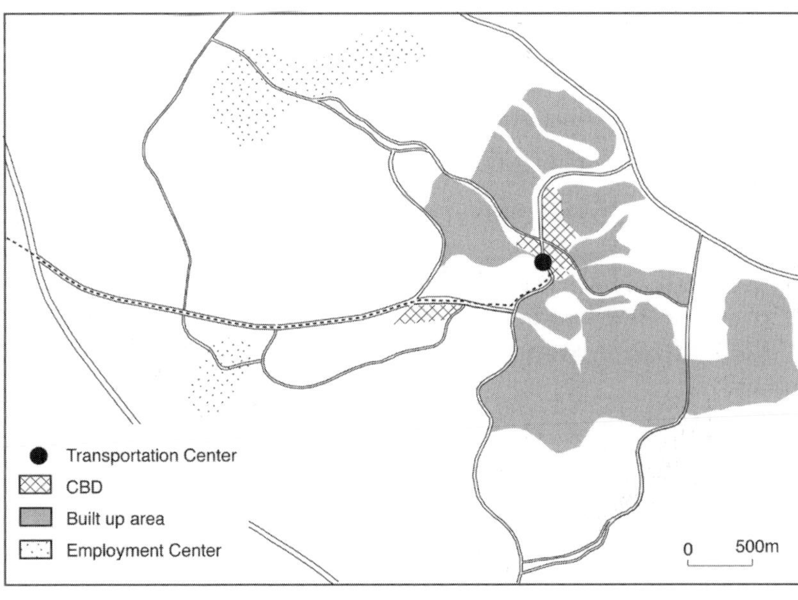

Figure 17.4 The Modiin plan
Source: a processed scheme from MD 20/20 plan as published in Modiin Municipality, 2008b
Site address: http://www.modiin.muni.il/modiinwebsite/ChannelArticle.aspx?PageID=242_138

4. The development of Modiin to date

The plans of Modiin were ratified in the mid-1990s, and by 1996 the first residents began moving in. At the end of 2007, 12 years on, the city had some 67,000 residents. These residents live in a series of neighborhoods built according to Safdie's plan. At present several additional neighborhoods are being built, and are expected to raise the city's population to approximately 80,000 residents in the coming years. The approved city plan is designed to accommodate 120,000 inhabitants, although the transport rights of way maintained in the plan can accommodate a much larger population.

The development of the employment centers was slower. While the infrastructure for these centers has been built, only a few large employers have moved in to date. Hence, the vast majority of residents commute either westward to the Tel Aviv metropolitan area or eastward to Jerusalem.

The internal road system has been built in conjunction with the building of the different neighborhoods. However, the trunk roads have not been built to full capacity, as they are planned to a higher capacity than was seen as necessary for a population of 60,000. But as the level of motorization is higher than was forecasted, congestion issues have arisen within the city. The regional transport system, which is not under the authority of the MHB and is not financed by it, has been slower to develop. As a result congestion developed on the roads leading from the city and to it, mainly during the morning and afternoon peaks.

The rail connection was built by the Israel Rail company according to its development plan. At first the rail extended to a station outside the town. However, as parking demand exceeded the parking provided near the station, spillover parking along the road leading to the station became a daily occurrence. In April 2008 the extension of the rail line to the transport center was completed, and the center opened. Within a very short period about 100,000 trips a month were recorded using this link.

The city's transport center, however, was not developed according to plan. Due to disagreements between the MHB and the Ministry of Transport (MOT) the central bus station was not built, and it is still unclear when it will be built. At present it is not funded. Moreover, the traffic circle was only partly developed. As a result, at present there is an improvized interface between temporary bus stops and the train station, and congestion has developed in the area. This is exacerbated by the lack of parking, or sites for kiss and ride in the vicinity of the transport center. The central rail station did not include parking for the commuters, as the station is located at the city center, and the plan sought to reduce private car use by reducing parking availability where there was to be a high level of service by public transport. The only parking provided in the vicinity of the transport center is at the nearby shopping mall. However, this parking is controlled by the mall owners, who plan to charge for it in order to deter all-day parking by commuters in favor of shoppers.

After years of complaints and discussions an organized intra-city bus service is operating, using a temporary operational terminal. Although it provides service to all parts of the city all day long, the bus system is under-utilized. The occupancy of the buses is low, serving mainly pupils. The bus services have so far failed to capture substantive numbers of commuters or shoppers within the city.

Pedestrian paths were built as planned. But anecdotal evidence suggests that many people abstain from walking due to the hilly topography. A few bicycle routes are planned. Some have already been paved. However, at present cycling is not a substantive mode of travel in Modiin. Thus, the dominant transportation mode for commuting and intra-city trips is the private car. More importantly for the purpose of this chapter – the planned integration between modes failed to materialize.

5. The impediments to integration: an analysis

Based on the brief literature review in the introduction above, several factors can be hypothesized as affecting the degree of integration in practice. These include the urban layout; the relationship between transport systems and land use (for details on this subject see Chapter 3 in this volume); and the institutional structure and interagency relationships (which in turn are a function of social capital – for details on this subject see Chapter 4 in this volume). The urban context and the transport/land use interfaces were essentially predetermined in the case of Modiin, a centrally planned new city, by the plan that was formulated and the way it was implemented. The institutional structure and interagency relationships can be expected to affect both the plan implementation and the way the public transport system operates. Hence, in order to assess the importance of the hypothesized factors in impeding the integration of transport systems in Modiin, five possible sources of failure are analyzed:

1. The plan – it may be that deficiencies in the original plan of the city impeded integration, as they created an unfavorable urban context.
2. Implementation of the plan – it is conceivable that certain crucial elements that were planned were not built, or not built in time, thereby precluding the transport/land use integration as was originally planned. Hence, the fault is not in the plan but rather in its implementation process.
3. The operations of the public transport system – it is possible that even when the infrastructure that is necessary for integration was planned and put in place in time, the operation modes of the public transport system did not utilize them in an integrative manner.
4. Institutional factors – it is also possible that lack of coordination either in infrastructure provision or in public transport operation (or both) is an outcome of institutional infighting or simple lack of cooperation and coordination between critical agencies. That is, different institutions with authority over different parts of the transport and land use systems and their interfaces failed to cooperate and coordinate their actions.
5. External factors – it is also conceivable that factors beyond control of the agencies involved, and beyond the city-scale, impeded integration. That is, that there are additional factors not noted in the literature that impeded integration in this particular case.

To identify the factors that impeded integration in Modiin these potential sources of failure are analyzed in this section one by one.

5.1 The plan

Modiin is being developed according to a regional outline plan for a city of up to 120,000 inhabitants (with the possibility of expanding it in the more distant future to 250,000 people). Detailed plans for the city's neighborhoods are derived out of this plan. The plan directs the sequence of such neighborhood development according to preconceived phases. Infrastructure, including transport infrastructure, has to be developed according to these phases, differentiated by the number of dwelling units built. To this end a table that requires the development of inter-urban transport infrastructure elements as a condition for further dwelling development was included in the plan. The conditions refer to inter-urban roads and a heavy rail connection westward toward the Tel Aviv metropolitan area (see Figure 17.2). The plan also keeps a 100 m-wide strip for a future mass transit system and sets the main city roads' right of way at 36 m. Thus, while the plan took into account the need to coordinate the city's development with the wider transport system, within the city limits it refers primarily to private car infrastructure, and less so to other modes.

The detailed plans located the neighborhood facilities (schools, local shopping, community centers, etc.) along or in the middle of the main valley roads. These locations proved to be problematic as they necessitated road-crossing from bus stations to schools and public facilities, and required almost all trips to these facilities to traverse down and up the hill slopes. The neighborhood streets follow the topography, thereby reducing the visual impacts of the roads. However, these winding and rather steep streets proved problematic for public transport and for pedestrian and bicycle use.

The desire to maintain the shape of the terrain, to keep valleys open and to limit building height to below future tree-cover constrained residential densities. As a result, most services could not be justified within walking distance from most residences. This impediment to non-motorized travel is aggravated by the location of services in the valleys – making most trips to them an uphill-downhill trip. The plan thus gave precedence to overall design principles (of roads, densities and location) over trip convenience, particularly in non-motorized modes.

The plan did not take into account the needs of intra-city buses. Hence it did not allocate land for an operational depot, and did not include bus ways or any other element that is needed for an effective operation of a city bus system. Hence, buses often have to travel through circuitous local roads or remain along distant trunk roads in the valleys. Another factor that reduces the level of service of buses is the maintenance of very wide rights of way (in order to be able to serve a population of 250,000 residents in the future). Such wide rights of way increase the distance between bus stops along the roadway and the buildings, thereby increasing the access time and effort to the buses along the main corridors.

An additional factor in the plan that has deleterious implications for transport integration was the lack of parking near the transport center. As noted above, the planners sought to induce the use of public transport by reducing parking at the city center (along the lines of the Dutch ABC system).[1] However, the result is that there are no park-and-ride facilities near the railway station, except within the shopping mall, where they are priced.

5.2 Implementation of the plan

The implementation of the city's plan was directed by the MHB, according to detailed plans authorized by the planning authorities. Thus dwelling units and infrastructure for employment centers were developed according to the phases outlined in the regional scheme. Local and intra-city trunk roads were also developed according to these phases. However, the regional transport infrastructure was not developed in tandem with the city's development, as the budgeting and development of the regional transport system is outside the jurisdiction of the MHB. Thus, despite the aforementioned conditions set in the city plan, the city's residential development advanced irrespectively of the regional transport system.

The main discrepancy between the city's development and the wider transport system pertains to the timing of the rail development. The national rail company built the rail extension to Modiin only after the city was populated. It completed the central rail station only after the city had over 60,000 inhabitants. As a result, many of the city's residents had already established car-dependent travel patterns by the time this station opened.

The completion of the central rail station was not accompanied by the building of the central bus station, as was planned. The MOT did not see urgency in building the central bus station, and consequently did not allocate budgets for this station. Thus, there is no central bus station in Modiin at the time of writing. Moreover, as the rail station was designed as part of the transport center, there are no interchange facilities in place between it and the adjacent temporary bus stations outside. Thus, the interchange between the local bus system and the train station continues to be user unfriendly.

While the infrastructure for the employment centers was built, they failed to attract a significant amount of jobs. Thus, in contrast to the plan, Modiin remains a commuter town, where almost all residents have to commute out to either the Tel Aviv or Jerusalem metropolitan centers.

5.3 Public transport operations

In the second half of the 1990s the MOT advanced a public transport reform, the crux of which was to introduce competition into the bus system, which until that time operated as a duopoly of two large cooperatives. As Modiin was a new city the MOT saw it as an appropriate site to implement the reform. Thus, Modiin was the first town in Israel in which the public transport system was tendered out (a tender from which the two large cooperatives were barred). This tender was won by a small company named Margalit. Margalit operated three lines that served the city. The lines were intercity lines coming from Tel Aviv or Jerusalem and entering the town's neighborhoods.

The operator did not have an operational depot within the city and thus had to park and maintain its bus fleet outside the city. At the time parts of the city were still under construction, with partially paved roads and temporary bus stops. As a result of these circumstances service was unreliable, and the level of service offered to intra-city travelers was low. But due to the rigidity of the contract between the state and the operator, the operator was unable to introduce changes. In an effort to solve the problems, two different public transport plans were prepared, concurrently, but none of them was fully implemented.

In 2006, a new public transport tender was issued. This time it allowed greater flexibility regarding route and service changes. Margalit, the previous operator won the bid again, and operated five circular lines, each operating in one direction. In August 2007, Veolia Transport (under the brand name Connex) bought Margalit's activities in Modiin, and today it is the public transport operator in the city. The new operator prepared a new operation plan along with the municipality and the MOT, which was to come into effect in September 2007, the date when the new rail connection to the city center was scheduled to open. However, the new operation plan was flawed, and thus failed on its first day – leading the mayor, who faced a strong public outcry, to demand a return to the previous scheme. The difficult entry of the new operator and the failure of the new operation mode in the first days of operation further diminished the public's view of Modiin's public transport system. In subsequent months the new operator began to reorganize the service. They conducted new demand surveys, built new bus stations and worked closely with the municipality and the MOT. During 2008 a temporary bus terminal was opened. Currently the company is operating 15 lines within the city and 11 intercity lines. The large number of lines being offered and the high frequency of services are an attempt to respond to public complaints and to supply both door-to-door services within the city and a good connection to other cities. The outcome is a large number of lines, long and circuitous routes, and buses that run empty most of the day, thereby incurring large expenses to the operator and subsequently heavy subsidies (these are determined on the basis of the contract between the operator and the MOT).

As a result of the travails of the operators, no coordination was established between the city bus operations and the operations of other modes, or between the bus operators and the physical planners of the various neighborhoods. Most importantly, there is scant coordination between the city buses and rail timetables. The timetable of the trains is determined by Israel Rail according to the national rail constraints. Thus, the trains' timetable is not well integrated with bus schedules and household travel patterns. Particularly, the rail timetable (which has very few departures from Modiin in the morning hours) does not allow many parents to get their young children to day care or primary school and still make it in time to the train by bus.

5.4 Institutional structure and policy

The planning and development of Modiin was led and financed by the MHB, which viewed the city as one of its major projects throughout the 1990s. However, the operation of the transport systems, and the public transport systems in particular, as well as the development of public transport infrastructure, are largely under the authority of the MOT.

The MOT is a highly centralized ministry. Essentially any public transport element, from the building of a bus stop, or the authorization of a local bus route, to the choice of public transport operators, falls within its purview, and requires explicit consent by the MOT. As a result, the MOT is inundated by various requests which require its attention. From the perspective of the MOT, Modiin was largely viewed as a small town, without any particular priority. Thus, with the exception of the first public transport tender, the MOT did not accord to Modiin any particular attention. This was manifest in the lack of MOT funding for the transport center and its slow responses to the public transport failures noted above. Thus, while the MOT could have conceivably advanced integration, given its wide-ranging power, in practice it did not provide Modiin with the attention needed to advance such integration.

The public transport system in the country as a whole is fragmented. Thus the rail company builds the rail tracks and stations, but it is not involved in the planning or development beyond the station's platforms. The lack of adequate parking, or connections to the intra-city public transport system in Modiin was not viewed by Israel Rail as part of its purview, and it was not involved in planning or discussions of these issues. The rail company also did not try to accommodate the local concerns within its scheduling, and there was no party that accosted it on this issue.

A third party that could have advanced integration, at least conceivably, is the municipality. However, being a new city, which initially was largely dependent on MHB budgets, the municipality did not have the power or capacity to criticize the MHB planning, or to advance its needs vis-à-vis the MOT. Thus, despite the fact that it was the agency which bore the brunt of citizen complaints regarding the public transport system, it had only very limited effect on the operation and planning of these systems. Therefore, it proved incapable of advancing the integration of the various transport modes.

5.5 Exogenous factors

The factors that affected transport development in Modiin, and the integration, or lack thereof, between the various transport modes and between transport and land use development are not limited to the city itself. Actually, several exogenous factors had an effect on the degree to which the transport system was integrated.

One such exogenous factor is the population composition. Most of the new residents in the city are households consisting of young parents with children. This has ramifications for the travel patterns of the households, as noted above. In particular, school hours constrain the parents' time-space prism. Many of these new households are middle class, many working in Israel's thriving high-tech industry and in professions relating to it. One of the ways firms in these businesses lure employees is by providing them with private vehicles, either directly or through leasing firms (Cohen-Blankshtain, 2008). Thus, many of the residents of Modiin do not pay the full price of their motoring, and private vehicle use is subsidized over public transport.

Another exogenous factor that may have affected the degree of transport integration in Modiin is the attitude of the city planners and developers in the field (i.e. those who planned and developed the various neighborhoods). In particular, the planners (most of whom are architects) and the developers viewed the transport system as a means to facilitate the sale of apartments – that is, as serving the residential development. Thus their main concern was to schedule the development of the transport system in tandem with the residential development, rather than the integration among transport modes or between public transport and development. Transport integration was thus not a priority to developers or planners in the field, most of whom were employed by the MHB, which also did not view transport integration as part of its purview.

6. The lessons

Many of the issues highlighted in this book as the necessary conditions to facilitate integrated transport policy in practice were present in Modiin, but integration has not been achieved to date.

The relatively small number of parties involved in the planning and development of Modiin and the centralized decision making in the planning and implementation stages could be expected to result in low transaction cost for integration. But in practice, the door-to-door transition costs within Modiin are substantial, implying that the transaction costs of integration were actually high.

The high transition costs identified in Modiin can be traced, at least in part, to all four factors identified earlier: the physical outlay of the city; the relationship between transport systems and land use; the institutional structure; and the interagency relations. Of these it seems that the institutional structure and lack of coordination between the agencies, particularly the MHB and the MOT, played the central role. Still, when looking at the five possible sources that impeded integration (the plan, its implementation, the operation of the public transport, institutional structure and policy, and exogenous variables), all five had some contribution – although again, the institutional structure and policy seems most central. However, if we go beyond this analysis, and ask why has the institutional structure been non-conducive for integration, it seems that the underlying reason for the lack of integration, despite the seemingly auspicious setting, was the fact that none of the actors really viewed the reduction of total transition costs as a primary concern, and none had the on-ground expertise necessary to reduce these costs.

If this is the case, several lessons can be gleaned from the case of Modiin. The first is that integration needs to be viewed from the bottom up. That is, in the planning and operation of public transport systems and their interfaces with land uses, it is necessary to analyze the movements of potential users from their perspectives. In the case of Modiin no one analyzed the city plan or the transport systems from the viewpoint of the two-worker household with two or more small children (which is typical in the city), who have to get out to school or day care before the parents can commute to a workplace that is often more than 30 minutes' drive away (either westward toward Tel Aviv or eastward toward Jerusalem). Only from this viewpoint can planners assess how and where public transport is a viable alternative within the time-space constraints of the household. In Modiin these constraints were not taken into account, therefore making public transport an infeasible alternative for many households. While it can be argued that it may be difficult to forecast the exact composition of the future population (although in the case of Modiin these forecasts were actually quite accurate), there were only scant attempts to incorporate behavioral insights into the public transport operations or to modify the implementation of the plan to take into account the behavioral facets of the population that moved in.

A second lesson is that the implementation time dimension is important. As households moved in they established travel patterns in accordance with the options that were available at the point in time when they moved. Once established, these may prove difficult to modify. Thus, the lack of suitable parking, public transport interfaces and kiss-and-ride options near the central rail station at the time it opened, as well as its late opening, may have had detrimental effects on the level of usage. Whether this is indeed the case requires further work, however.

A third, and perhaps more important, lesson is that integration requires specialization. Even if the planners and transport consultants are highly experienced and well aware of state-of-the-art techniques and discourses, integration may not materialize if none of them sees it as a primary priority and actually strives to implement it. As most planners and transport experts have not conducted projects whose explicit goal is to foster integration, they may pay lip service to it, and even design specific interfaces to facilitate it, but they may not be aware of the full range of issues they need to consider in order to make integration a reality.

A fourth lesson is the importance of institutional structure and cultures. In Modiin the MHB took the lead, and did not bother to bring anybody else in. Thus, from the perspective of all other actors, and in particular the MOT, Modiin was the MHB's problem, or baby, and not theirs. As the institutional culture in Israel is non-cooperative the result was that no other agency was ready to make an extra effort to ensure that the transport system in Modiin would indeed be integrative. Moreover, as the MHB's planning branch is dominated by architects, and has no transport specialists, there was nobody in the MHB who was placed in a position to criticize the plan and its implementation from a transport integration perspective.

A fifth, critical, lesson is that no amount of planning can assure that integration will actually materialize. It is essential that the integration ideal be pursued during the implementation phase. Yet, as Bardach (1977) cautioned long ago, it is more likely that a program (or plan) will not be implemented as foreseen than it will be. Thus, it is necessary to have a "fixer" (to use Bardach's terminology) who will help overcome the almost inevitable institutional discrepancies, oversights and the outright opposition of vested interests. In other words, it is imperative that there will be a person who will serve as an "integration agent" and will see transport system integration as his/her major goal. As most of the transport systems are run by multiple operators, and are separated from land use authorities, it may be difficult to identify the agency that will employ such an agent. Perhaps it can be suggested that such an agent should be locally based; that is, with the local authority that is the closest to the residents who use the system, and thus better aware of the discrepancies between the various systems.

Finally, it can be seen that partial integration or coordination has very limited benefits. If only some of the physical features are planned and built, the contribution to public transport patronage might be very limited. As noted at the outset of this section, integration has to be viewed and analyzed from the perspective of the user. The user is not interested in the institutional boundaries or in part of the trip, but only in the end result – the fulfillment of his/her desires. The ability of the transport and urban system to address these desires on the basis of public transport use is contingent on all the elements being in place, from the house's doorstep to the traveler's destination.

In the case of Modiin, the various elements needed to better integrate the transport system and the land use layout are slowly being put in place, sequentially. Thus, eventually the central bus terminal will be built, more parking will be provided near the rail station, a bus depot will be made available and bus schedules are already better integrated with those of other activities. To what extent these will affect the travel patterns of residents remains an open question at this point.

In summary, transport integration is not something that will simply happen, or be advanced successfully in a very incremental manner by haphazard actors. It has to be planned for, it needs people who specialize in such integration to plan it, and it requires an "integration agent" who will see it through. If these prerequisites are not present then even in highly propitious circumstances, such as were available in Modiin, integration may fail.

Acknowledgements

The authors wish to thank Mr Zvi Yeres, Ms Rakefet Tibi and architect Meiron Cohen for their help.

Note

1 The Dutch ABC system is a land use parking policy that restricts the provision of parking for businesses as a function of public transport provision, whereby the best-served nodes are allowed virtually no on-site parking.

References

Bardach E., 1977, *The Implementation Game: What Happens after a Bill Becomes a Law?*, MIT Press, Cambridge, MA.

Cohen-Blankshtain G., 2008, "Institutional constraints on transport policymaking: the case of company cars in Israel," *Transportation* 53, 411–424.

Furubotn E.G. and Richter R., 2000, *Institutions and Economic Theory*, University of Michigan Press, Anne Arbor, MI.

Geerling H. and Stead D., 2003, "The integration of land use planning, transport and environment in European policy and research," *Transport Policy* 10, 187–196.

Hull A., 2008, "Policy integration: what will it take to achieve more sustainable transport solutions in cities?," *Transport Policy* 15, 94–103.

Keil R. and Young D., 2008, "Transportation: the bottleneck of regional competitiveness in Toronto," *Environment and Planning C: Government and Policy* 26, 728–751.

Laws D. and Hajer M., 2006, "Policy in practice," in: Moran M., Rein M., Goodin R. (eds) *The Oxford Handbook of Public Policy*, Oxford University Press, New York.

Macario R., 2007, "Integration: an instrument for sustainability of urban mobility," in: Rietveld P. and Stough R. (eds) *Institutions and Sustainable Transport: Regulatory Reform in Advanced Economies*, Edward Elgar, Cheltenham, UK and Northampton, MA.

Modiin Muncipality (2008a) [Title: Planned transport centre of Modiin?]. Available online at: http://www.modiin.muni.il/modiinwebsite/ArticlePage.aspx?PageID=370_266 (accessed 14 April 2010).

—— (2008b) [Title: 20/20 plan?]. Available online at: http://www.modiin.muni.il/modiinwebsite/ChannelArticle.aspx?PageID=242_138 (accessed 14 April 2010).

Potter S. and Skinner M.J., 2000, "On transport integration: a contribution to better understanding," *Futures* 32, 275–287.

Safdie M., 1997, *The City After the Automobile: An Architect's Vision*, Westview.

Viegas J.M., 2005, "Integrated transport systems: public-private interfaces," in: Button K. and Hensher D. (eds) *Handbook of Transport Strategy Policy and Institutions*, Elsevier.

Chapter 18

Integrating public transport management in France

How to manage gaps between mono-scale policies

Pierre Zembri

1. Introduction

The integration of public transport policies, especially where several decision levels are operating within a given urban territory, may provide the best guarantee for efficient intermodality. In Great Britain, this requirement was underlined during a conference given at the Royal Society of Arts in November 1973 by Sir Colin Buchanan and Geoffrey Crow (Buchanan and Crow, 1974). Since then a large number of contributions have reinforced this idea, and a systemic approach has been developed to analyse travel movements and location decisions. More recently, a large number of authors have integrated the environmental aspect into this approach (in particular, see Hull, 2007; Geerlings and Stead, 2003, Bertolini *et al.*, 2008; May *et al.*, 2006), given that transport is now a principal source of pollutants and CO_2 emissions in the urban environment. Public authorities acknowledged this movement (European Conference of Ministers of Transport, 1998), noting a growing social demand for a better quality of urban life, and this in turn means a better quality of transport services (Hine and Scott, 2000).

The recent energy crisis provides an opportunity to promote best practices in terms of organizing modes of transport of a sufficient quality to replace private cars. Reconfiguration of urban form, to locate housing and workplaces closer to means of public transport is a logical step forward, although it requires a long lead time, and similarly improving the efficiency of existing networks has now become essential. The recent growth in public transport ridership is unprecedented, even in countries where the private car plays a dominant role, such as the United States.[1] In the specific case of France, the most recent surveys carried out in large cities such as Lyon, Rennes, Rouen and Lille reveal that public transport has begun winning back some market shares following a long decline (Quételard, 2008). Between October 2007 and June 2008, the regions and SNCF (French railways) engaged in a forward-thinking exercise to examine the consequences of quadrupling regional traffic (TER[2] × 4) by 2030. This was a response to the current annual ridership growth rate which was between 7 per cent and 13 per cent in 2007.

But the French case, with the exception of Île-de-France, is also characterized by a considerable dispersal of responsibilities in terms of transport organization, given that it is subject to four decision-making levels (Table 18.1). Each network operating in a given territory has an 'organizing authority'[3] that has no obligation to coordinate its actions with other ones. At each level, the various policies that were supposed to be jointly applied are not necessarily run by the same departments and the financing systems differ from one mode to another. This makes integration difficult and makes for a lower level of overall efficiency in the public transport service, and in the organization of alternative individual modes of transport such as walking and cycling.

This chapter proposes placing the current unfavourable situation in an international perspective by examining the configurations to be found in other European countries. The extent to which transport policies are integrated appears to depend on both the institutional framework and the nature of the relations between transport operators and organizing authorities.

In the next section a description of the French institutional framework is given, together with some details about the functioning of the French public transport market, which is characterized by an open market for urban and inter-urban networks and by a monopoly for the regional rail market. This is followed by an analysis of the quality of intermodality resulting from the organization already described. The chapter ends with an international comparison showing that the number of decision levels is not the only criterion to take into consideration.

2. Description of the French institutional framework

The National Transport Framework Act (Loi d'orientation des transports intérieurs – LOTI), dated 30 December 1982, provided the four political decision levels for the three transport markets (Table 18.1). While the distribution of competences initially appears logical, a large number of overlapping functions have subsequently revealed themselves.

It is at the level of large towns that the situation is most complicated, as there is a real interlocking of competences (Zembri, 1999, p. 57). Although the four categories of services are present on the same territory (long-distance trains, regional trains and coaches, départemental coaches and urban services), there is no obligation to coordinate the services. There is little integrated fare pricing, given that the level of integration depends on the willingness of the various institutional partners to work together and to transfer resources from one mode to another. It was only imposed in the Paris region (Île-de-France) in 1971, with the creation of the Région des Transports Parisiens.

The differences are equally important in terms of financing, making barriers harder to overcome. There is just one market (urban transport) that benefits from a specific resource called the Versement Transport[4] (VT), which was created in 1971 exclusively for the Paris region. It takes the form of a tax on the payroll of companies with over nine employees located in the territory served by urban transport networks. Its level, voted by the local authority, depends on the density of the networks, with the maximum only being charged in Paris and the Hauts-de-Seine Département, which incorporates the La Défense business centre. From 1973, VT collection was extended to other large cities and then to smaller towns. Since 2000, it has been applicable to any urban centre with over 10,000 residents. The development of the Metro, tramway and other public transport systems owes a considerable amount to this vital resource given that it now represents 40 per cent of the sector's income (Figure 18.1). The levying rates also take the existence of guided public transport projects into account (Table 18.2), and most recently they have included the integration level of the organizing authorities. For example, the choice of Federation of Municipalities status (Communauté d'Agglomération[5]) provides a 0.05 per cent 'bonus' over the maximum rates.

On the basis of this data and on the fact that the unitary expenses are

Table 18.1 Modes, markets and organizing authorities in France

Decision-making level	*Urban transport*	*Inter-urban bus transport*	*Rail transport*
State		Interregional	Interregional
Region		Inter-départemental	Regional
Département[1]		School buses and intra-départemental	
Local authorities (in the framework of an urban transport perimeter – PTU)	Bus, light rail and Metro		

Note:
1 Level of administration analogous to the British county or the Italian province. We use the French term and the corresponding adjective 'départemental' in order to avoid confusions with 'department' as a part of an institution.

Table 18.2 Transport levy rates applicable since 2006

Category	*Maximum levying rate on payroll*
Towns with between 10,000 and 100,000 residents	0.55 %
Towns of over 100,000 residents outside Île-de-France	1.00 %
Towns of over 100.000 residents outside Île-de-France with guided public transport project (tramway, Metro, etc.)	1.75 %
City of Paris and the Hauts-de-Seine Département (La Défense)	2.50 %
Seine-Saint-Denis and Val-de-Marne Départements	1.60 %
Other Île-de-France Département (outer ring)	1.30 %

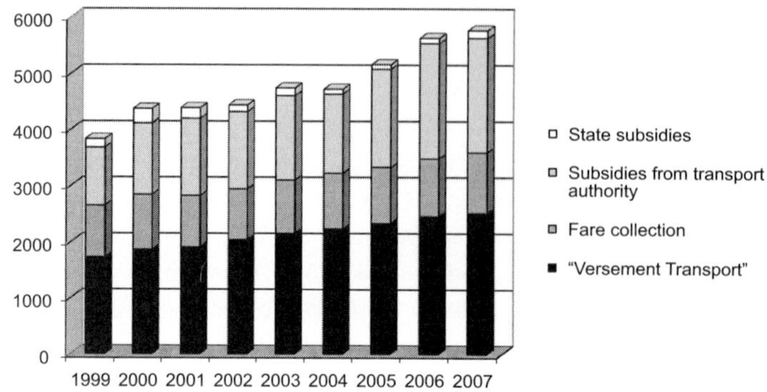

Figure 18.1 Sources of urban public transport financing outside Île-de-France and their increases between 1999 and 2007 (in millions of Euros)
Source: GART, 2008

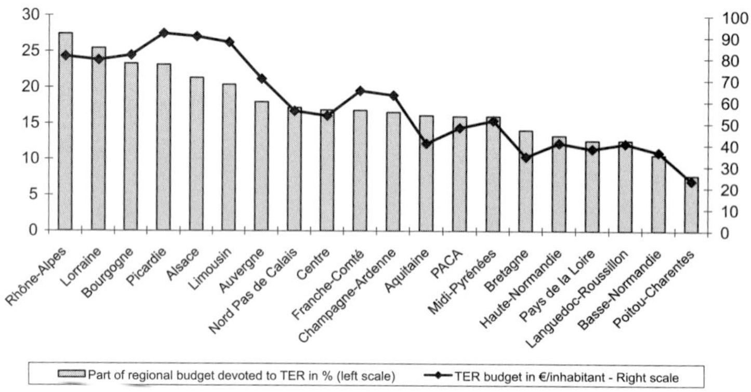

Figure 18.2 Measurement of the regional budgetary effort using two indicators (2005)
Source: Rail et Transport, *2006*

under control, it can be concluded that the development of the urban public transport sector is strongly supported. The progressive financial withdrawal of the State has not resulted in a loss of income. In contradiction with the earlier trend, the recent national 'Grenelle Environment Round Table' (Grenelle de l'Environnement) held in November 2007 saw the State promising to finance new tramway projects.

Regional rail transport also profits from abundant financing, but this is essentially State funding. The transfer of responsibility for these services to the 20 concerned regions (Île-de-France and Corsica being subject to a specific plan) took place between 1995 and 2002 and saw the State make a considerable budgetary transfer. Its participation in the cost of fast-growing TER services increased from €598 million in 1994 to €1.7 billion in 2004. The budgetary effort made by the regions was equally important, although unequal due to considerable disparities in their resources (Figure 18.2). Fare income represents about 30 per cent of operational costs (Association des Régions de France, 2007). The TER is not eligible for the VT, even for lines included in urban transport perimeters (PTUs).

Inter-urban transport by buses (excluding school transport which is fully covered by public funds) is the sector least well equipped in operational terms. Investments made by the Départements are generally low and rather patchy. Most operators only have fare income to balance their operating costs given that public subsidies give precedence to renewing the rolling stock. Investment in the 85 surveyed Départements represented a total of €36 million in 2003,[6] and 39 per cent of the Départements have done nothing to harmonize price setting between the different sub-networks.

At this point, the three existing transport markets (urban, inter-urban buses and regional rail) have little in common, either in terms of the means of financing or the involvement of the organizing authorities. In the next section discussion focuses on the methods used to operate the services.

3. The operation of the French public transport market: historically open to competition

Compared with a large number of European countries (Spain, Italy, Switzerland, etc.) that have remained loyal to the municipal urban transport company model, or its German multi-service variant called Stadtwerke, urban transport in France (with the exception of Île-de-France) was opened up for competition through a tendering process. This took place in the 1960s, during a period of underinvestment and decline in use of public transport. The initial long-term concessions (up to 99 years) given to contractors that had built a large proportion of the 101 tramway networks were gradually ended. These concessions took the form of models similar to those of the current build, operate and transfer (BOT) contracts or municipal ownership systems set up to achieve the same end but operated exclusively through public funds. The main reason for the decline in ridership was the competition from the car, a lack of investment in new rolling stock, and an outdated infrastructure that needed replacement. In all cases, with the exception of three lines in Lille, Marseilles and Saint-Etienne, the change in system coincided with the winding up of the old tramway networks that took place from 1961.

The adopted principle remained innovative for quite some time, and covers short-term operating contracts (3 to 8 years) based on public tenders, with payment of a subsidy calculated according to criteria that can be different between one town and another (lump sum, capped, by kilometre produced, etc.). It also addresses the ownership of the infrastructure, machinery and equipment, as these remain the property of the local authority making the investments, but the personnel remain attached to the operation and are transferred from one contractor to another should the service provider be changed.

The result is the emergence of pure operation contractors that did not have to make major investments and made profits on the basis of increased productivity or a greater level of clientele using the networks. During the 1970s and 1980s, there was considerable consolidation which resulted in the emergence of three large groups, a few private independent or semi-private companies,[7] and a small number of direct operations by local authorities with their own staff and vehicles (Figure 18.3).

The inter-urban sector was even more fragmented and generally placed the operations of services in the hands of small- and medium-sized local and regional companies. The market was considered to be open to competition, with the exception of 16 Départements which continued to run the systems themselves (self-production). The three largest groups continued their development by successive purchases, with Veolia Transport taking the leading role. In 2002, Veolia Transport

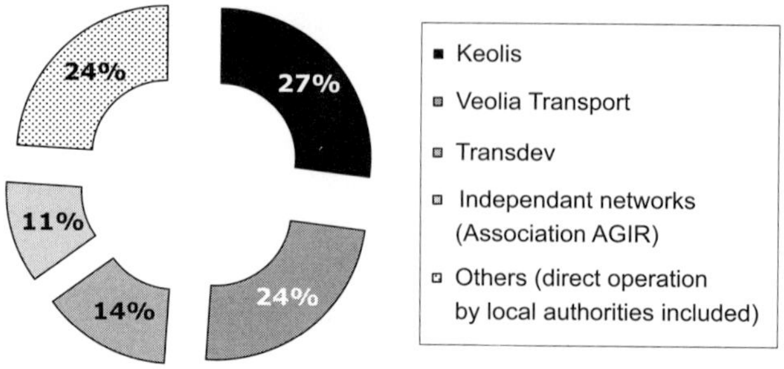

Figure 18.3 Typology of urban transport operators in France in 2007
Source: GART, 2008

bought Verney, the largest remaining independent family-owned group. Operating on both urban and inter-urban markets where work organization, the status of employees, and salaries were different, the groups were quick to use their urban operations to subcontract services to their local inter-urban subsidiaries.

In 2008, the three main French groups having profited considerably from the calls for tender across the world using the French model were respectively, 1st (Veolia Transport), 3rd (Transdev) and 4th (Keolis), on a world scale based on their turnover.[8] Veolia Transport (€6,080 billion turnover) is the only group present on all five continents.

Regional rail transport continued to be a monopoly, as the market had not been opened as in Germany and Sweden. The regions could only contract with SNCF and, even then, the contracts were for relatively short periods (5 to 10 years) and became increasingly complex when the reform based on the Solidarity and Urban Renewal Act dated 13 December 2000 came into force on 1 January 2002. 'Bonus-malus' clauses[9] were added, especially insofar as service quality was concerned. Major disputes developed as SNCF did not provide all the means necessary to meet the level of offer to which it had committed itself. However, to develop its regional sector, SNCF reorganized and incorporated itself into a new management structure (created in 2005) called Proximités, integrating Keolis, which became a subsidiary, in order to constitute a multimodal operator.[10] Currently only one region (Alsace) envisages opening its network to competition as from 2010. This is the first year that the market can be opened to local authorities.

While the players are close to one another on urban and inter-urban markets, their different statuses make it difficult to integrate operations. In addition, the growing power of the large groups is perceived as a potential danger by urban transport authorities (Zembri, 2007). These are now asking themselves whether there should be an 'allotment' of networks. This means breaking them into sub-networks whose size would make it possible to attract smaller or more specialized contractors, for sectors such as on demand responsive public transport or the transport of disabled persons. SNCF, which took control of Keolis, is attempting to impose 'packages' that incorporate the urban network and extensions on the national rail network by the use of hybrid light rail networks, and SNCF is trying to control the multimodal information and fare pricing. All this is causing some concern. However, the carriers themselves are not effectively grouping operations together and such effective grouping cannot be done by the organizing authorities under current regulations.

4. A difficult integration

It was demonstrated (Zembri 1999, p. 58) that, with the exception of the Île-de-France, the negative effects of the French institutional configuration take two forms.

First, there are the *barrier effects* between urban transport perimeters (PTUs), which generally favoured lower fares, higher levels of service and the adjoining territories. These effects can be even more prejudicial given that the PTU does not necessarily cover the entire urbanized territory, and the inclusion of the communes is not obligatory. For reasons of political opposition to the main central commune or for fiscal selfishness (refusal to share the income from local taxes on businesses), certain communes have refused to join their PTU, and nobody can oblige them to do so. In some large towns (Tours, Rennes, Nancy) there are even two separately managed PTUs.

Second, there are the *difficulties in providing integrated railway network* services within large towns. Many public transport authorities cannot directly intervene in the management of railway services, even for lines completely located within their territorial jurisdiction, and these lines cannot claim any urban transport financing (such as income from the transport payment). The only responsible body is the regional government and this implies agreements between the regional and the urban authorities, and the transfer of resources (especially those coming from the VT).

These problems were highlighted when hybrid light rail projects were developed during the 1990s. According to the model developed in Karlsruhe and then in Kassel (Germany), the task was to interconnect urban tramway networks and the regional rail network in a way that could compensate for the outside-the-city-centre locations of certain central stations. To date, only one project has been translated into reality. This is the network introduced in Mulhouse, but even here there is no certainty that the project will be fully completed. This simultaneously raises the issue of the nature of the authority responsible for the service (regional government, local government, or both together through a common structure), as well as the nature of the operator (SNCF, urban operator, or both at the same time). The problem is that the projected lines go beyond PTUs. If they use routes already served by regional trains, then the regional government has no choice but to be a stakeholder. If the approach taken is to recuperate former railway lines that are no longer used, the Département is also involved as it is responsible for the inter-urban bus services that will be affected by the project.

Another cause of friction between public transport players is the creation of mobility service centres in application of Article 113 of the Solidarity and Urban Renewal Act (Loi Solidarité et Renouvellement Urbain – SRU), dated 13 December 2000. This requires all towns with over 100,000 residents to have a multimodal information service available to users. In addition, the requirement that public transport networks will be fully accessible (imposed by the 2005 law on equal opportunities) demands that public transport services be adapted to be fully accessible to the disabled (transport des personnes à mobilité réduite – TPMR) or allow advanced bookings by the disabled in addition to regular accessible lines. Finally, on demand responsive public transport (transport à la demande – TAD) is being developed in most large towns, both around the outskirts as well as in certain highly populated districts.

This has led a growing number of transport authorities to prepare specific calls to tender for the distribution of information to passengers, with the management of TAD/TPMR bookings. The contractors submitting tenders included specialized service providers as well as known transport players that wanted to add a new business to their carrier activity. However, certain candidatures have an ulterior motive. For instance, the Clermont-Ferrand mobility service centre was won by Veolia Transport despite the fact that the group was not present on the local urban network. Similarly, SNCF and its Effia subsidiary consider that intermodality should be constructed on the basis of a framework formed by TER lines on a regional scale and that they need to control multimodal information.

The SRU Act also attempted to facilitate hybrid projects by providing the possibility of creating multilevel transport authorities incorporating several local authority levels that are able to charge the transport payment in their particular area of jurisdiction. Eight years later, it can be seen that this initiative has not been particularly successful, as only eight companies have been created, despite over 100 towns being eligible. The distribution of responsibilities within these types of structures is problematic, with the root problem being that they create an additional management level without in any way altering the existing stacking of competences. To quote Francis Beaucire (1997): 'The complete separation of design and decision between the organizing authorities, whether due to indifference or, more seriously, rivalry, has until now completely prevented the potential development of all the various forms of intermodality'.

5. Generally unfavourable international comparisons

Figure 18.4 compares the way public transport is organized in the European Union countries. The comparison is largely based on the number of decision-making levels which range from a single level in Ireland and in Ulster, to four in France. The other countries have two or three organizing authority levels. A first conclusion would be that the main source of the French problem lies in this overly high number of decision-making levels, which other comparable countries do not have. Even within France, the Parisian case, where there is only a single level (the Syndicat des Transports d'Île-de-France, dominated by the regional authority) and a high level of integration (financing methods, fare levels, etc.), would be a good example to follow, but is considered as a very costly one.

But the reality is not that simple. The number of levels must be cross-referenced with market organization methods, including the relevance of scales and whether or not partitions exist between the various networks. Taking, for example, countries with only two decision-making levels, a number of highly differentiated situations in terms of the integration of network and transport policies are revealed.

In the case of the Scandinavian countries (Sweden, Norway, Denmark and Finland) and Switzerland, Belgium and the Netherlands,[11] there are two levels, the State and the regions. The former is responsible for national railways while the latter manage all the rest, including the local (urban and rural) and regional rail and

Figure 18.4 Transport authorities in Europe: number of decision-making levels and urban regions with a particular status
Source: Ollier and Pico, 2006

bus services. The scale of the regions means that the vast majority of standard travel movements (home–work, home–studies) are included. Tariff integration is complete, no matter whether there are one or more operators active in the region. In the Netherlands, fare prices are organized on a national level using numbered zones. Travelling from one town to another using the national railway network means purchasing a multimodal transport service from the departure zone to the arrival zone and this includes local transport, both upstream and downstream from the railway route. It is possible, within a given zone, to travel on any of the transport modes and use any of the transport operators present. In a small country like Denmark, there are only five transport authorities and these correspond to the main towns and their zones of influence.

In the case of the United Kingdom (with the exception of Greater London and Northern Ireland), there are also two decision-making levels, namely the State and the towns. However, the organization of the market since the introduction of the 1985 Transport Act means that it is no longer possible to apply a network integration policy on the level of the towns. This is despite the existence of the Public Transport Executive (PTE), which only has a very limited role, mainly infrastructure planning (e.g. tramways) and the financing of social services when the private operator offering is considered to be incomplete. In practice, there is no tariff integration, as each transport operator is free to set the prices it wants, and it is rare that the tickets issued by one operator are accepted by the others. Wanting to harmonize fare rates would interfere with the free competition principle, an absolute concept enacted by the Transport Act. In the case where a local authority constructs a tramway network, the transport operators present in the town can retain parallel itineraries and thus be in competition with the new service.

The German and Austrian cases reveal that having three decision-making levels (State, Länder[12] and towns) is not incompatible with a high level of networks integration. In practice, they work in partnership within pricing communities (*Verkehrsverbund*) that correspond to the zone of influence of a town or a group of towns. The issue of urban growth is resolved by the inclusion in the community of the concerned communes step by step. Generally speaking, pricing communities are based on a company whose shareholders are both the various levels of transport authorities (with the exception of the State) and the public and private operators, including national railways, municipal transport companies, and other road or rail transport operators. The company acts as a clearing house, distributing the income and subsidies between the various operators. The setting of fare rates is integrated and based on 'honeycomb' zones. Principles of continuity are negotiated between neighbouring pricing communities for cross-border routes. The result is a very high performance 'network of networks' that is completely seamless for the user despite the presence of a number of operators and several decision-making levels. The system is highly resistant to the opening of the regional railway markets to competition now taking place in Germany. New operators having obtained a public service designation are invited to join the community and fare setting remains in the hands of the organizing authorities.

This comparison concludes by noting two contradictory logical approaches that can currently be observed across the world. First, there is the fragmentation of networks into sectors that are subject to separate calls for tenders; and second, there is the full integration of certain contracts by combining various operational modes and scales.

In the former case, the organizing authority assumes a growing responsibility in terms of the design and the commercialization of the service, with the transport operator simply becoming a service provider, a mechanism in a system within which it has no power. Its role is essentially to produce within the best economic conditions while fully respecting the specifications. It is possible to break up the networks, subcontract the management, the distribution of information, and even the control of the execution and the quality of the services. This prospect is particularly appealing to transport authorities concerned with being overly dependent on the small number of large operators present in the market. This kind of management makes possible the use of services of smaller or more specialized contractors (such as those specialized in the on demand responsive public transport sector).

In the latter case, the transport authority places all the industrial and commercial risks in the hands of the chosen transport operator and only retains organizational and supervision functions. The highest level of simplicity was attained in 2006 in the province of Limburg (the Netherlands) where all transport networks (regional rail, urban and inter-urban bus, school buses, on-demand services) were grouped together and tendered as a single block. Tenderers were invited, for a given subsidy level, to suggest the best possible network configuration able to meet the needs of the organizing authority. Veolia Transport won the competition by offering the best level of service for the largest part of the region (with an extended use of on-demand services).

6. Conclusions

Making changes to the French system based on good practice observed elsewhere means working in two directions. First, there are the defining perimeters of competence without any preliminary territorial or modal constraint. This should be viewed at a scale allowing the extended zones of attraction of towns to be taken into consideration. Integration requires unitary management, financing and pricing, as well as the multiplication of connections between networks. The systematic association of a scale (or a mode) and an organizing authority has revealed its limits.

Second, there is the smoothing out of the regulatory differences between the various modes of public transport (particularly urban and inter-urban services) so that, if required, calls for tender for large multimodal networks can be made. This would make it possible to derive the best possible use of the existence of large groups, incorporating all the skills being sought and the opening up of the last few remaining protected markets (e.g. regional railways).

In all cases, the transport authority must retain control over the configuration of the networks (infrastructures, services), pricing, information and the quality of the service. Experience has proven that a successfully integrated network is one where the public authority is powerful and exercises a continuous control over the execution of the service and the use of public funds, no matter what the nature or number of operators.

It would be possible in France to retain the existing number of decision-making levels, despite the fact that it is the only country in Europe with so many. But this outcome would require the pooling of services and financing within the framework of a German or Austrian type of pricing community and the geographical regions would need to be much larger than those of the existing PTUs in France.

Notes

1 As an illustration, here are the titles of a few recent articles printed in *The New York Times*: 'Gas prices send surge of riders to mass transit' (10 May 2008), 'Travellers shift to rail as cost of fuel rises' (21 June 2008), 'Politics failed, but fuel prices cut congestion' (3 July 2008), 'Fuel prices shift math for life in far suburbs' (25 June 2008).

2 TER means 'regional express rail services' (Transport Express Regional).

3 This is the literal translation of the French official term 'Autorités organisatrices' for transport authorities. The organizational tasks include network design, tariff setting, collection and redistribution of resources, contracting with operators, etc.

4 A literal English translation would be 'transport payment'.

5 This is a public establishment responsible for cooperation between municipalities. It brings

together a large number of competences (particularly town planning, transport, road management and station management) which, at least in theory, allows it to implement fully integrated policies.

6 Figure from GART (2007).
7 These associate a private operator (large group or independent) owning less than 50 per cent of the shares and the organizing authority owning the rest.
8 Figures from GART (2008). This ranking only concerns service provider groups with no particular territorial attachment and which, consequently, are not 'in-house' within the meaning of the 2007 European regulation concerning public service obligations.
9 When the fixed objectives of performance are exceeded, the operator profits from additional money given by the public authority, as a reward for its efficiency (the 'bonus'); and, when the objectives are not achieved, the operator is penalized and receives less money from the authority (the 'malus').
10 Keolis was not competing with SNCF on the rails. In the French market, this company only operates urban public transport and inter-urban bus networks. SNCF just wanted to become a 'global' operator through external growth.
11 In the specific case of the Netherlands, three large towns provide the exception to the rule: Rotterdam, Amsterdam and The Hague, which all make use of a third level.
12 The German administrative unit equivalent to regions and counties.

References

Association des Régions de France (2007). Presentations available online at: http://www.arf.asso.fr/index.php/documents/congres_de_l_arf/3e_congres_de_l_arf (accessed 10 April 2010).

Beaucire, F. (1997) 'La famille des "Inter": sous des termes "mode", de nouvelles façons de voir', *Transports Urbains*, 97, October: 3–4.

Bertolini, L., Le Clercq, F. and Straatemeier, F. (2008) 'Urban transportation planning in transition', *Transport Policy*, 15, 2: 69–72.

Buchanan, C. and Crow, G. (1974) 'An integrated transport system', *Journal of the Royal Society of Arts*, 2, February: 117–128.

European Conference of Ministers of Transport (1998) *Urban Travel and Sustainable Development*, Paris: ECMT.

GART (2007) *L'année 2006 des transports urbains en France.*

—— (2008) *L'année 2007 des transports urbains en France.*

Geerlings, H. and Stead, D. (2003) 'The integration of land use planning, transport and environment in European policy and research', *Transport Policy*, 10, 3: 187–196.

Hine, J. and Scott, J. (2000) 'Seamless, accessible travel: users' views on the public transport journey and interchange', *Transport Policy*, 7, 3: 217–226.

Hull, A., (2007) 'Policy integration: what will it take to achieve more sustainable transport solutions in cities?', *Transport Policy*, 15, 2: 94–103.

May, A., Kelly, C. and Shepherd, S. (2006) 'The principles of integration in urban transport strategies', *Transport Policy*, 13, 4: 319–327.

Ollier, B. and Pico, F. (2006) *Organisation and Major Players of Short Distance Public Transport*, Brussels: UITP.

Quételard, B. (2008). 'Du nouveau dans le "partage modal"?', *Transports Urbains*, 112, March: 9–12.

Rail et Transport magazine (2006), 'Palmarès des Régions 2006' 11/08/2006, pp. 46–59.

Zembri, P. (1999). 'Mutations de la mobilité et rigidité des périmètres de gestion des transports publics en France: un divorce croissant entre deux territorialités', *Rivista Geografica Italiana*, 2: 55–72.

— (2007) *Pour une approche géographique de la déréglementation des transports*, Vol. 2, unpublished Habilitation Thesis, University of Paris 1, December.

Chapter 19

Intermodalism in the US

Issues and prospects

Joseph S. Szyliowicz

1. Introduction

Transportation has emerged as an important item in practically every country's political agenda as existing systems are increasingly unable to deal with new demands and expectations for both passengers and freight. Problems of urban congestion, environmental pollution and bottlenecks that impede the smooth flow of goods internally and internationally are commonplace. Planners attempting to ameliorate such problems have increasingly come to recognize the degree to which the traditional approach, one that deals separately with the individual modes (road, air, rail, water), no longer suffices. Accordingly a new vision of transportation has emerged, one that holds that it is essential to view transportation from a system perspective and to develop policies and projects that recognize the interrelatedness of the modes and their particular characteristics. In the United States this approach is known as intermodalism.

Although there is general agreement on the importance of this approach, what it actually means remains unclear. For example, it has been defined as: "the coordinated passage of goods and people by way of two or more of the primary modes of transport (sea, air, rail, road) from origin to destination as defined by the passenger or the shipper and consignee, with a single travel directive bill of lading or ticket and a single price covering the entire trip" (Alt *et al.*, 1997, p. 34).

Such a definition captures the integration dimension well but it fails to incorporate such critical elements as choice and inclusiveness. Thus, intermodalism has also been defined as: "a system that is both safe and efficient and productive and flexible in responding to the needs for good movements and ... offer(s) people choices and flexibility in their personal movements." This system must also be "international, intelligent and inclusive" (Jeff, 1998, p. 13). Yet even this definition is inadequate because it fails to recognize explicitly the externalities of transportation.

It is possible to develop an integrated system that is safe, efficient, flexible, intelligent, international, and inclusive but which continues to pollute the environment and waste energy. Accordingly, the following definition is proposed:

> An intermodal system is one in which the individual modes are linked, governed, and managed in a manner that creates a seamless and sustainable transportation system. Such a system should be economically efficient, environmentally sound, safe and secure and ethically based.

Implicit in such a perspective is the idea that each mode should be utilized for the purpose for which it is best suited in terms of these considerations. Thus, as many containers as possible should move by rail and not by road, and the aviation mode should only be used for high value long-distance and trans-oceanic trips. Furthermore, every effort should be made to minimize the negative impacts that are inherent in each mode. Such a system maximizes efficiency, offers more choices for personal and freight mobility, and minimizes environmental impacts and the use of energy – a critical point given transportation's heavy reliance on petroleum and its contribution to global warming (an interesting case in relation to the global warming effects of transport is the related legislation in California; see Chapter 4 in this volume).

Creating such a system has proven to be a complex and difficult challenge in the US owing to numerous technical and institutional barriers and obstacles (these barriers are also common to other places; see for example Chapters 17 and 18 in this volume for specific examples and Chapter 2 for a more general discussion in a European context). However, powerful forces have made the need for such changes ever more evident and today many voices are calling for profound changes to the existing transportation systems.

2. The pressures for intermodalism

To begin with, the globalizing forces that are shaping the contemporary world have important implications for transportation, which itself has played a major role in spurring these forces. Indeed some argue that intermodalism was born with the development of the container; although its roots can be traced back to the 1950s when trains began to carry trailers. But the real revolution started in 1956 when a ship called *Ideal X* sailed from New Jersey to Texas with a load of containers. The visionary Malcolm McLean had devised a technology that permitted a qualitative change in the way that freight was handled. Until then, ships were loaded and offloaded much as they had been for centuries; now goods could be shipped across the globe more cheaply than ever before, with profound consequences for international trade and investment patterns.

The container also affected the road and rail modes. It provided new business for the American railroads, which possessed under-utilized capacity. Deregulation accelerated this process by eliminating numerous rules and regulations that had prevented carriers from innovating and competitors from entering the market. Imaginative entrepreneurs, who recognized the economic advantages of shipping freight in an integrated way through different modes, were able to seize market opportunities. New patterns of cooperation and competition emerged as a host of new actors became active and founded corporations such as UPS, FedEx, and J.B. Hunt, which have since become major global actors. Concomitantly, important technological innovations, such as double stack trains, further spurred the revolution whereby air, ship, rail and truck became intertwined, so that today the US is able to transport large numbers of containers intermodally.

Within a few decades, global supply chains linked resources and markets and existing consumption patterns were transformed, as what had once been exotic goods became readily available. Production patterns changed as well, not only geographically but through such techniques as "Just in Time" production which further reduced costs by increasing productivity and minimizing inefficiencies. Today national development and well-being no longer depends solely on the ways in which the productive sectors operate but also on the ways in which their products are distributed rapidly and efficiently to international markets. This requires integrated national transportation systems with sophisticated ports and other infrastructures.

The US has attempted to move in this direction but there is widespread agreement among policy makers, shippers, operators, and many other groups that the existing transportation system no longer meets current needs, let alone those of the future. Since the legislative framework, SAFETEA-LU (the Safe, Accountable, Flexible and Efficient Transportation Equity Act: A Legacy for Users) will expire in 2009, an important debate has begun on the kinds of changes that are necessary. It has been driven by a growing recognition of the sorry state of the existing system, which has reached its capacity limits and a large percentage of its components – urban roads, bridges and rail – are in poor condition (Brookings Institution, 2008, ch. 3). Freight congestion issues have gained particular attention for, as a recent headline in my local newspaper stated, "Supply Chain near Stopping in Its Tracks" (Tarm, 2008).

Moreover, this situation will only deteriorate further; although, because of the global recession, at a slower rate than has been projected. The number of containers that the US imports, which stood at 8 million in 2000, for example, was expected to reach 28 million by 2015 (National Surface Transportation Policy and Revenue Study Commission, 2008 (hereafter cited as "NSTPRSC"), vol. II, ch. 2, p. 12), but this estimate has recently been revised sharply in 2010 down to 18.1 million. Thus, such issues as the ability of US ports to handle container freight adequately (Pisano, 2006, p. 5), or the investment needs of the railroads in Southern California where the major container ports are located, an estimated $150 billion over the next three decades (Pisano, 2006, p. 7) are no longer as acute. The existing system, however, already imposes long delays and thus heavy costs on all the freight companies, an estimated $7.8 billion in 2005 (Pisano, 2006, p. 7). Hence, the problem identified by the following forecast will have to be resolved, although the timeframe can, most likely, be extended: "Without this investment ($148 billion), 30 percent of the rail miles in the primary corridors will be operating above capacity by 2035, causing severe congestion that will affect every region of the country and potentially shift freight to an already heavily congested highway system" (American Association of Railroads, 2007).

Human lives are also at stake – those delays cause increased diesel particulate emissions. These emissions, although less damaging than other pollutants, are largely responsible for 2,400 deaths annually in California alone (American Association of Railroads, 2007).

Americans are increasingly concerned with such environmental impacts, as well as with the social costs of transportation, including disruptive freight movements through their communities and associated safety and security issues (Szyliowicz, 2009). As a result, transportation projects require detailed environmental impact evaluations that can serve to delay their implementation by years.

At the same time, people are demanding ever faster, more reliable, and convenient travel services and their numbers are rising rapidly. Demographics – the size, age, gender of the population–shapes not only demand for goods but individual mobility as well. The US population is projected to reach 364 million by 2030 and 420 million by 2050 (NSTPRSC, 2008, vol. II, ch. 2, p. 4), an increase which will place additional demands upon an already stressed passenger system. Deregulation produced an explosive growth in air travel and tourism, which created the well-known problems of air traffic congestion and limited access to airports. The latter situation has been aggravated by the fact that, unlike Europe where most airports have direct links to city and inter-urban rail systems, many American airports lack a transit link. However, here too, the global recession has inexorably led to a sharp drop in demand, the duration and impact of which remains unclear.

Essentially, passenger transportation remains highly dependent on the private automobile and public highways and, from an intermodal perspective, lags well behind freight, which is commonly shipped in containers on an efficient rail system. The limited development of rapid transit as an attractive alternative to the automobile imposes heavy costs upon society in terms of adverse environmental, health, energy use and safety impacts, and social equity since many groups such as the poor, the elderly, and the handicapped do not have access to a car. Nor should one overlook the opportunity costs in the use of resources for overly intense auto production versus investing in other modes and intermodal connections.

The financial costs alone are staggering. In 2005 urban congestion cost about $78 billion (based on wasted time and fuel), an increase of almost $5 billion from 2004. The amount of wasted fuel was about 3 billion gallons of fuel – the equivalent of 290,000 gasoline tank trucks (Texas Transportation Institute, 2007). To this must be added the cost of automobile accidents that has been estimated at $164.2 billion a year (American Automobile Association, 2008).

However, as noted above, transportation is now widely viewed not merely in traditional economic terms but in terms of its sustainability. Because of such concerns, today it is practically impossible to envisage, in the US, a program to greatly expand the highway system. Nor is it likely that the severely constrained airport capacity will soon be eased to any significant degree despite the increased use of slack terminal resources with growth in the use of secondary airports. Accordingly, over the past decade or so, increasing numbers of public interest groups and policy makers in the public and private sector have come to recognize the need to change this situation, and have been paying renewed attention to the development of rapid transit in urban areas and intercity rail travel because of their many advantages in terms of fuel efficiency, pollution, and other environmental impacts. However formidable barriers will have to be overcome before the goal of a truly intermodal passenger system is achieved.

3. Development of intermodalism

By the end of the twentieth century, political leaders had come to recognize that traditional transportation policies and practices no longer sufficed to provide the country with the kind of transportation system that could meet national goals. The rapidly changing situation required a new vision and in 1991 the US made a definite commitment to intermodalism. That year Congress enacted the landmark Intermodal Surface Transportation Efficiency Act (ISTEA) that moved policy away from the age-old emphasis on specific modes, notably the highways, towards intermodalism. The ISTEA legislation signaled the beginning of a new era in transportation policy and planning through its explicit use of the word "intermodal" in the title. For the first time, federal legislation recognized the constraints and negative consequences imposed by traditional modal policies and the need for a new approach that emphasized flexibility, innovation, and greater public involvement. Its purpose was explicitly stated in the 1991 Intermodal Surface Transportation Efficiency Act's "Declaration of Policy" as follows:

> It is the policy of the United States to develop a National Intermodal Transportation System that is economically efficient and environmentally sound, provides the foundation for the Nation to compete in the global economy, and will move people and goods in an energy efficient manner.
>
> (USDOT, 1991)

Such a vision could obviously not be achieved easily or quickly even though it is limited to the transport sector and does not include important related elements such as land use. One important gap that was immediately apparent was the limited understanding of the actual state of intermodalism and the role of various actors, as well as the barriers and obstacles, which hindered its development. Accordingly, the National Research Council organized The National Commission on Intermodal Transportation (1994), and the Transportation Research Board convened several important conferences – the National Conference on Intermodalism: Making the Case, Making it Happen (1994); the National Conference on Setting an Intermodal Transportation Research Framework (1996); Intermodal Transportation Education and Training (1997); and the Committee for the Study of Policy Options to Address Intermodal Freight Transportation (1998) – to study and discuss such issues.

Their very titles suggest the problem areas that concerned government officials and other experts interested in the growth of intermodalism – the status of the existing system, regulatory and institutional issues, educational and research questions, and the special problems of freight. Suffice it to say that there was strong agreement that considerable progress was being made, but also great concern with the barriers and obstacles that had to be overcome if the vision of a true intermodal system were to be realized. The passage of new legislation, the Transportation Equity Act for 21st Century (TEA-21) in 1998 reinforced the principles of ISTEA, as did its 2005 successor, SAFETEA-LU. This bill had as its goal to address critical transportation challenges, "challenges such as improving safety, reducing traffic congestion, improving efficiency in freight movement, increasing intermodal connectivity, and protecting the environment – as well as laying the groundwork for addressing future challenges" (USDOT, 2005). However, as is apparent from the previous discussion, these legislative efforts have failed to transform an aging and inadequate system.

Since SAFETEA-LU was scheduled to expire in September 2009, two national commissions, the National Transportation Planning and Revenue Commission (NTPRC) and the National Surface Transportation and Infrastructure Financing Commission (NSTIFC) have been convened to analyze the existing situation and to make recommendations for change. The NTPRC report, *Transportation for Tomorrow*, reached a dramatic conclusion that: "the current Federal surface transportation programs should not be 're-authorized' in their current form. We must begin anew. This New Beginning is the dawn of the third era in the modern history of the Federal surface transportation program."

Whether a "New Beginning" will be achieved remains to be seen but difficult issues, many of which have contributed greatly to the present state of the transportation system, will have to be dealt with before an intermodal system that meets the needs of the future emerges.

4. Critical issues

4.1 National priorities

Obviously, the US still lacks a clear-cut vision of the national surface transportation system it should create, let alone of the role of intermodalism therein. As a result, it is difficult to identify a national transport policy – the policy chapter in a recent report was titled: "Federal Transportation Policy is Absent, Outdated and Underperforming" (Brookings, 2008, p. v). Essentially, policy consisted of narrowly focused programs that never constituted a coherent whole. SAFETEA-LU, for example, contains extensive lists of highly specific highway and transit projects and programs, a large number of which are "earmarks" explicitly directed to a congressman's state or district. The number of such "pork barrel" projects increased from ten in 1982 to a staggering 6,300 in SAFETEA-LU (NSTPRSC, 2008, vol. I, p. 6). The result has been scathingly described as follows:

> Lost in a morass of pet project pork and politics, American Transportation policy today is an unaccountable free for all, geared more to building bridges to nowhere than maintaining the ones we have, developing world class transit or unblocking the movement of freight at our sea, rail, and air hubs. Federal transportation expenditures are neither evidence-based nor outcome-oriented nor performance-measured, leading to politically driven, rather than market strengthening investments.
>
> (NSTPRSC, 2008, vol. I, p. 12)

In the absence of an explicit vision and a non-politicized process, it is obviously not possible to develop clear-cut goals, a coherent strategy to achieve them, or cost-effective policies, but the difficulties involved in radically changing this situation should not be underestimated.

4.2 Technology

Americans have great faith that technology can solve problems so it is not surprising that the US has been making great efforts to develop new technologies to ameliorate the numerous physical and institutional impediments that constrain the seamless flow of passengers and freight, and to increase safety. Though pressure to ease the problem by building new roads remains, these efforts are severely constrained by the limited amount of available land and the growing realization that it is simply not possible to build one's way out of congestion. Thus, it is not surprising that there is widespread support for the development and implementation of new technological applications.

However, much remains to be done. For passenger intermodalism to succeed, for example, it is not sufficient to build transit systems; an appropriate information infrastructure that facilitates a seamless journey is required. Someone using public transit has to identify transfer points and check one or more schedules. There is often a lack of clear and precise information on how to shift from one mode to another. New developments in communication and information hold great promise for facilitating transit and intermodal journeys by integrating information and ticketing systems but many issues of standardization and coordination still have to be overcome.

On the freight side too, problems remain. Despite significant technological advances by the private sector, the ability to employ electronic data interchange (EDI) remains inadequate, primarily because of the limited coordination between modes and the problems posed by a lack of common standards. Nor can one overlook the need for persons with the requisite skills to deal with these new technologies and their ability to help develop a new culture within transportation organizations, one that is hospitable to innovation.

4.3 Education and training

Such an effort clearly involves the educational system. Concerns about the degree to which future transportation professionals were receiving adequate training became increasingly widespread in the 1990s, owing to the rapid technological changes that were taking place; the growing attention being accorded by the public and by policy makers to the negative social and environmental impacts of transportation; and especially, the growing shift towards intermodalism.

Accordingly, several organizations sponsored a series of conferences and studies to define the kind of training that future transportation professionals required.[1] Essentially, a consensus emerged that existing programs, most of which are modally structured and heavily oriented towards technical issues, did not pay adequate attention to logistics and intermodalism or to the management of technology and innovation. A specific study on intermodalism (APEC, 2003) concluded that transportation curricula seldom incorporated all the knowledge and skill sets that were required and that in-service training was inadequate.

Nor should one ignore public education in its many dimensions. It is not feasible to expect that an intermodal transportation system can be realized unless the public recognizes the promise of intermodalism and accepts the importance of engaging in the kinds of behaviors that are required. Decisions concerning transportation projects and policies are influenced by many factors, such as the incentives that are associated with particular transportation options, but public knowledge does play an important role.

4.4. Coordination and integration

Besides skilled professionals on both the passenger and freight sides, the effective use of many new technologies requires a high degree of coordination and integration of resources. Achieving coordination is never easy but it is a sine qua non for intermodalism because of the variety of actors who are involved in developing and implementing intermodal policies and projects.

The actors can be divided into three general groups – regulators, operators, and users. The first category includes not only the federal body, USDOT, but also all the State Departments of Transportation (SDOTs), the Metropolitan Planning Organizations (MPOs), city councils, and numerous other local government structures such as regional transportation districts. The second includes the shippers and the modal carriers. The third is of increasing relevance for there is growing recognition that the public should be involved in transportation decisions in a meaningful way. Bringing all these groups together so as to achieve a consensus on projects and policies has proven to be no easy matter because of the existing institutional arrangements which represent the "primary barriers to sustainable transport" (Rietveld and Stough, 2007, p. 1).

The most important issue is the continuing dominance of the individual modes, the "stovepipes." Until passage of the ISTEA act, national and state policy was unimodal; all agencies were organized around modes so that policies and resource allocations never cut across modes. That structure did not change quickly. As the National Commission on Intermodal Transportation (1994) noted, "planning and policies, particularly at the Federal level, do not encourage and accommodate intermodalism ... Federal government institutions are organized along modal lines, which inhibits planning and developing an intermodal transportation system."

The Commission's report led to an attempt, in 1995, to restructure the USDOT so as to facilitate the development and implementation of intermodalism but this effort was rejected by Congress, a reflection of the high degree of politicization and the power of modal interests. Over time each mode has developed its own organizations, cultures, constituencies and powerful interests who benefit from a modal focus. Congress and its committees reflect these modal interests and are responsive to these groups. State/local units also are linked to modal actors. Hence, while there are powerful forces supporting each mode, there are few politically effective intermodal groups and the balance of power remains heavily biased in favor of the specific modes, especially highways. As a result, "Intermodal plans [which] by their nature tend to cut across different modal areas … frequently suffer from the lack of an internal champion to advance them within those different areas" (NSTPRSC, 2008, vol. II, ch. 6, p. 7), as well as the absence of incentives to enhance intermodalism.

Even today, the USDOT is still essentially organized by modes and lacks the organizational capacity to promote the development of intermodalism. A recent study of its operations concluded: "While DOT has taken actions to address intermodal barriers … no one office is coordinating these actions across the department. The Office of Intermodalism, which has responsibility for initiating and coordinating federal intermodal policy, is primarily focused on research and analysis." (GAO, 2007).

Many proposals have been advanced over the years on how to turn the USDOT into a more effective organization, one that can develop and implement a coherent vision for a national transportation policy. These range from radical restructuring to more modest proposals but this topic should be at the center of any future debate involving the surface transportation system. It is particularly important if one is thinking in intermodal terms because not only is the transportation sector involved but many issues require the cooperation of other sectoral actors. For example, any attempt to deal with energy use or environmental impacts necessarily involves other cabinet-level departments. Only a respected, effective and efficient USDOT pursuing a particular vision can hope to develop the required cooperative arrangements as well as the support of key congressional leaders for its policies (Hall and Sussman, 2006).

This too is a difficult challenge, given the nature of the American political system. At the national level it politicizes transportation policy in many ways (such as "earmarks"). Though this technique is often used in a wasteful manner, it is also the only way to promote regional projects. Structurally, the problem is that numerous House and Senate Committees, many of which are modally oriented, have jurisdiction over issues that are incorporated in transportation legislation. Reform of this pluralistic system, which clearly complicates any systemic approach, is unlikely.

It also devolves a large responsibility for transportation to the states. Essentially, the federal government disburses grants to the states for particular programs and oversees their implementation in terms of programmatic goals. Thus one very contentious issue that will have to be resolved in any attempt to create a national transportation system is the relative roles of the federal government and the states (Lockwood, 2008, p. 31). The significance, complexity, and intractability of this issue cannot be overstated.

Many of the problems evident at the federal level are also commonplace at the state level, particularly interagency cooperation. The major actors here are the SDOTs and MPOs. ISTEA increased the power of MPOs but these organizations often possess inadequate staff, tools, and experience to promote intermodalism, even though they usually understand the necessity for such a system in their regions. Their concerns are with urban issues (traditionally passenger systems), but SDOTs, on the other hand, are still largely highway-focused. They are staffed largely by highway engineers and most funding is still directed to the highway mode.

Creating genuine partnerships between these agencies has proven very difficult because each has its own interests, culture, resources, goals, and political alliances. Nor can one overlook the degree to which cooperative planning efforts are an innovation, so that those seeking to work together have limited experiences to draw upon. As a result, conflicts between MPOs and SDOTs over the priorities that should be allocated to various projects are commonplace, as is the outcome – the SDOTs control most of the funding and thus tend to emerge victorious. Moreover, land use and transportation are closely related (see Chapter 3 in this volume) but different agencies have jurisdiction over the former so that even more actors have to be involved in any meaningful attempt to achieve an integrated system at any level. Nor can one overlook the role that non-governmental actors, such as business interests, public interest groups, and the public at large, are now playing in transportation planning. The freight industry, despite its importance and needs, tends to have a low priority in state transportation planning and plays only a minor role in most states.

A recent study sought to identify the degree to which SDOTs had adjusted to this new environment, specifically the degree to which SDOTs have actually adopted an intermodal approach as mandated by ISTEA, TEA-21 and SAFETEA-LU. It found that only limited progress has actually occurred (Goetz *et al.*, 2007). While each SDOT has changed its organizational structures to reflect a more intermodal approach, there are questions as to whether institutional cultures have kept pace with these changes. On the one hand, an increasing number of intermodal projects are identified in comprehensive plans, an increasing number of specific plans are produced on intermodal aspects of transportation, and an increasing number of intermodal projects have been built in virtually every state. However, despite these efforts, knowledgeable respondents generally felt that more could, and should, be done to improve intermodal planning. Perhaps predictably, a major concern focused on the lack of funding for intermodal projects. Essentially, highway interests remain dominant and an intermodal mindset has not permeated the entire transportation policy community – state transportation commission, state legislature, SDOT leadership, SDOT staff – charged with transportation decision making and planning. Furthermore, many intermodal policies and projects are regional in scope and cover several states or urban areas, usually requiring extensive negotiations between many governmental actors.

Aggravating the problem of modal relations are the differing private/public sector perspectives and public attitudes towards new transportation projects. Not only is the private sector fragmented and competitive, it possesses a different planning perspective from that of the public sector, being oriented towards the short term rather than the long term. Nor can one ignore the position of labor unions, which often view intermodalism negatively because the new arrangements may mean the loss of jobs.

The need is obvious – to move towards cooperative arrangements and partnerships. But this is easier said than done given the historic conflictual relationship between the private and public sectors and between labor and management, and, indeed, between governmental institutions as well. Incentives for cooperation and coordination between various actors and for people to interact intermodally would be very helpful but, at present, there are, at best, only limited rewards for such behavior.

4.5 Laws and regulations

Although these can be considered to be a sub-set of institutional arrangements, they play such an important role that they deserve separate recognition. There is no doubt that they have profoundly influenced the development of intermodalism and will also shape its future. Historically they reinforced modal power and a great degree of deregulation was required before intermodalism could flourish. But many regulations and laws still hinder the development and implementation of transportation policies and projects. Numerous unnecessary, inconsistent, and complex regulations are in force not only at the federal level but also at the state and local levels as well. The result is that "it takes too long and costs too much to deliver transportation projects" (NSTPRSC, 2008, vol. II, ch. 6, p. 10). While the process is generally dysfunctional, some laws and regulations are particularly onerous in terms of the development of intermodal projects. For example, even if favorably disposed towards intermodalism, SDOTs are often restricted by regulations which prevent spending on non-highway projects. Furthermore, laws governing labor and liability differ among the modes, thus creating unnecessary complexities and difficulties for shippers and carriers who wish to operate intermodally even within the US, let alone internationally. As one expert has noted: "streamlining of regulatory responsibilities and rules across modes will do much to promote the seamless intermodalism for which the nation should strive" (Dempsey, 2000).

4.6 Infrastructure

That the US infrastructure is in poor condition and needs repair is obvious. If a better intermodal system is to be created, attention must be paid to several issues. First, serious maintenance problems such as upgrading the condition of existing bridges, highways, and transit systems have to be dealt with. Second, various elements, especially transit facilities, require expansion and the issue of high-speed intercity rail has to be tackled. Finally, particular attention must be paid to intermodal facilities. The most important of these involves eliminating existing bottlenecks and building linkages across modes in the form of the nodes (where the modes come together, namely freight and passenger terminals). The capacity of the existing freight terminals is in many cases strained and new ones are urgently required. Building such facilities, however, is no easy matter due to the kinds of constraints that have already been discussed – the number and variety of actors, the tendency by MPOs to favor passenger projects, the public concern with externalities, and difficulties in securing the necessary financing.

4.7 Financing

This is always a significant bottleneck but the existing funding system is essentially broken. The national finance commission (NSTFC), in its preliminary report, identified several basic problems: existing revenues are inadequate to maintain the existing system and to improve it; costs and demand are growing faster than the revenues provided by current funding mechanisms; system demands are outpacing investments; system maintenance costs are competing with the necessary expansion of the system; and the fuel tax, which has been the key federal funding source for our system, is no longer sufficient at current rates. Ineffective investment decisions aggravate this already desperate situation due to such factors as the lack of performance standards, "earmarking," a lack of investment analyses, and inflexible current funding arrangements (NSTPRSC, 2008, vol. II, ch. 6, p. 7). The gap between current sustainable revenues and estimated needs for all modes through 2020 is between $155 and $200 billion (NSTPRSC, 2008, vol. I, p. 6). How this money will be raised will be much, and heatedly, discussed.

Even if significant new resources become available, how intermodal projects will fare is not clear for it has traditionally proven difficult to find funding for such projects. As the GAO recently noted, intermodalism suffers from "limited federal funding ... in part due to statutory requirements" (GAO, 2007, p. 1) Even with the passage of the SAFETEA-LU legislation, funds have to be shifted from the highway mode to others, such as transit. Under these conditions it is not surprising that there is so much opposition to spending on intermodal freight and passenger projects. As the NTPRC report noted: "All too often, investment decisions for Different Asset categories are made within agency 'stovepipes,' with a focus on minimizing near term agency costs as opposed to maximizing the long-term benefits to system users and society at large" (NSTPRSC, 2008, vol. II, ch. 6, p. 6).

Intermodal freight projects must overcome another barrier – these are expensive projects that the private sector cannot finance by itself. Accordingly, the question of the role of government in financing infrastructure development is an important and complex one that forces us to rethink the relationship between the government and the private sector and raises many complex questions. These include: 1) How to identify the private sector activities that deserve governmental support, for many possibilities exist ranging from rail facilities to roads leading to terminals; and 2) How should the funding be arranged? Here too many possibilities can be identified ranging from tolls to user charges, to limited partnerships to the issuance of tax-exempt bonds, to the establishment of trust funds.

5. Conclusion

Hopefully, new policies will soon be adopted to deal with the many shortcomings that characterize today's US transportation system. There are some promising signs, notably the widespread recognition of the need for change and the important reforms suggested by the National Commission. These include consolidating the present scattershot federal projects and programs into 10; the first and most fundamental, entitled "Rebuilding America," calls for developing and maintaining the nation's infrastructure. The others focus on specific areas such as Freight Transportation; Metropolitan Mobility; Safe Mobility; Rural Mobility; Intercity Passenger Rail; Energy Use; Environmental Impacts and Research; Development, and Technology programs.

What will happen to these recommendations remains to be seen but certain problems are already evident. First, many of these areas are interrelated and it is not clear how they will be linked functionally and institutionally. The Commission recognizes this issue, stating:

> While the 10 programs identified above represent 10 distinct areas of Federal interest, individual projects may contribute to achieving goals in multiple areas, and thus the programs cannot be considered completely independent. The Commission believes that coordination among the planning activities required for each of the programs will be essential. Coordination should begin as plans are developed at the local, State, and regional level, but the USDOT will need to take an active role in consolidating these separate plans into a national strategic plan.
>
> (NSTPRSC, 2008, vol. II, ch. 6, p. 28)

Whether the USDOT, in its present configuration, can perform such a function must be questioned, as well as the degree of coordination that can be achieved, given the numerous barriers identified above. Accordingly, attention should be given to its reorganization as well as to various mechanisms that would facilitate cooperation. Particular attention should also be accorded to potential technological "fixes."

A second more fundamental problem, however, is the lack of a common perspective as underlined by the fact that three of the twelve Commissioners signed a minority report. They argued that the majority had failed "to develop a policy framework" to deal with the transportation crisis. Specifically, they objected to the role assigned to the federal government, the imposition of new federal regulations, inadequate attention to pricing mechanisms, and to the role of the private sector. In their words:

> Our approach would sustain current gasoline and diesel tax levels and refocus Federal efforts on (a) maintaining the Interstate Highway System; (b) alleviating freight-related bottlenecks that impede the flow of commerce and goods; and (c) providing States with appropriate analysis, incentives, and flexibility regarding the adoption of market-based reforms to their highway systems. Every effort should be made to facilitate the application of tolling and congestion pricing to the transportation system.
>
> (NSTPRSC, 2008, vol. III, p. 60)

Clearly, profound philosophical, value, and institutional differences remain to be resolved. These include the balance between federal and local interests and priorities; the ways in which funds will be raised and distributed; the role of the private sector; and the degree to which users will be expected to pay through tolls, congestion charges and the like. On each of these, as well as on many others, heated debates can be anticipated, the outcomes of which will profoundly shape the character of the transportation system.

Third, even the majority report leaves much to be desired from an intermodal perspective. It does touch upon various intermodal dimensions such as the need to strengthen network connections between modes; implement freight projects "that are beyond the traditional modal and governmental orientations"; develop high-speed intercity rail corridors; and increase transit capacity. These are important projects and their successful implementation would greatly enhance the intermodal character of the surface transportation system. However, the report fails to accept intermodalism as a critical point of focus. And, educational issues are notably absent.

The lack of an intermodal vision is a significant lacuna. As I have suggested above, this is an essential prerequisite for the creation of a truly intermodal system. In its absence, the outcome will probably be limited to some improvements in such areas as funding, infrastructure improvements, laws and regulations, and the deployment of technology. However, the modal orientations and the institutional structures which complicate coordination and integration will probably remain powerful forces and will continue to influence every dimension of the surface transportation system. For a genuine intermodal system to emerge, powerful leadership that shares that vision, and is willing to struggle to implement it, is necessary. Any call for leadership is often a substitute for action. However, in this case, it should be clear that an elite consensus must be forged on a common intermodal vision and that this consensus must be supplemented by widespread support for the vision from all relevant publics, including local officials, community leaders, the private sector, public interest groups, the media and the public at large. Doing so will clearly be a difficult challenge but one that has to be realized if America's future transportation needs are to be met in a sustainable manner.

Acknowledgements

I wish to thank two anonymous reviewers for their detailed and insightful comments on the original draft. In preparing this paper I have drawn upon some of my previous work, notably "Intermodalism: The Challenge and the Promise," in *Transportation Law Journal*, Vol. 27, No. 3, September 2000; reprinted in *Intermodal Transportation: Selected Essays* (National Center for Intermodal Transportation, 2000).

Note

1 Recent examples include "The Education of Transportation Planning Professionals" (Handy *et al.*, 2002) and "Urban Transportation Planning Education Revisited" (Khisty and Kikuchi, 2003).

References

Alt, R., Forster, P.W. and King, J.L. (1997) "The Great Reversal: Information and Transportation Infrastructure in the Intermodal Vision," in *National Conference on Intermodal Transportation Research Framework*, Washington, DC Transportation Research Board.

American Association of Railroads (2007) "Landmark Study Puts Rail Infrastructure Needs at $148 Billion." Available online at: http://dev.aar.org/AAR/IndustryInformation/National_Capacity_Study/NatRailFreightInfrastructureStudy.aspx (accessed October 3, 2009).

American Automobile Association (2008) "Crashes vs. Congestion – What's the Cost to Society?" Available online at: http://www.aaanewsroom.net/Assets/Files/20083591910.CrashesVsCongestionFullReport2.28.08.pdf (accessed October 3, 2009).

APEC Transportation Working Group (2003) "Identification of Needed Intermodal Skills and Development of Required Training Programs." Available online at: http://www.apec-tptwg.org.cn/new/Steering-Committees/Competitive/intermodal-task-force/tptwg-18-final-papers/apec-training.pdf (accessed October 3, 2009).

Brookings Institution (2008) *A Bridge to Somewhere: Rethinking American Transportation for the 21st Century*, Washington, DC. Available online at: http://www.brookings.edu/reports/2008/06_transportation_puentes.aspx (accessed October 3, 2009).

Dempsey, P.S. (2000) "The Law of Intermodal Transportation," *Transportation Law Journal*, No. 27: 367–417.

GAO (2007) "DOT Could Take Further Actions to Address Intermodal Barriers, Highlights," Washington, DC. Available online at: http://www.gao.gov/new.items/d07718.pdf (accessed October 3, 2009).

Goetz, A.R., Szyliowicz, J.S., Vowles, T.M. and Taylor, G.S. (2007) "Assessing Intermodal Transportation Planning at State Departments of Transportation," *World Review of Intermodal Transportation Research*, Vol. 1, No. 2: 119–145.

Hall, R.P. and Sussman, J.M. (2006), "Promoting the Concept of Sustainable Transportation Within the Federal System – the Need to Reinvent the U.S. DOT," Boston: MIT Working Paper Series, ESD-WP-2006–13. Available online at: http://esd.mit.edu/staging/wps/esd-wp-2006–13.pdf (accessed October 3, 2009).

Handy, S., Weston, L., Song, J., Lane, K.M.D. (October 2002), "The Education of Transportation Planning Professionals," Southwest Region University Transportation Center, Research Report SWUTC/02/167522.

Jeff, G. (1998) "Welcoming Remarks," in *Intermodal Education and Training*, Washington, DC.: National Academy Press.

Khisty, S.J. and Kikuchi, S. (2003) "Urban Transportation Planning Education Revisited," *Transportation Research Record*, No. 1848: 57–62.

Lockwood, Stephen (2008) "National Transportation Policy Options: A Time for Choice," *ITE Journal*. Available online at: http://findarticles.com/p/articles/mi_qa3734/is_200808/ai_n29492040/?tag=content;col1 (accessed October 3, 2009).

National Commission on Intermodal Transportation (1994) *Toward a National Intermodal Transportation System – Final Report*, Washington, DC. Available online at: http://ntl.bts.gov/DOCS/325TAN.html (accessed October 3, 2009).

National Surface Transportation Policy and Revenue Study Commission (NSTPRSC) (2008) *Final Report*, Washington, D.C. Available online at: http://transportationfortomorrow.org/fi nal_report/ (accessed April 14, 2010; site due to be shut down on June, 30, 3010).

Pisano, M. (2006) "Southern California Goods Movement Action Plan," *Efficient Goods Movement and the Environment Symposium Series*, Washington, DC.: Eno Transportation Foundation.

Rietveld, P. and Stough, R. (eds) (2007) *Institutions and Sustainable Transport: Regulatory Reform in Advanced Economies*, Cheltenham, UK: Edward Elgar.

Szyliowicz, J. (2009) "Terrorism, Mobility, and Transportation Security," in M. Tahmisoglu and C. Ozen (eds) *Transportation Security Against Terrorism*, NATO Science for Peace and Security Series E, Human and Societal Dynamics, Vol. 54, Amsterdam: IOS Press.

Tarm, M. (May 30, 2008) "Supply Chain Near Stopping in its Tracks," *The Denver Post*.

Texas Transportation Institute (2007) *Urban Mobility Report*. Available online at: http://mobility.tamu.edu/ums/ (accessed 14 April 2010).

USDOT, Federal Highway Administration (1991) *Intermodal Surface Transportation Efficiency Act of 1991*. Available online at: http://ntl.bts.gov/DOCS/istea.html (accessed October 3, 2009).

—— (2005) "SAFETEA-LU Overview." Available online at: http://www.fhwa.dot.gov/safetealu/summary.htm (accessed October 3, 2009).

Chapter 20

The pursuit of integration

How far and what next?

David Banister and Moshe Givoni

1. Introduction

The chapters in this book have all taken different perspectives on the notion of integration in transport, and they have all emphasized the importance of it as a concept. Yet it has proved to be very elusive, both in terms of definition and in terms of the means by which it can be investigated. Furthermore, many of the chapters also demonstrated how difficult it is to achieve integrated transport in practice.

Several authors have used the following definition given by NEA *et al.* (2003), which was also given in the introduction to this book:

> The organizational process through which the planning and delivery of elements of the transport system are brought together, across modes, sectors, operators and institutions, with the aim of increasing net social benefits.

This definition provides a useful starting point (even if it lacks direct reference to the environmental aspect, as noted by Anderton in Chapter 4), as it is explicitly related to transport, can be applied to include freight as well as passenger transport, and as it covers the range of actors and modes with the overall objective of increasing net social benefits. Much of the literature on transport integration is embedded within this rather narrow framework that sees the transport system as a closed system, and where integration is seen as promoting a better transport system through a better travel experience on one mode or between different modes. Several of the chapters in this book reflect this approach to integration, often referred to as integration at the physical level.

At the same time, other chapters have taken a much wider interpretation of integration, placing it in the context of decision making and policy integration.

Here, it is suggested that there may be five types of policy integration (Stead, Chapter 2), that cover horizontal (actors, agencies and sectors), vertical (levels of decision making from EU to local), spatial, temporal and modal (transport specific) aspects, and that transport integration only addresses two of these five categories (namely the horizontal and modal dimensions).

This distinction, between transport integration (or integrated transport) and policy integration, is important, as the two are closely interlinked and understanding and progressing both is essential to the improvement of the transport system and services it provides. Thus, to the many challenges and elements of integration already addressed in this book we will add the need to link between integrated policy and integrated transport. The fact that most of the chapters did not attempt to make the link but rather concentrated on one of these is probably the best evidence for this need. The chapters in Part 4, which take the wider perspective, have demonstrated how difficult it is to make this link in practice.

In this concluding chapter, we aim to distil the main messages from the book, highlighting some of the more theoretical and conceptual challenges that need to be addressed, as well as identifying some of the difficult issues that have been presented in the previous 18 substantive contributions. We then build on this foundation to suggest the way forward in the pursuit of integrated policy and (also as an outcome of it) integrated transport, and we also put forward a wider agenda that links transport integration more effectively with policy integration.

Given the evidence provided in the previous chapters, and the short summary above, this final chapter focuses on bringing together the arguments and the means by which integration can achieve goals in transport and policy more widely. This suggests that different levels and elements of integration should be pursued depending on the particular situation, but equally there should be clear patterns and opportunities that can allow more general conclusions to be drawn.

2. Theory and integration

Integration has become a central theme in transport policy for the last 10 years, but it has remained an elusive concept. Traditionally, it was used in a rather narrow context, as defined above, but policy integration has widened the debate to cover integration between sectors and to address some of the organizational factors. However, there now needs to be a wider debate about the role of integration, what purposes it should serve and in what way. The difficulties in achieving integration also suggest that it is a broader theoretical concept that can never be really achieved, and that the discussion should be more modest, and refer instead to terms like 'coordination' and 'cooperation', as these sometimes might suffice to deliver many of the benefits of full integration.

The horizontal coordination of sector-based policies is central to the working of the public sector (Peters, 1998), but it is increasingly difficult to achieve because of the growth in the number of actors involved, the competing priorities of central government, the decentralization and division of responsibilities, and the reluctance to make difficult decisions. The vertical coordination element only complicates the picture as it adds several scales and levels of implementation. Coordination may be appropriate where the public sector dominates in the provision of services, but increasingly decisions have become more market-based, making it more difficult for coordination. At its minimal level, cooperation implies dialogue and information (Stead, 2008). At the higher level, coordination suggests that there is a broader policy framework, together with clearly defined goals for integration. This in turn implies greater transparency and an understanding of the different priorities among the actors and agencies, but most importantly that there are means by which conflicts can be resolved. The highest level is policy integration, where that understanding moves towards a common agenda where the different actors and agencies actually work together to achieve real outcomes that have benefits for all parties. This is what Stead (2008) has called a hierarchy of policy integration (Figure 20.1).

In transport, the debate should perhaps not be about the highest level of integration, but more about coordination and cooperation. Policy-making integration can be pursued to a high level of interaction and interdependency. Thus a clear distinction is needed between integrating institutions on the one hand and the physical integration of the different elements of the transport system on the other.

Referring back to the Integration Ladder[1] (Hull, 2005) described in Table 1.1 in Chapter 1, we argue that only the lowest two levels of integration (that of 'Physical and operational integration of public transport' and 'Modal integration') are really pursued in transport and have been partly achieved. While the focus on achieving integration at these levels is important to advance a better transport system and possibly a more sustainable one, it is problematic in excluding other sectors from the debate and the drivers of demand for transport. In this respect, transport is increasingly seen as a *product* to be produced (in higher volume and quality) rather than a service to be provided, and there are subtle differences stemming from these different approaches. It is perhaps one of the main reasons that in most cases (public) transport services are provided as a separate bundle of trips rather than a door-to-door service. For transport to be considered as a service, it is necessary for it to be considered in the context of how other sectors of society and the economy need and use it. Full integration at the policy level is required for transport to be considered as a means to achieve broader goals (e.g. social cohesion, economic development, and environmental sustainability) rather than merely getting people and goods from A to B. In this respect, it is important to remember that for the majority of activities the demand for transport is a derived demand.

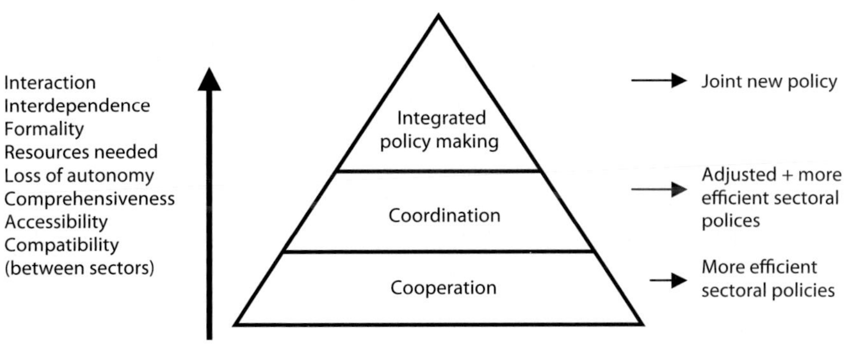

Figure 20.1 Integrated policy making, policy coordination and cooperation
Source: Stead, 2008, p. 141

In the Integration Ladder, Hull (2005) suggests that there is some horizontal integration (Levels 1, 2 and 3), but that vertical integration (Level 4 and above) has been much harder to achieve as there is too much fragmentation between the different actors and agencies. This partial success at the lower rungs of the ladder is probably the result of considering transport only within the 'transport' sphere, when working to improve the transport system. Efforts to achieve a sustainable transport system require transport to be viewed as a service, and a better transport service will require higher levels of integration than are currently achieved at the horizontal level. To make real progress towards such a transport system, integrated transport policy (integration at the vertical level) is essential, and in such a policy context the role of transport in achieving certain goals (e.g. economic and/or environmental) must be understood and accounted for. When integrated transport policy is achieved, then partial integration at the physical level (or cooperation and coordination) might suffice.

Considering the above, this chapter now brings together the evidence from the different chapters of this book to outline the main integration challenges and to identify the next steps.

3. Integration challenges

3.1 Policy integration

The policy integration agenda would place transport within a longer-term, more radical perspective that questions the role that transport can play in achieving sustainable development. This is asking a more fundamental question about transport integration. Current approaches to integration, as in most of the analysis related to integration, look at the means by which integration can lead to more efficient transport. This in turn would probably lead to more travel, so the question becomes one of how efficiency in the passenger and freight transport sectors can be increased through better integration, but at the same time looking at how the total amount of travel can be decreased (Hickman *et al.*, Chapter 3). This is where cross-sector analysis becomes important; for example to locate various activities at reasonable distances from each other and from home/work locations (Susilo, Chapter 8), as well as understanding the appropriate level at which decisions should be taken.

It is clear that the effectiveness of policy implementation depends on the structures through which they are delivered, and this in turn requires appropriate levels of integration (Anderton, Chapter 4). This is well illustrated by the cross-cutting agenda related to the climate change debate. The question here concerns the best decision-making level for better policy integration, whether it is at the EU or federal level, at the national or state level, or at the individual city or neighbourhood level – this is a form of subsidiarity. In the Californian example, it is argued that the sub-national (state) level has made clear progress in setting up the necessary policy frameworks for integration in the transport sector (Anderton, Chapter 4); while in France it has been argued that there are too many levels of government (Zembri, Chapter 18), and it is only in Paris (sub-regional level) that there is a simple integrated organization for the provision of public transport with a high level of integration. As in Paris, London is considered to perform better than other UK cities with respect to the provision of quality public transport because transport responsibilities are mainly with the mayor and a single authority (Transport for London).

A slightly different argument is presented at the city level in the case of Modiin (Feitelson and Gamlieli, Chapter 17). In essence, all (planning and most implementation) responsibilities for creating a new city were in the hands of one organization, the Ministry of Housing and Building (MHB). As such, transport was not really considered as an essential element, partly because some responsibilities for this aspect were allocated to another body, the Ministry of Transport. While, essentially, integration was only needed between the two organizations, this proved difficult to achieve. This supports Hull's (2008) argument about the importance of the relationships between the different players. The quality of the linkages between agents is crucial to integration and for effective implementation of complex decisions. Included here are concepts of whether there is a history of cooperation between the agents, as well as openness and trust between them, and this relates to the culture of the organizations involved. There is sometimes a high social cost associated with these linkages, but their importance should not be underestimated.

Education for 'integration' appears to be crucial at all levels to reduce 'transaction costs' associated with integration, and efforts in this direction in the US (Szyliowicz, Chapter 19) seem to be justified. The need to educate for integration is vital to bridge not only across organizations but also across different professions (differences in approach, methods, and culture) involved in the policy making and implementation process. For example, Parkin (Chapter 9) illustrates the need for planners and engineers to understand each other's approach when planning walking and cycling networks. Feitelson and Gamlieli (Chapter 17) suggest that if the architects employed by the MHB had more understanding and knowledge of transport, outcomes would have been different. Thus they make a case for specialization to achieve integration. This builds on Bardach's (1977) idea of 'fixers'. The role of these integrators is to explicitly encourage integration through building up human capital.

The spatial and temporal dimensions of integration mentioned by Stead (Chapter 2) may be instrumental here, as policy and transport integration may depend on decisions being made at the appropriate spatial level (subsidiarity). Integration in this respect is also required since most of the actual actions to advance integrated transport need to take place at the local level, while policy, and often the financial resources required to achieve integrated transport, are decided and allocated at much higher levels. Time is central to the creation of linkages as well as openness and trust between the agents involved in those decisions (Hansen, 2006). Fostering human capital is essential in achieving integration at high (policy) levels.

A central question arising from this discussion is what tools can be used to increase policy integration. Hull (2008) argues for new planning authority structures to overcome these barriers to integration, as they are also acting as a limiting factor in moving towards sustainable development. Feitelson and Gamlieli (Chapter 17) suggest that there should be a person tasked with being responsible for integration (a 'fixer' as noted above). Obviously, the success of this approach depends on the powers and responsibilities given to such organizations and individuals, but it must be recognized that at the same time such an approach simply introduces additional elements and levels that need to be integrated and therefore might be counterproductive with respect to the outcome.

Feitelson and Gamlieli (Chapter 17) note that one of the explanations for why integrated transport was not achieved in Modiin, despite the conductive background, is that none of the actors really viewed the reduction of total transition costs (related to integration) as a primary concern, and none had the on-ground expertise necessary to reduce these costs. A lesson they draw is that institutional responsibility, in this case for (integrated) transport, needs to be shared. Thus, if transport is not the concern/responsibility of the sectors using it, integrated policy and integrated transport (systems) are not likely to result. Shared responsibility (for transport) requires making transport a matter for consideration of any actor generating demand for it. Bringing transport into non-transport institutions (that generate demand for transport) is the way forward. Creating another level of institutions where transport is considered together with other non-transport matter is probably not the most effective solution. Only the former approach is likely to result in integration being properly addressed and realized throughout the chain from policy setting through to policy implementation, and finally the operation of the transport system.

3.2 Transport integration

Within transport, integration can be seen as a response to market failure (Preston, Chapter 12), as the classic neoliberal argument would see all necessary coordination being provided by competition. However, the reality suggests that the market does not provide information on all the services being provided, and there is wasteful competition, network failures and a significant underestimation of the externalities. In particular, markets were never seen as being capable of or designed for deciding and directing policy. At the same time, market mechanisms (namely some level of competition) are important to preserve, as they contribute to efficiency and to a better quality of service.

Thus, there seems to be a basic dilemma that has not yet been resolved between the freedom and benefits that the market provides, and the need to correct or intervene where there is evidence of failure. This relates very much to the respective roles of the private and public sectors in service provision, and whether they can both work together to provide the best service for users. There may be irreconcilable differences between the profit-maximizing interests of the private sector operators and the more social welfare objectives of the public sector. Both have an important role in providing integrated transport services, but the key question here is whether they can actually work effectively together. The challenge is in identifying the most appropriate levels of competition for the provision of transport services so that both welfare and integration objectives can be achieved. This would include decisions on the type of competition (e.g. competition in the market or for the market), the form of integration and the importance of different policy goals. Given that markets were never seen as being capable of directing policy, the role of the market in promoting integration seems to be at the service provision level.

This is very much the case in many situations, for example, where bus and rail services are provided and run by the private sector in competition with other service providers (often the competition is at the stage of bidding to provide the service), but there still is no integration. The root of this problem is probably in the unimodal approach and the narrow focus within the organizations which were set up to govern the transport system. This constraint is in addition to the lack of integrated policy, as discussed above.

If in formulating transport policy a door-to-door journey was considered and public bodies responsible for transport had tendered a door-to-door service, the market or private operators would have delivered just that. In this kind of environment, the efficiency of transport services would be increased through the consolidation of loads and the raising of occupancy levels, as opposed to the current situation where much transport operates with empty or almost empty vehicles. These empty vehicles are evidence of one of the problems of transport (that demand is usually in one direction) but also that the pricing structures are not working, that people and businesses are not aware of their real costs, and in general, that the market is not working 'properly'.

New information technology provides important means by which real-time information can be utilized to make better use of existing capacity and in general has great potential to improve the quality of service, specifically for multimodal transport. The use that is currently being made of it is another example for what hinders integration from taking place. There is a clear, ongoing shift in information and communications technologies (ICT) from a capital-intensive resource owned and maintained by different organizations and individual users to a commodity that is purchased and used according to need (Hoose, Chapter 10). The question here is how the available information can best be organized and used, both to manage the transport system most efficiently and to determine how individual users (and firms) can best access that information. In other words, ICT provides an opportunity to better match demand with supply and to improve efficiency through route guidance, and other forms of information dissemination.

However, at present ICT is not properly used to advance 'seamless travel' because of the lack of integration, and this lack of integration is not related to technology but to organizational separation between the different elements responsible for managing and operating the transport system. To some extent, issues of competition (between operators) or lack of interest/motivation to share information hinders best use of available capacity through better harnessing of technology (Hoose, Chapter 10). An example would be the organizational separation between the fleet operator and the road network manager. Another example is the incompatibility of the London smart card system (OYSTER) with the cards passengers use on coach or rail services into London. It is the institutional organization of the transport system that prevents integration. Hoose (Chapter 10) in this respect concludes: 'Just as transport demand is a "derived" demand to enable people to go about their daily lives so technology can be seen as a second order-derived demand to enable transport systems to deliver more effectively' (p. 181–82).

At the heart of addressing the current fragmentation in the approach to transport, and the key challenge in advancing an integrated transport system, is the need to move away from the primary concern over cost minimization and speed towards the 'total travel experience'. One of the main deterrents to using public transport is not the quality of the main part of the journey, but the difficulties of getting to and from the public transport system. It is well known that the value of time increases for the inconvenience of interchanges and access (Iseki and Taylor, 2009), and this may deter many people from taking complex multimodal trips, as levels of satisfaction are significantly reduced (Givoni and Rietveld, Chapter 11). Many of the chapters in this book comment on and provide analysis of some of these issues that may deter users. Examples include the quality of information and supply chain complexities (Allen *et al.*, Chapter 5), service quality and reliability (Zamparini and Reggiani, Chapter 6), and the comfort and transfers between different transport modes (Holvad, Chapter 14; Sabir *et al.*, Chapter 16).

The attractiveness of door-to-door travel is clear, and it has been seen as a major factor in the increasing dominance of private transport (Parkin, Chapter 9). Public transport is only really competitive in congested conditions (when it has priority) or over long distances, with the car (or truck) having clear advantages in most other situations. Public transport integration needs to address the weakest-link argument and examine the whole journey concept more thoroughly (including the potential for walking and cycling), as these factors are as important as speed and frequency of public transport services in determining choices (Rietveld, 2000).

Better use of capacity can be made, for example, by better using available technology, and there are no technical or even real physical barriers to providing a door-to-door service, but the current fragmentation of the transport system prevents this from taking place. There is no doubt that integration at the physical level is essential to advance sustainable transport, for example, through modal shift of passenger and freight from road to rail transport. Yet, for this to take place it is probably sufficient for actual integration to take place at higher levels than at the operational level. This would start with integrating policy objectives and continuing with integrating the institutions governing transport so that these consider a door-to-door journey as one unit of travel rather than looking separately at the trip within the city, and then between cities. If this takes place, it is likely that actual integration would not be necessary at the operation level, and competition could be used at this level to enhance seamless travel with individual service providers competing for providing a door-to-door service. In this respect, it is important to account for the economies of scale in an integrated system, as compared with each public transport mode taken separately (Bakker *et al.*, Chapter 7). Under competitive conditions service providers would benefit from such a system.

One lesson which seems to arise from current progress on integration of transport services is that once the physical infrastructure for integration is provided by the public sector, the private sector is likely to provide more integrated transport services. The best evidence probably relates to the provision of high-speed rail infrastructure at major airports and how this results in airlines opting to substitute their aircraft services with rail services, and in general in air-rail integration (Givoni and Banister, 2006). The creation of Railteam (Holvad, Chapter 14) is another example where the integration was brought about after regulatory reforms and EU efforts to advance interoperability.

3.3 The integration challenge

Means and objectives can be mixed up in the name of sustainability and an example of the above conflict between policy integration and integrated transport (and the wish to promote public transport) was nicely illustrated by Bakker *et al.* (Chapter 7) in their discussion of the evaluation of sprawl policies in the Netherlands. In that example, they show that contrary to what might be expected, allowing more sprawl, rather than increasing urban densities and increasing public transport provision, shows more benefits than costs. They claim that this illustrates how means (public transport) turn into goals, on the implicit assumption that public transport is a good thing. Such an approach, they argue, mixes up political choices (to support public transport) with policy objectives (in general to increase welfare). They argue this proposition to emphasize the importance of a rigorous evaluation before deciding on any policy (the importance of rigorous evaluation was also underlined by Punzo *et al.*, Chapter 15). But the argument also demonstrates why transport should not be considered on its own, nor based on the premise that public transport promotion is beneficial.

It is clear that an integrated transport system will result in a more efficient system and a better service to users. When aiming to improve the transport system through better integration of its elements, the role of competition is not clear. While competition in passenger services seems to hinder integration and efficiency, it is a major element in promoting integration and efficiency in freight services. For example, the competitive forces which operate within the logistics industry are responsible for a much better use of ICT to achieve integration and thus a better and more efficient service. As noted earlier, in perfect competitive markets there should be no need to address integration, as the market forces will bring about all the necessary integration. But, transport markets are not perfect and so interventions are needed to address market failures, distortions, externalities and social inequalities. The solution lies in not letting the market be part of setting the policies but only in delivering them. There is a lot to learn from the realization of logistic companies that they need to provide a reliable door-to-door service and from the way the organizations governing freight transport have provided the necessary environment to achieve this. At the same time, there is also a need to realize that one of the outcomes of the free market element in freight transport means that there may be too much reliance on road transport, when better alternatives, in sustainability terms, are available, such as rail and waterway transport (Macharis *et al.*, Chapter 13).

The difference between the passenger and freight sectors highlights one level of integration which has not been discussed so far in this book, namely the possible benefits of integrating passenger and freight transport. To a large extent, certainly in research terms, these are seen as two separate transport systems, where the other network is considered to a limited extent, mainly to avoid conflicts rather than to look for synergies (Allen *et al.*, Chapter 5). The benefits of integration here can be substantial.

The air transport network seems to be one part of the transport system that is quite successful in achieving integration (but only within its boundaries), and this is achieved through what seems to be the right balance between public bodies governing it and private commercial actors competing for and running it. Perhaps due to the background of full public and state ownership, the air transport industry is currently one of the most competitive industries, which over the years has managed to expand substantially, reduce the costs to the users and maintain a relatively good level of service, while keeping high standards of safety. Furthermore, particularly before the phenomenon of airline alliances and before internet booking became widespread, buying a ticket from any airport in the world to another airport was possible, even if this involved competing airlines. This level of integration is considered unimaginable in many other transport networks. Another interesting example is that only the air transport industry fully integrates freight transport services with the passenger ones, allowing airlines to achieve a much better utilization of their fleet.[2] In air transport it appears that the right elements of the system were deregulated while others were not. If there is a downside to the way the air transport industry is presently run it is in the lack of integration when it comes to setting the policy, which currently is not accounting for wider societal goals and is too focused on transport.

As means and objectives can be mixed up in the name of sustainability, the current pursuit of integrated transport as a goal might only promote more travel and would not advance any other goals. There is a realization in the US that it needs to better follow the European experience when it comes to providing urban transport passenger services (namely by means of transit systems). It also needs to show a willingness to allocate resources for better services. Insights from European cities might promote integrated transport but not sustainability if there is no policy integration that puts sustainability high on the agenda.[3]

To summarize, if integrated policy can be advanced one of the outcomes of this will be more integrated transport services, and this is essential for achieving sustainability goals. At the operational level (the integrated transport level), the challenge is to move away from efficiency in vehicles towards efficiency of movement. But this shift needs to be initiated from above (the integrated policy level), not at the operational level which basically reacts to goals, targets and priorities set by those governing the transport system.

4. The way forward

For systems to work together, integration is needed at all levels and for all types of activities, and this is crucial for the transport system. Thus, the Sisyphean[4] efforts to achieve integrated transport are not in vain, they are important! Yet, the ample evidence of failure to achieve real progress alongside the relative dearth of success stories (which do exist) suggests the concept of and approach to achieving integrated transport must be carefully rethought, but not abandoned.

Integrated transport emerges from this book as a part-whole problem where the parts do not add up to the whole, as the countless elements of integrated transport are addressed almost independently of each other in the various chapters of this book. This has not resulted in a clear concept of integration, let alone a road map of how to get there. In fact, many of the chapters have found it difficult to directly relate their material to the concept of integration. Perhaps integration is a chimera. The same conclusion may hold for integrated transport in practice, where the transport system is governed and provided by the public and private sectors. The importance of integration is recognized and actions to promote it are made, with some local success. Yet, we are still far from having an integrated transport system or an integrated transport policy. Thus, there needs to be much more clarity about the different concepts of integration and what can actually be achieved in reality. More specifically, what should be integrated and how still remains somewhat open. Some points of direction were offered above and these, as well as the general message from this book, pose questions for a new research agenda on transport and policy integration.

Policy decisions in transport and other sectors are becoming more complex, with increasing numbers of actors and agencies involved, and ever more complicated processes necessary for implementation. Complexity may be increasing at a faster rate than our knowledge of the means to address integration, thus making the task more difficult or even impossible. But integration is not about finding ideal solutions to complex problems. It is more about achieving effective outcomes that satisfy the objectives of different agencies and sectors, and in the transport context, providing a better quality of service to the user.

There is a sense that integration became a goal in itself. Since integrated policy and integrated transport are so important, naturally setting a goal to 'integrate' seems logical, but this probably only serves to make the task even harder. The goal from a policy perspective should be sustainability (again it is sufficient here to define it as an increase in socio-economic benefits with the same or lower environmental cost) and from a transport provision perspective a door-to-door service. This door-to-door service should be reliable, convenient, accessible and reasonably/sufficiently fast, and meet the sustainability definition. Integration is probably the most important means to achieve these objectives. But we also need to recognize the built-in conflict between the two levels of integration. A fully integrated transport system (one that provides a seamless door-to-door service) will surely result in increased mobility and a higher demand for transport which runs counter to the integrated policy objectives.

Therefore, a hierarchy between the two levels is needed. At the higher level and with greater priority, policy needs to be integrated across sectors and departments of government, so that policy with respect to economic and social decisions and sectors like energy, industry, education, etc. will take full account of the transport considerations (and transport implications). Transport and transport integration need to be placed within the wider debates on climate change, energy futures, sustainable development and social cohesion. This means that the traditional imperative of planning for an increasing level of mobility needs to be moderated with a concern over whether the same levels of activity can be supported by less travel rather than more. This may be the pathway to true transport integration. At the lower level, transport policy should be directed by policy goals with respect to the above debates and it should not be considered on its own, as a closed system. The principle, that transport is only to serve other needs and to fulfil other objectives, needs to be recognized. Integrated transport is only second to integrated policy.

Notes

1 Hull's (2005) Integration Ladder consists of eight rugs, which include (starting from the highest level of integration): 8. Integration of policy measures; 7. Integration of policy sectors; 6. Institutional and administrative integration; 5. The integration of environmental issues in transport policy making; 4. Integration with social objectives; 3. Integration with market needs; 2. Modal integration; 1. Physical and operational integration of public transport.

2 An aircraft flying with almost no passengers might still be fully loaded with freight and therefore generate revenues. Interestingly, perhaps the long tradition of separately considering freight and passenger transport on the railways prevented the idea of passenger trains pulling freight carriages behind from happening. Furthermore, evolution in the design of passenger trains means that, currently, new passenger trains are built as one unit (and not as a separate locomotive and carriages, which is still the standard in freight trains) and therefore the idea of transporting freight together with passenger is no longer feasible.

3 We thank Joseph Szyliowicz for these comments.

4 Sisyphus was a king of Corinth whose punishment in Hades was to roll a heavy boulder up a hill which as it reached the top rolled down the other side; he had to repeat this throughout eternity – an example of fruitless toil.

References

Bardach, E (1977) *The Implementation Game: What Happens after a Bill Becomes Law?*, Cambridge, MA: MIT Press.

Givoni, M and Banister, D (2006) Airline and railway integration, *Transport Policy* 13, pp. 386–397.

Hansen, CA (2006) Urban transport, the environment and deliberative governance: A role of interdependence and trust, *Journal of Environmental Policy and Planning* 8(2), pp. 159–179.

Hull, A (2005) Integrated transport planning in the UK: From concept to reality, *Journal of Transport Geography* 13(4), pp. 318–328.

Hull, A (2008) Policy integration: What will it take to achieve more sustainable transport solutions in cities?, *Transport Policy* 15(2), pp. 94–103.

Iseki, H and Taylor, BD (2009) Not all transfers are created equal: Towards a framework relating transfer connectivity to travel behaviour, *Transport Reviews* 29(6), pp. 777–800.

NEA, OGM and TSU (2003) *Integration and Regulatory Structures in Public Transport*, Final Report, DGTREN, Brussels, November.

Peters, BG (1998) Managing horizontal governance: The politics of coordination, *Public Administration* 76(2), pp. 295–311.

Rietveld, P (2000) Non-motorised modes in transport systems: A multimodal chain perspective for the Netherlands, *Transportation Research D* 5(1), pp. 31–36.

Stead, D (2008) Institutional aspects of integrating transport, environment and health policies, *Transport Policy* 15(2), pp. 139–148.

Index